Drying in the Dairy Industry

Advances in Drying Science and Technology

Series Editor: *Arun S. Mujumdar*

McGill University, Quebec, Canada

Handbook of Industrial Drying, Fourth Edition
Arun S. Mujumdar

Advances in Heat Pump-Assisted Drying Technology
Vasile Minea

Computational Fluid Dynamics Simulation of Spray Dryers: An Engineer's Guide
Meng Wai Woo

Handbook of Drying of Vegetables and Vegetable Products
Min Zhang, Bhesh Bhandari, and Zhongxiang Fang

Intermittent and Nonstationary Drying Technologies: Principles and Applications
Azharul Karim and Chung-Lim Law

Thermal and Nonthermal Encapsulation Methods
Magdalini Krokida

Industrial Heat Pump-Assisted Wood Drying
Vasile Minea

Intelligent Control in Drying
Alex Martynenko and Andreas Bück

Drying of Biomass, Biosolids, and Coal: For Efficient Energy Supply and Environmental Benefits
Shusheng Pang, Sankar Bhattacharya, Junjie Yan

Drying and Roasting of Cocoa and Coffee
Ching Lik Hii and Flavio Meira Borem

Heat and Mass Transfer in Drying of Porous Media
Peng Xu, Agus P. Sasmito, and Arun S. Mujumdar

Freeze Drying of Pharmaceutical Products
Davide Fissore, Roberto Pisano, and Antonello Barresi

Frontiers in Spray Drying
Nan Fu, Jie Xiao, Meng Wai Woo, Xiao Dong Chen

Drying in the Dairy Industry
Cécile Le Floch-Fouéré, Pierre Schuck, Gaëlle Tanguy, Luca Lanotte, Romain Jeantet

For more information about this series, please visit: www.crcpress.com/Advances-in-Drying-Science-and-Technology/book-series/CRCADVSCITEC

Drying in the Dairy Industry

From Established Technologies to Advanced Innovations

Edited by
Cécile Le Floch-Fouéré
Pierre Schuck
Gaëlle Tanguy
Luca Lanotte
Romain Jeantet

CRC Press
Taylor & Francis Group
Boca Raton London New York

CRC Press is an imprint of the
Taylor & Francis Group, an **informa** business

First edition published 2021
by CRC Press
6000 Broken Sound Parkway NW, Suite 300, Boca Raton, FL 33487-2742

and by CRC Press
2 Park Square, Milton Park, Abingdon, Oxon, OX14 4RN

CRC Press is an imprint of Taylor & Francis Group, LLC

Library of Congress Cataloging-in-Publication Data

Names: Le Floch-Fouéré, Cécile, editor.
Title: Drying in the dairy industry : from established technologies to advanced innovations / edited by Cécile Le Floch-Fouéré, Pierre Schuck, Gaëlle Tanguy, Luca Lanotte, Romain Jeantet.
Description: Boca Raton : CRC Press, 2021. | Series: Advances in drying science & technologies | Includes bibliographical references and index.
Identifiers: LCCN 2020027410 (print) | LCCN 2020027411 (ebook) | ISBN 9780815359982 (hardback) | ISBN 9781351119504 (ebook)
Subjects: LCSH: Dairy products--Drying. | Dried milk. | Infant formula industry--Technological innovations. | Dairy products industry--Technological innovations. | Dairy processing--Technological innovations. | Dried milk industry--Technological innovations.
Classification: LCC SF259 .D79 2021 (print) | LCC SF259 (ebook) | DDC 637/.1--dc23
LC record available at https://lccn.loc.gov/2020027410
LC ebook record available at https://lccn.loc.gov/2020027411

ISBN: 978-0-8153-5998-2 (hbk)
ISBN: 978-1-351-11950-4 (ebk)

Typeset in Times
by Deanta Global Publishing Services, Chennai, India

Contents

Advances in drying science and technology

It is well-known that the unit operation of drying is a highly energy-intensive process encountered in diverse industrial sectors ranging from agricultural processing, to ceramics, chemicals, minerals processing, pulp and paper, pharmaceuticals, coal polymer, food and forest products industries as well as waste management. Drying also determines the quality of the final products. Making drying technologies sustainable and cost effective via the application of modern scientific techniques is the goal of academic as well as industrial R&D activities around the world.

Drying is a truly multi- and interdisciplinary area. Over the last four decades the scientific and technical literature on drying has seen exponential growth. The continuously rising interest in this field is also evident from the success of numerous international conferences devoted to drying science and technology.

The establishment of this new series of books entitled Advances in Drying Science and Technology is designed to provide authoritative and critical reviews and monographs focusing on current developments as well as future needs. It is expected that books in this series will be valuable to academic researchers as well as to industry personnel involved in any aspect of drying and dewatering.

The series will also encompass themes and topics closely associated with drying operations, e.g., mechanical dewatering, energy savings in drying, environmental aspects, life cycle analysis, technoeconomics of drying, electrotechnologies, control and safety aspects and so on.

Arun S. Mujumdar
Series Editor
McGill University, Quebec, Canada

Preface

A universal food *par excellence* and of great nutritional value, milk has been the subject, over the centuries, of fundamental discoveries to improve its shelf life, to allow its transport and to guarantee its quality, e.g., milk dehydration aimed at stabilizing the surplus coming from the dairy industry and ensuring storage. Applied first on products long considered as by-products (whey, etc.), the resulting powders were mainly intended for animal feed and had a low added-value. But dairy powders are nowadays used as food ingredients and are produced on a large scale. Indeed, the global market for dairy powders has undergone two major transitions over the past five decades, first by integrating scientific and technological knowledge, and second due to the growing demand of emerging markets from the late 2000s. Worldwide demand is rising at present, thus increasing the volume and the value of the dairy trade supplied by the industry at a global scale. The challenge represented by this segment makes now dried dairy products a relevant axis of development for the industry. Indeed, by considering dairy powders as key ingredients in the formulation of a wide variety of food products, the control of their functional and nutritional properties, which determine their value and uses, is a main concern. This is the case for whey products which have been increasingly studied over the past decade, as well as the rising of the infant formulae market. Moreover, growing interest is expected in the future for specifically designed nutritional properties, including health benefits from the presence of live probiotics in products. In the case of dried products with a long shelf life, this specific point remains particularly challenging.

This trend stresses the need for the development of alternative approaches to experiments at the industrial scale for setting optimized operating conditions, as the management of these latter is always costly and difficult given the complexity of industrial-scale drying equipment, the diversity of formulations and the limited control of the process parameters *in situ*. Controlling the drying process therefore requires progress on fundamental questions of major interest (such as the physical approach to the transformation of droplets into particles, etc.) in the dairy industry.

This handbook, *Drying in the Dairy Industry: From Established Technologies to Advanced Innovations* aims to focus on the advances and innovations in the drying of dairy products by highlighting the interactions between product and process. The book is divided into four sections.

In the first section, some basic fundamentals of the spray-drying operation, the key processing steps and its application for the production of various dried dairy products are presented, with a focus on the composition of these latter.

In the second section, the key operations in view of modulating and controlling dairy products are presented. It includes membrane processes, homogenization, concentration by vacuum evaporation, agglomeration and fluidization.

In the third section, different powder properties and their influencing factors, such as water activity, caking and physical and microbial properties, are detailed.

In the fourth section, some recent innovations and prospects are discussed in terms of products (infant and follow-on formulae, lactose hydrolyzed or camel milk powders) and processes (production of dairy powders without the use of a drying chamber, spray drying of probiotics).

Lastly, a general conclusion will allow the topic of the drying of milk at the laboratory scale to be broached, thanks to an overview of the more recent pioneering works performed on the evaporation of dairy solutions using mainly microscopy and microfluidic approaches.

Dr Cécile Le Floch-Fouéré and co-workers

Series editor

Dr. Arun S. Mujumdar is an internationally acclaimed expert in drying science and technologies. He was the Founding Chair in 1978 of the International Drying Symposium (IDS) series and has been Editor-in-Chief of *Drying Technology: An International Journal* since 1988. The 4th enhanced edition of his *Handbook of Industrial Drying* published by CRC Press has just appeared. He has been the recipient of numerous international awards including honorary doctorates from Lodz Technical University, Poland, and University of Lyon, France.

Please visit www.arunmujumdar.com for further details.

Editors

Cécile Le Floch-Fouéré has been an assistant professor in the Food Science and Engineering Department Agrocampus Ouest, Rennes, France, since 2008. Her primary area of expertise is process engineering and dairy technology at the research unit Science and Technology of Milk and Egg in Rennes. Cécile received her PhD in 2008 in Biophysics from the University of Rennes 1. Since 2017, she has been the leader of the Spray-Drying – Concentrated Matrices – Functionalities team. Her current research focuses on the understanding of the interface formation mechanism during the drying of food powders. To date, Cécile has participated in the supervision of 6 MSc and 11 PhD students and has co-authored about 30 publications in peer-reviewed journals, 67 conferences and posters and 2 books including 3 chapters.

Pierre Schuck has a Master's degree in Food Science and Technology and obtained his PhD in Physicochemistry and Quality of Bio-Products in 1999. He is a research engineer at the research unit Science and Technology of Milk and Egg at INRAE at the French National Research Institute for Agriculture, Food and Environment (INRAE) in Rennes. His main interest is the spray drying of dairy products with a particular interest in the physical mechanisms of water transfer and in the influence of the physico-chemical parameters before, during and after spray drying on the properties of the dairy powders. His expertise is well-recognized in the field, and he is in great demand as a consultant in many dairy companies around the world. His scientific contributions have led to 116 papers in peer-reviewed journals, 112 conferences and posters, 7 patents, 37 licenses, 11 books and 26 book chapters.

Gaëlle Tanguy works as a research engineer in the dairy processing team of the research unit Science and Technology of Milk and Egg at INRAE in Rennes. Her current research focuses on the concentration by vacuum evaporation of dairy products, and more especially on the behavior of dairy components during concentration in relation to the operation and the fouling of falling-film evaporators. She has co-authored about 14 publications in peer-reviewed journals.

Luca Lanotte is a researcher at INRAE. His primary area of expertise is linked to the physics and the rheology of dairy biocolloids at the research unit Science and Technology of Milk and Egg in Rennes (France). Luca obtained his PhD in Physics and Chemical Engineering in 2013. The PhD project (Vinci Project 2009) was realized in collaboration with the University Federico II of Napoli (Italy) and the University Joseph Fourier of Grenoble (France). Currently, his research activity focuses on the physico-chemical mechanisms characterizing the drying in mixes of dairy proteins. To date, Luca has co-mentored six MSc students and one PhD student. He has co-authored 15 publications in peer-reviewed journals and his works have been presented in 28 international/national conferences and meetings.

Romain Jeantet is currently a food engineering professor and deputy director of Research at Agrocampus Ouest (Rennes, France), an adjunct professor of chemical engineering at Suzhou University (China), and, since 2009, deputy director of the research unit Science and Technology of Milk and Egg in Rennes, which represents 83 permanent staff and 25 PhD students. During his PhD and until 2001, he was first involved in research topics dealing with nanofiltration, and then pulsed electric fields applied to dairy and egg products until 2004. Since 2003, his research activity has been focused on the concentration, spray drying, storage and rehydration of dairy products with a particular interest in the control of the heat and mass transfer involved and in the influence of the process parameters on the final functional properties of dairy products. His research has led to 113 papers in peer-reviewed journals, 120 conferences and posters, 2 software programs (37 licenses) and 5 patents, 13 books including 81 chapters and 10 other book chapters.

Contributors

Zahra Afrassiabian
Sorbonne University, Université de Technologie de Compiègne, ESCOM
Centre de recherche
Compiègne cedex, France

Alexia Audebert
Royal BEL Leerdamer
Utrecht, The Netherlands

Nidhi Bansal
School of Agriculture and Food Sciences
The University of Queensland
Brisbane, QLD, Australia

Laurent Bazinet
Institute of Nutrition and Functional Foods (INAF), Dairy Research Center (STELA), Department of Food Sciences and Laboratory of Food Processing and ElectroMembrane Processes (LTAPEM)
Université Laval
Rue de l'Université, Quebec, Canada

Bhesh Bhandari
School of Agriculture and Food Sciences
The University of Queensland
Brisbane, QLD, Australia

Jennifer Burgain
Université de Lorraine, LIBio
Vandœuvre-lès-Nancy, France

Antônio Fernandes de Carvalho
INOVALEITE – Departamento de Tecnologia de Alimentos (DTA)
Universidade Federal de Viçosa (UFV)
Viçosa, MG, Brasil

Xiao Dong Chen
Suzhou Key Laboratory of Green Chemical Engineering
School of Chemical and Environmental Engineering
College of Chemistry, Chemical Engineering and Materials Science
Soochow University
Suzhou, China

Thomas Croguennec
STLO, INRAE, Institut Agro
Rennes, France

Bernard Cuq
IATE, INRAE, l'Institut Agro - Montpellier SupAgro
Université Montpellier
CIRAD, France

Anne Dolivet
STLO, INRAE, Institut Agro
Rennes, France

Marie-Hélène Famelart
STLO, INRAE, Institut Agro
Rennes, France

Mark A. Fenelon
Teagasc Food Research Centre
Moorepark, Fermoy, Co. Cork, Ireland

Tatiana Lopes Fialho
Universidade Federal de Viçosa (UFV)
Viçosa, MG, Brasil

Tristan Fournaise
Université de Lorraine, LIBio
Vandœuvre-lès-Nancy, France

Claire Gaiani
Université de Lorraine, LIBio
Vandœuvre-lès-Nancy, France

Muhammad Gulzar
Glanbia
Ballyragget, Co. Kilkenny, Ireland

Thao Minh Ho
School of Agriculture and Food Sciences
The University of Queensland
Brisbane, QLD, Australia

and

Department of Food and Nutrition
The University of Helsinki
Helsinki, Finland

Song Huang
DuPont Nutrition & Health
Changning District
Shanghai, China

Gwénaël Jan
STLO, INRAE, Institut Agro
Rennes, France

Romain Jeantet
STLO, INRAE, Institut Agro
Rennes, France

Luca Lanotte
STLO, INRAE, Institut Agro
Rennes, France

Cécile Le Floch-Fouéré
STLO, INRAE, Institut Agro
Rennes, France

Christelle Lopez
STLO, INRAE, Institut Agro
Rennes, France

Solimar Gonçalves Machado
INOVALEITE – Departamento de Tecnologia de Alimentos (DTA)
Universidade Federal de Viçosa (UFV)
Viçosa, MG, Brasil

Evandro Martins
INOVALEITE – Departamento de Tecnologia de Alimentos (DTA)
Universidade Federal de Viçosa (UFV)
Viçosa, MG, Brasil

Serge Méjean
STLO, INRAE, Institut Agro
Rennes, France

Eoin G. Murphy
Teagasc Food Research Centre
Moorepark, Fermoy, Co. Cork, Ireland

Ítalo Tuler Perrone
INOVALEITE – Departamento de Ciências Farmacêuticas
Universidade Federal de Juiz de Fora
Juiz de Fora – MG, Brasil

Mathieu Persico
Institute of Nutrition and Functional Foods (INAF)
Dairy Research Center (STELA)
Department of Food Sciences and Laboratory of Food Processing and ElectroMembrane Processes (LTAPEM)
Université Laval
Rue de l'Université, Quebec, Canada

Mathieu Person
ARMOR PROTEINES/SOFIVO
St-Brice-en-Coglès, France

Jérémy Petit
Université de Lorraine, LIBio
Vandœuvre-lès-Nancy, France

Ramila Cristiane Rodrigues
INOVALEITE – Departamento de Tecnologia de Alimentos (DTA)
Universidade Federal de Viçosa (UFV)
Viçosa, MG, Brasil

Yrjö H. Roos
University College Cork
Cork, Ireland

Khashayar Saleh
Sorbonne University, Université de Technologie de Compiègne, ESCOM
Centre de recherche
Compiègne cedex, France

Joël Scher
Université de Lorraine, LIBio
Vandœuvre-lès-Nancy, France

Pierre Schuck
STLO, INRAE, Institut Agro
Rennes, France

Cordelia Selomulya
School of Chemical Engineering
UNSW
Sydney, Australia

Rodrigo Stephani
INOVALEITE – Departamento de Química
Universidade Federal de Juiz de Fora (UFJF)
Juiz de Fora, MG, Brasil

Gaëlle Tanguy
STLO, INRAE, Institut Agro
Rennes, France

Meng Wai Woo
Department of Chemical & Materials Engineering
Faculty of Engineering
University of Auckland
Auckland, New Zealand

Dr. Nima Yazdanpanah
Procegence
Washington DC, USA

Zhengzheng Zou
School of Agriculture and Food Sciences
The University of Queensland
Brisbane, QLD, Australia

chapter 1

Spray-dried dairy product categories

Thao Minh Ho, Nidhi Bansal and Bhesh Bhandari

Contents

1.1 Introduction

Spray drying is a common technique used to convert a liquid, a slurry, an emulsion, a suspension or a low viscosity paste into dried powder. This process involves atomizing (or spraying) the feed material into very small droplets which subsequently come into contact with a high-temperature air-drying medium in a drying chamber. The large surface area of the atomized droplets of the feed material imparts a very fast drying rate (Woo, 2017). Due to the unique characteristics of the spray-drying process, especially its short drying time and rapid water evaporation, it has been used not only in the food and dairy industries, but also in many other industrial sectors such as pharmaceuticals, agrochemicals, chemicals, detergents, pigments, biotechnology and ceramics (Broadhead et al., 1992; de Souza et al., 2015; Cao et al., 2000; Patel et al., 2014a). In these fields, spray drying has been employed for three main purposes: (i) to dehydrate products for improving preservation, ease of handling, transport and storage; (ii) to encapsulate bioactive compounds within a protective matrix; and (iii) to induce the structural (phase) transformation of solid materials to obtain dried powders with distinct properties (Ho et al., 2017). However, in the scope of this chapter, only specific applications of spray drying for the production of dairy products are addressed. Although spray drying was introduced to the market in the 1920s and became popular in the dairy industry in the 1970s, with recent technical advances and innovations, the number of dairy products produced by spray drying has increased exponentially (Pisecky, 1985; Schuck, 2002). In the production of dairy products, spray drying is not limited to conventional milk powders (skim milk powder [SMP] and whole milk powder [WMP]), but has expanded to high protein and lactose powders (caseinate, milk protein concentrates [MPC] and whey protein powders), high fat powders (cream powder and butter powder) and even encapsulated powders (Table 1.1). The final dried products can be free-flowing powders of individual particles, agglomerates or granules depending on the properties of the feed materials and the conditions of spray drying. The physiochemical and functional properties of these products are greatly influenced by the spray-drying conditions and the post-spray-drying treatment of powders. In this chapter, the basic fundamentals of a spray-drying operation unit, and the compositions and key processing steps of such dried dairy products are discussed.

1.2 The fundamentals of the spray-drying process in the production of dairy products

Many aspects of the spray-drying unit operation relating to the production of food powders, such as heat and mass transfer, drying kinetics and the characteristics of the different components in a typical spray-drying system, the different types of spray dryers and so on, have been well described elsewhere (Masters, 1985; Anandharamakrishnan and Ishwarya, 2015; Bhandari et al., 2008; Filkova and Mujumdar, 1995; Woo, 2017; Woo and Bhandari, 2013). Therefore, only the basic fundamentals of the spray-drying process

Table 1.1 Applications of spray drying in the production of dairy powders

Applications	Typical products
Dehydration	*Conventional milk powders* (regular, instant or agglomerated forms): • Skim milk powders (SMP) • Whole milk powders (WMP) • Infant formula milk powders *High protein and lactose powders:* • Caseinate powders • Milk protein concentrates (MPC) • Milk protein isolates (MPI) • Whey protein powders (sweet whey powder, acid whey powder, demineralized whey powder, delactosed whey powder) • Whey protein concentrates (WPC) • Whey protein isolates (WPI) *High fat powders* (fat content ~40–88%): • Cream powders • Butter powders • Cheese powders
Encapsulation matrix	• Oils (flavor oils, essential oils, butter oil and fish oil) • Useful bacteria and probiotics • Food colors and vitamins

Source: Refstrup (1995).

are given in this section. Due to the high energy consumption during the spray-drying operation, the removal of water from the feed prior to spray drying is typically required to increase the solid concentration to a certain level via alternative energy efficiency approaches such as reversed filtration and evaporation to keep energy costs to a minimum. For example, in the production of SMP and whey powders, milk and whey feeds are typically concentrated to 25% and 42–60% total solids, respectively, via the reverse-osmosis membrane and vacuum evaporation systems before they are subjected to spray drying (Masters, 1985; Roy et al., 2017). The energy required to evaporate 1 kg of water during spray drying is three to six times higher than that of a double-effect vacuum evaporator (Carić, 1994) and more than double that in the membrane concentration. Moreover, the preconcentration of the feed for spray drying enables the acquisition of the dried powders' desirable properties. The milk powders produced from preconcentrated milk have a large particle size, a small amount of entrapped air and a long shelf-life due to the increase in viscosity and to the possible crystallization of lactose during preconcentration. Hence, the preconcentration of the feed can be considered an essential pre-treatment step in the spray drying of dairy powders.

The essential parts of a typical spray drier include the feed pump, atomizer, drying chamber, air heating and dispersing system, a product recovery system (cyclone) and a process control system. Figure 1.1 illustrates typical spray-drying equipment and the associated processing conditions that greatly affect the properties of the dried product. In terms

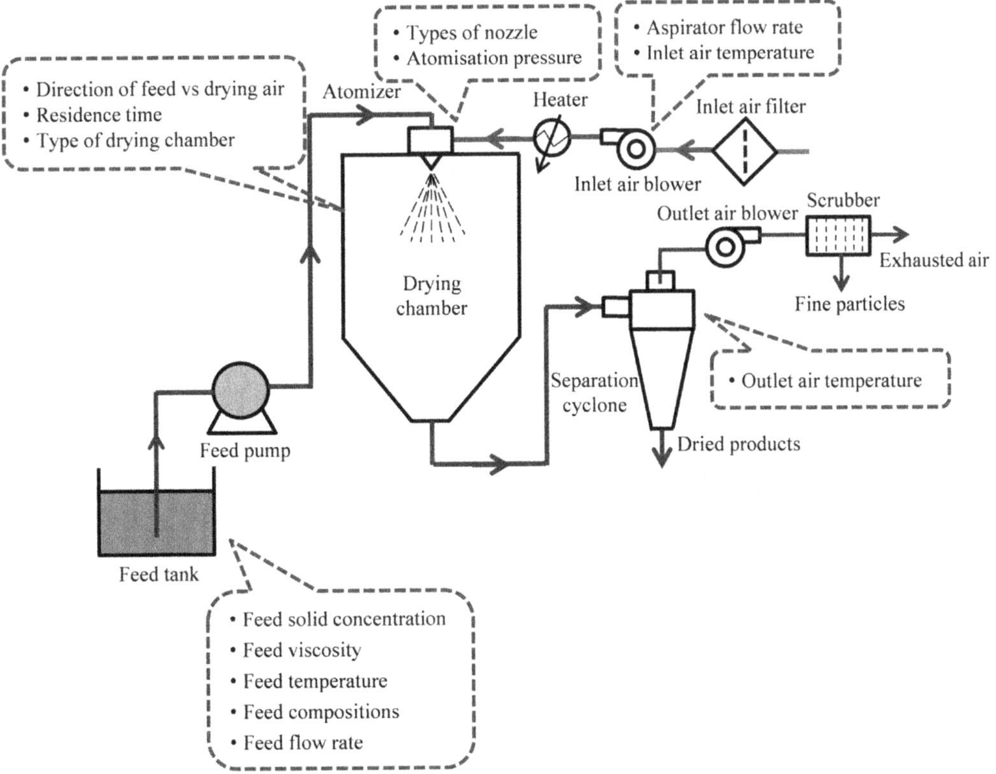

Figure 1.1 A diagram of a typical spray dryer indicating the main steps in the spray-drying process and associated processing conditions affecting the dried product properties. Adapted with permission from Ho et al. (2017).

of operation, spray drying is a continuous multi-stage unit involving (i) the atomization of the feed solution, (ii) mixing of the sprayed droplets and hot air, (iii) the evaporation of water and (iv) the separation of dried products from exhausted air (Anandharamakrishnan and Ishwarya, 2015). The atomization stage is the most important operation in spray drying because it determines the trajectory, speed, shape, structure and size distribution of the droplets, and thus the design of the drying chamber and the particle size of dried products (Ho et al., 2017). The feed is supplied to an atomizer located on the top of the spray dryer via a feed pump. Typically, centrifugal pumps, positive displacement pumps or peristaltic pumps are the most commonly used types in the spray drying of dairy products depending on the feed rate, the viscosity of the feeds and the type of atomizer used. However, the feed pump is dependent on the feed properties (e.g. viscosity, surface tension, temperature and concentration), the type of atomizer and the feed amounts to be pumped. Under atomization, the feed is sprayed into micron-size droplets with a very large surface area. In thermodynamics, the rate of heat and mass transfer is directly proportional to the effective surface area. Thus, such a large surface area of atomized droplets facilitates rapid water evaporation and a short drying time. Several common types of atomizers are used to disintegrate the droplets from the feed, such as a rotary atomizer, a pressure nozzle atomizer, a two-fluid nozzle atomizer, an ultrasonic atomizer or an electrohydrodynamic atomizer. Each atomizer is different in terms of its energy requirements. The characteristics and operating principles of these atomizers were documented by Masters (1985), Anandharamakrishnan (2015) and Bhandari et al. (2008). The selection of a suitable atomizer is based on the physical properties of the feed, the design of the drying chamber and the desired final properties of the dried powders.

Upon leaving the atomizer, the droplets come into contact with hot drying air in three basic ways: (i) concurrent flow (e.g. the droplets and the hot drying air travel in the same direction); (ii) countercurrent flow (e.g. the droplets and the hot drying air travel in the opposite direction); and (iii) mixed flow (e.g. the droplets and the hot drying air travel in random direction providing higher levels of mixing and longer residence time). In the production of dairy products, the concurrent flow is the most operated mode due to the short residence time of particles that allows the dried particles to avoid passing through the high-temperature drying zone. As a result, undesirable changes in many bioactive components present in the materials are prevented during spray drying (Bhandari et al., 2008). In the drying chamber, due to the huge surface area of the atomized droplets, water evaporation from the atomized droplets occurs within a fraction of a second and within a very short travelling distance. The formation of a crust (e.g. drying boundary layer) on the surface of the powder particles occurs so fast that the spray-drying process can be considered to include only a falling rate period, especially the spray drying of small droplets that have a high amount of dissolved solids (Anandharamakrishnan, 2015). The dried powders are typically collected from the bottom of the drying chamber and discharged. Depending on the type of spray dryer, the operating conditions and the material properties, the exhausted air in the spray-drying system can carry about 10–50% of the total powder (Pısecký, 1997). Separating the powder from the exhausted air not only improves the drying yield but also minimizes air pollution. This can be done using either cyclones or a combination of cyclones and bag filters (even wet scrubbers). Beside the limitations on the cost-effectiveness and energy efficiency of spray drying, in comparison with other drying techniques used in the production of dairy powder, spray drying offers many advantages as illustrated in Table 1.2.

One of the major problems associated with spray drying most dairy powders is the stickiness and caking of the powder particles. This results from the collision of insufficiently powder dried particles with other particles (cohesion forces) and/or with drying

Table 1.2 Advantages and disadvantages of spray drying in the production of dairy powders

Advantages	Disadvantages
• Continuous operation • Flexible production rates and capacity design • Wide range of operating temperatures • Short residence time and rapid drying • Suitable for both heat-sensitive and heat-resistant materials • Possibility for automatic operation and maintenance • Production of powdered products with predetermined characteristics and without requiring further grinding	• Expensive with high installation costs • High energy requirement • Heat degradation possibility due to high temperature of air drying • Stickiness of dried powders in the drying chamber resulting in congestion, low yield and difficulties in cleaning

walls (adhesion forces) (Boonyai et al., 2004). The stickiness of the powders during spray drying leads to many operating problems (difficulties in cleaning, possibility of blockage in the ducts or cyclone and fire hazards), a deterioration in product quality (possibility of overheating resulting in unpleasant sensory characteristics and degradation) and a low product yield due to wall deposition of the dried powders (Papadakis and Bahu, 1992).

Two mechanisms can explain the stickiness between the powder particles in spray drying dairy products. The first mechanism is related to the changes in the thermodynamic state of amorphous components. During spray drying, water evaporation occurs so fast that the solutes do not have sufficient time to rearrange into an ordered structure (Ho et al., 2017). The main amorphous components present in spray-dried dairy powders are proteins and lactose. However, as compared to the proteins, the lactose dominates the changes in the physical properties observed in dairy powders (Roos, 2002). Thus, the stickiness of spray-dried powder particles is usually associated with the lactose components. During spray drying, as the water content is reduced, the temperature of the atomized droplets increases. At glass transition temperature (T_g), the amorphous lactose transforms from an immobile glass state to a rubbery state. A decrease in the viscosity on the particle surface in the rubbery phase leads to an increase in the mobility of the lactose molecules which results in the formation of rubbery bridges between adjacent particles. If the amorphous lactose remains above T_g for sufficient time, the crystallization of the amorphous lactose and the formation of solid bridges between the powder particles occur (Foster, 2002). It was reported that signs of stickiness of dairy powders occur at a surface viscosity of about 10^6–10^8 Pa.s (Downton et al., 1982). The stickiness of dairy powders caused by the amorphous components can be avoided by spray drying at temperatures lower than 20°C above their T_g (Bhandari et al., 1997). It is observed that the mobility of the amorphous molecules is greatly affected by the moisture content, temperature and relative humidity to which the materials are exposed. An increase in the moisture content reduces T_g and the activation energy to rearrange the molecules, enhancing the stickiness (Roos and Karel, 1992; Wang and Langrish, 2007). The second mechanism associated with the stickiness of spray-dried dairy powders relates to the melting of fat. In the spray drying of high fat milk powders, as the temperature of the drying medium is much higher than the melting temperature of fat, most fat transforms into a fluid-like state and tends to migrate over the particle surface due to its hydrophobic nature. The melting of fat facilitates the formation of liquid bridges between the powder particles. As the temperature drops below the melting temperature of fat, these bridges partially solidify, resulting in stickiness and caking of the powders (Foster, 2002). Moreover, the crystallization of lactose can stress the fat droplets inside the powder particles and force the fat to spread over the powder surface, imparting the stickiness between the powder particles (Fäldt and Bergenståhl, 1996). However, the contribution of fat to the stickiness of the dairy powders is only significant when it is present in amounts large enough to prevent encapsulation by proteins (O'Callaghan and Hogan, 2013).

As mentioned, the stickiness of the powders in the spray-drying process is described by both cohesion and adhesion phenomena. While the adhesion phenomenon is governed by viscoelastic behavior, phase transformation and melting of the surface fat, the cohesion phenomenon is determined by differences in the surface energy between the dried products and the equipment. The stickiness caused by the adhesion of the powder to the dryer wall plays a role in the early stage of the drying process. When the equipment wall is completely covered with the powders, stickiness is then governed by the cohesive forces between the particles. Thus, it is important to avoid the early stage of adhesion which can act as a seed for further accumulations of sticky particles (Bhandari and Howes, 2005). In order to minimize and/or avoid stickiness of the dried dairy powders, many approaches have been reported; however, some of them are not readily available commercially because of the incomplete understanding about new systems (Bhandari and Howes, 2005; Bhandari et al., 1997; Woo, 2017), such as:

a) Spray drying at low drying temperatures and low humidity conditions, and the introduction of cold air with controlled relative humidity at the end of the drying chamber or cooling the drying chamber wall to minimize the phase transformation and the melting of fat.
b) Use of a two-stage spray-drying process in which fluidized bed drying at mild temperatures is integrated into the bottom of the spray-drying chamber to provide further drying of partially dried powder particles in the spray-drying stage, to induce partial crystallization of the lactose or to act as an agglomeration unit. Due to the low drying temperature during fluidized bed drying and the short residence time in a spray-drying section in the two-stage spray-drying system, thermal degradation of the products can be avoided.
c) Intermittent sweeping of the drying chamber wall with dehumidified cold air to prevent the stickiness of the powder on the drying wall.
d) Spray drying in a conical drying chamber (to minimize the adhesion of the drying powders to the dryer wall) and a concurrent flow (to avoid overheating the drying powders).

In spray drying, the powder particles are usually small (<50 μm in diameter), resulting in poor flowability and slow reconstitution or lump formation during rehydration. Thus, the manipulation of the powder agglomeration can lead to an increase in the dissolution rate of the powder due to an increase in the particle size and changes to the powder structure. There are two common ways to produce agglomerated milk powders, known as agglomeration in a drying chamber and fluidized bed agglomeration. In the first method, the fine powder particles travelling in exhausted air are collected and recirculated back to the drying chamber. Once they are in contact with the drying air and the atomized milk droplets, their surfaces are humidified (by the evaporated water) and become sticky. The adhesion of other drying particles on such surfaces due to the mixing actions in the drying chamber results in the agglomeration of the particles. In a fluidized bed agglomeration, a vibrating multi-section fluid bed with a perforated bottom is connected to the discharge opening of the drying chamber. When the powder from the drying chamber enters the first section of the fluid bed, steam is introduced to humidify the powder particles. Then, the humidified powder particles are conveyed to the drying section with a decreasing temperature to induce the agglomeration of the powder particles and further the dehydration of the agglomerated powders (Tetra Pak, 2018). In many cases, food-grade surfactants (primarily lecithin) can be sprayed onto the particles within the fluidized bed to produce high fat instant milk powders (Woo, 2017).

1.3 Spray-dried dairy powder products

Many commercial dairy powder products such as whey protein powders and caseinate powders are produced either by spray drying or roller drying. The differences in the dehydration mechanism between spray drying and roller drying lead to dissimilarities in the properties and functionality of milk powders. An example for caseinate powders is illustrated in Table 1.3 (DMV, 2012). However, in the scope of this chapter only spray-dried dairy powders are introduced. The production steps for the different types of spray-dried dairy powders are shown in Figures 1.2 and 1.3.

Table 1.3 Differences in the properties of spray-dried and roller-dried caseinate powders

Properties	Roller drying	Spray drying
Particle size (μm)	<500	<800
Morphology	Spherical	Irregular
Bulk density (g/L)	350–400	470–520
Dispersibility	Relatively poor	Better
Flowability	Medium to poor	Good
Viscosity (15%, 20°C) (Pa.s)	2.5	3.0
Occluded air	A lot	A little

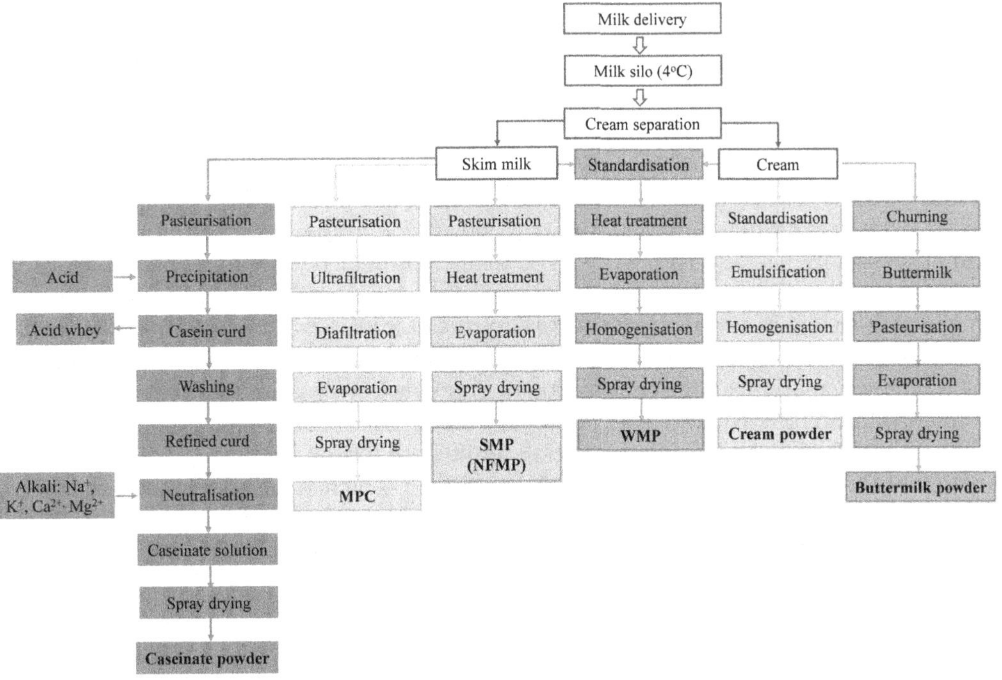

Figure 1.2 A chart illustrating the main steps in the production of different types of milk powders: caseinate powder, MPC – milk protein concentrate, SMP – skim milk powder, NFMP – non-fat milk powder, WMP – whole milk powder, cream powder and buttermilk powder.

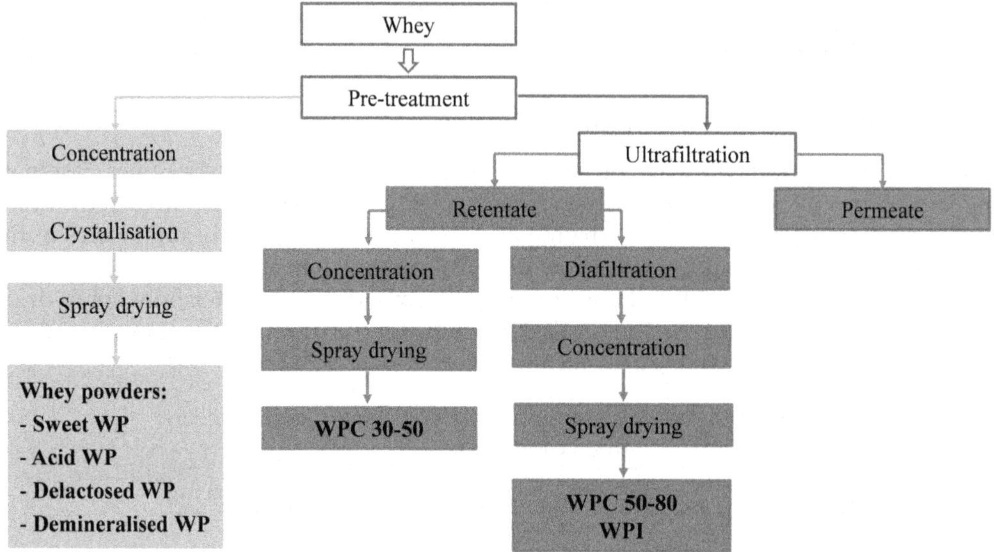

Figure 1.3 A chart illustrating the main steps in the production of different types of whey-based powders: WP – whey powder, WPC – whey protein powder, WPI – whey protein isolate.

1.3.1 *Conventional dairy powders*

Conventional spray-dried milk powders include non-fat milk powder (NFMP), SMP and WMP. They can be in either regular or agglomerated form.

1.3.1.1 *Skim milk and non-fat milk powders*

In terms of composition, NFMP and SMP are very similar, but different in their requirements for protein content. SMP must have a minimum protein content of 34% while NFMP does not have any standard for protein content (Tetra Pak, 2018). The main composition of SMP is shown in Table 1.4 (Deeth and Hartanto, 2009). Both SMP and NFMP are produced from skim milk using a similar procedure including heat treatment, evaporation and spray-drying stages (Figure 1.2). The heat treatment not only destroys bacteria and inactivates enzymes in milk but it also controls the denaturation degree of whey proteins. Depending on the heat treatment conditions, different types of SMP with different functional properties and applications are manufactured (Table 1.5). An increase in heat treatment intensity increases the denaturation degree of whey proteins, reducing the solubility of the produced SMP (Bhandari et al., 2008). Pasteurization at 70–80°C for 15–50 s is typically applied for low heat SMP production while both pasteurization and subsequent heat treatment at 90–100°C for 30–50 s and 115–125°C for 1–4 min are used for medium and high heat counterparts, respectively. Due to the low solid concentration of skim milk (12–14%, w/w), it is typically concentrated under a vacuum evaporator to 50–55% (w/w) total solids prior to spray drying with inlet and outlet air temperatures of 180–230°C and 70–95°C, respectively (Bhandari et al., 2008). Due to the high lactose content (49–52%, w/w) in amorphous structures induced by rapid water evaporation during spray drying, SMP is very hygroscopic and prone to sticking and caking during handling, transport and storage. The partial crystallization of lactose prior to spray drying can minimize these unwanted changes to skim milk powders.

Table 1.4 Composition of SMP and WMP

Composition (%, w/w)	SMP	WMP
Moisture content	3.0–5.0	2.0–4.0
Proteins	35.0–37.0	25.0–28.0
Fat	0.7–1.3	25.0–27.0
Lactose	49.0–52.0	37.0–38.5
Ash	8.2–8.6	6.0–7.0

Table 1.5 Different types of SMP

Differences	Low heat SMP	Medium heat SMP	High heat SMP
Heat treatment	70–80°C/15–50 s	90–100°C/30–50 s	115–125°C/1–4 min
WPNI* (mg WPN/g)	>6	1.5–1.9	<1.5
Solubility	Highest	Medium	Lowest
Applications	Reconstituted milk	Ice cream, sweetened condensed milk, confectionery	Bakery products

* WPNI is the whey protein nitrogen index.

1.3.1.2 Whole milk powders

The production of WMP follows a similar procedure to the manufacture of SMP (Figure 1.2). However, raw milk is standardized by mixing a proportion of cream to skim milk to give a minimum fat content of 25% (w/w) in the final powder. Moreover, a two-stage homogenization (15–20 MPa in the first stage and 3–5 MPa in the second stage) is normally carried out between the evaporation and spray-drying steps to prevent the cream from separating (Ranken et al., 1997). Due to the presence of a high fat content in standardized milk, spray drying is accomplished at a slightly lower air-drying temperature than that used for SMP production to prevent the stickiness of milk powder (Bhandari et al., 2008). A high fat free content on the particle surface makes WMP prone not only to stickiness and caking, but also to fat oxidation and difficulties in agglomeration and rehydration. Lecithin is usually added to the powder to improve its wettability during reconstitution.

1.3.2 High protein milk powders

1.3.2.1 Caseinate powders

The general process in the production of caseinate powders is shown in Figure 1.2. Caseinate powders are produced from either freshly precipitated acid casein curds (obtained by mixing pasteurized skim milk with food-grade acids until the isoelectric point of caseins is reached) or from dry acid caseins via a reaction with an alkali solution (Sarode et al., 2016). Depending on the type of alkali solution used to neutralize refined acid casein curds, which are initially washed several times in water to remove any residual whey, various types of caseinate powders with different functionalities (such as sodium caseinate, calcium caseinate, potassium caseinate and ammonium caseinate powders) are produced. Sodium caseinate is the most widely used caseinate in food processing due to its steric stabilization mechanism in oil-water emulsions, its strong and long-lasting electrostatic combination and its water absorption ability (Badem and Uçar, 2017). In the food industry, sodium caseinate is often used as a protein source, as an ingredient in milk or cream substitutes and as an emulsifier or a humectant in dairy and meat products (DMV, 2012). In terms of popularity, calcium caseinate follows sodium caseinate. It is used in both pharmaceutical preparations and food processing as a supply source for both calcium and protein. Sodium caseinate and calcium caseinate are quite similar in composition, but different in functional properties, as shown in Table 1.6 (DMV, 2012; Sarode et al., 2016).

1.3.2.2 *Milk protein concentrate and milk protein isolate (MPI) powders*

MPC and MPI are other typical types of high protein milk powders, together with caseinate powders. While MPCs refer to milk powders containing less than 90% protein (a dry basic, d.b.), MPIs are milk powders with higher than 90% protein (d.b.) (Meena et al., 2017). Unlike caseinate powders which contain only caseins, proteins in MPCs and MPIs contain both caseins and whey proteins at the same ratio as is naturally found in milk, and remain in an almost undenatured state due to the minimum heat load applied during processing. Thus, they preserve their native functional properties (Patel et al., 2014b). Depending on the processing conditions, especially the degree of ultrafiltration and diafiltration, several types of MPCs with different protein concentrations, ranging from 42 to 85% (d.b.) are produced and they are typically named according to their protein content, e.g. MPC 42, MPC 70, MPC 75, MPC 80 and MPC 85 containing about 42, 70, 75, 80 and 85% protein, respectively, as shown in Table 1.7 (Deeth and Hartanto, 2009; Meena et al., 2017). Due to their high protein content, MPCs have a very broad application in the food industry. MPCs are used as ingredients in and replacements for WMP, SMP and NFMP in food formulation, cheese production and health-related products (Kelly, 2011). Low protein MPCs (42–50% proteins) are often used in cheese, yogurt and soup applications while high protein MPCs (>70% proteins) are typically used in beverages, medical foods, enteral foods and protein bar applications (Patel et al., 2014b).

Table 1.6 Composition and functional properties of sodium caseinate and calcium caseinate powders

Properties	Sodium caseinate	Calcium caseinate
Moisture content (%)	3.8	3.8
Protein (%)	91.4	91.2
Lactose (%)	0.1	0.1
Fat (%)	1.1	1.1
Minerals (%)	3.8	3.6
pH	6.5–7	6.5–7
Molecular conformation in solution	Random coil	Micelle-like aggregates
Appearance of solution	Translucent	Opaque
Mouthfeel solution (2–5%)	Sticky	Milk-like
Dispersibility	Slight tendency for lumps	Good
Solubility	Good	Good
Viscosity (relative)	High	Low
Emulsifying property	Very good	Good
Salt tolerance	Very high	Limited
Heat stability	Very high	High

Table 1.7 Composition of different types of MPC and MPI powders

Milk powders	Composition (%, w/w)				
	Moisture content	Proteins	Fat	Lactose	Ash
MPC 35	3.4–3.6	35.4–35.7	0.5–3.5	49.6–53.0	7.7–8.1
MPC 42	3.5	42.0	1.0	46.0	7.5
MPC 50	3.8–4.5	49.8–49.9	0.53–2.7	35.8–38.0	7.7–7.8
MPC 56	3.8	57.1	1.3	30.1	7.7
MPC 60	3.9–4.0	56.4–60.8	0.51–3.0	19.5–24.0	7.8–8.1
MPC 70	4.2	70.0	1.4	16.2	8.2
MPC 75	5.0	75.0	1.5	10.9	7.6
MPC 80	3.9	80.0	1.8	4.1	7.4
MPC 85	4.9	85.0	1.6	1.0	7.1
MPC 88	5.3	88.0	2.27	0.74	7.05
MPI	<6.0	>89.5	<2.5	<5.0	<8.0

The production of MPCs is well described in the literature (Kelly, 2011; Meena et al., 2017; Patel et al., 2014b; Uluko et al., 2016). Typically, MPCs are obtained from pasteurized skim milk via advanced low heat separating technologies, i.e. membrane filtration processes (ultrafiltration with/without combination with diafiltration), followed by evaporation and spray drying (Figure 1.3). During the ultrafiltration of skim milk, high molecular weight substances such as proteins and residual fat are retained on the membrane and concentrated in the retentate, while low molecular weight substances such as water, lactose, soluble minerals and vitamins pass through the membrane with the permeate. In the production of high protein MPCs, e.g. MPC 80, diafiltration is commonly applied to further remove lactose and soluble minerals to increase the protein concentration (Meena et al., 2017; Patel et al., 2014b). Ultrafiltration alone allows the proteins to be enriched up to 35–42% (d.b.), and with the additional assistance of diafiltration, the obtained protein concentration increases up to 85% (d.b.) (Kelly, 2011). The presence of residual fat and micellar calcium phosphate is a limiting factor for increasing the protein concentration beyond 90%. Therefore, in the production of MPI, microfiltration is performed to remove any residual fat. During filtration, an increase in the retentate viscosity due to the enrichment of the protein concentration causes a rapid decline in the permeate flux through the membrane. Thus, an additional thermal concentration via vacuum evaporation is required to increase the solid concentration in the retentate prior to undergoing spray drying. Controlling the heating conditions during the thermal concentration and spray-drying process is essential to preserve the quality and functionality of the proteins in MPC and MPI powders.

1.3.3 *High fat dairy powders*

High fat dairy powders have a higher than 35% fat content (Varnam and Sutherland, 2001). However, according to the Codex Alimentarius standard (CODEX, 2011), they must contain at least 42% fat. Due to the unique functional properties of milk fat, high fat dairy powders have been used in the manufacture of many products such as ice cream, cheese, infant formula, chocolate, desserts, bakery products, pastry making and confectionery,

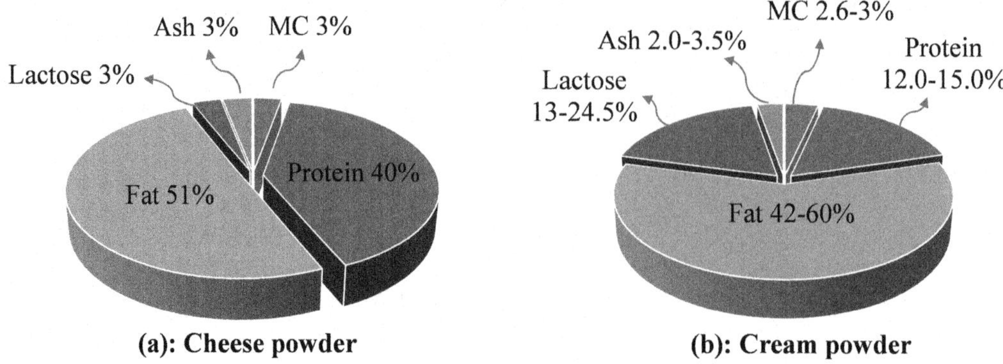

Figure 1.4 The composition of cheese powder (a) and cream powder (b).

dehydrated soups and sauces (Early, 1990; Himmetagaoglu and Erbay, 2019). Typical high fat dairy powders are cream powder and cheese powder. Their proximate compositions are shown in Figure 1.4 (Deeth and Hartanto, 2009; Pısecký, 1997).

1.3.3.1 Cream powders

The industrial process for cream powder production is similar to that for WMP; however, due to the high fat content in cream, the emulsification of cream is required to stabilize the fat. The most commonly used emulsifiers include caseinate, whey proteins, MPC or SMP depending on considerations related to the ability to stabilize fat and flavor (Havea et al., 2009). The high fat content of the cream emulsion also makes spray drying problematic because of the stickiness and caking of dried cream powder. It was reported that the stickiness of cream powder is a function of the air-drying temperature and air relative humidity (Paterson et al., 2007). An increase in the outlet air temperature from 30 to 50°C increases the stickiness of cream powder due to the fat melting, but a further increase to 80°C reduces the stickiness because the effects of a reduction in the surface viscosity on the stickiness are more profound than those of the melting fat. The cream powder contains a relatively high content of amorphous lactose (13–24.5%) (Deeth and Hartanto, 2009). Thus, the stickiness of cream powder in spray drying is also caused by the phase transformation (amorphous to crystalline) of lactose which is greatly affected by the relative humidity of air drying. After spray drying, the surface of the cream powder particles is predominantly fat with approximately 99.0% (Kim et al., 2002). The high surface fat content of spray-dried cream powder makes it prone to fat oxidation during storage (Andersson and Lingnert, 1997, 1998a, 1998b). Thus, oxygen-eliminating packaging of cream powder (e.g. oxygen impermeable bags and nitrogen flushing of bags) is an efficient solution to enhance its shelf life. Also, lecithin may be sprayed on cream powders to improve their wettability.

1.3.3.2 Cheese powders

Aged cheese is often used as a starting material in the production of cheese powder to compensate for the loss of aroma and flavor during spray drying (Havea et al., 2009; Písecký, 2005). Four main steps are involved in cheese powder production (Figure 1.5). In the first stage, aged cheese and ingredients such as emulsifying water, salts, color agents, flavor agents, filling materials (such as SMP, maltodextrin, gelatinized starch) and antioxidants are blended to produce a slurry with 35–40% of the total solid fraction. The mixture is then heated to 75–80°C with strong agitation until a hot molten slurry which is homogeneous in color and consistency and free of lumps or non-hydrated particles is obtained. The hot

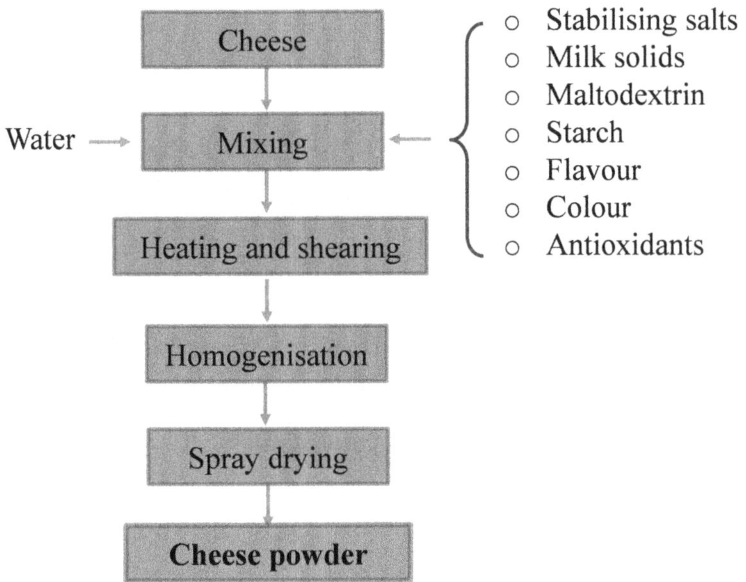

Figure 1.5 The general production procedure for cheese powder.

molten slurry is subsequently homogenized (15 MPa in the first stage and 5 MPa in the second stage) to ensure homogeneity and the absence of free fat. Spray drying of the hot molten slurry is typically performed with an inlet air temperature of 180–200°C and an outlet air temperature of 85–90°C in a two-stage drying system with an integrated fluidized bed allowing the powder to cool to ambient temperature and to agglomerate. In spray drying cheese powder, the outlet air temperature is the most important parameter and it must not exceed 95°C (Fox et al., 2017). A higher outlet air temperature results in a higher rate of browning and Maillard reactions, lower solubility of cheese powder, a higher fat free content and lower acceptability in sensory properties (Erbay et al., 2015; Koca et al., 2015).

1.3.4 Buttermilk powder

Buttermilk powder is produced from spray drying buttermilk, a by-product of butter production. Its main composition is similar to that of SMP (Figure 1.6). During butter churning, a large proportion of fat goes into the butter, leaving the buttermilk with a low fat content. However, buttermilk has a large proportion of fat globule membrane materials which are rich in phospholipids (Deeth and Hartanto, 2009). It was reported that the amount of phospholipids in buttermilk (0.72–0.88 mg/g) is four to seven times higher than that in whole milk (0.12–0.14 mg/g) (Christie et al., 1987; Elling et al., 1996). Buttermilk powder has been used as an ingredient or an additive in the production of confectionery, bakery ice cream, desserts, candy bars and chocolates. The production protocol of buttermilk powder (Figure 1.2) is similar to that of low heat SMP (Farkye, 2006).

1.3.5 Whey-based powders

The most common whey-based powders are composed of whey powders (e.g. sweet whey powder, acid whey powder, demineralized whey powder and delactosed whey powder) and whey protein powders (e.g. WPC and WPI). The approximate compositions and

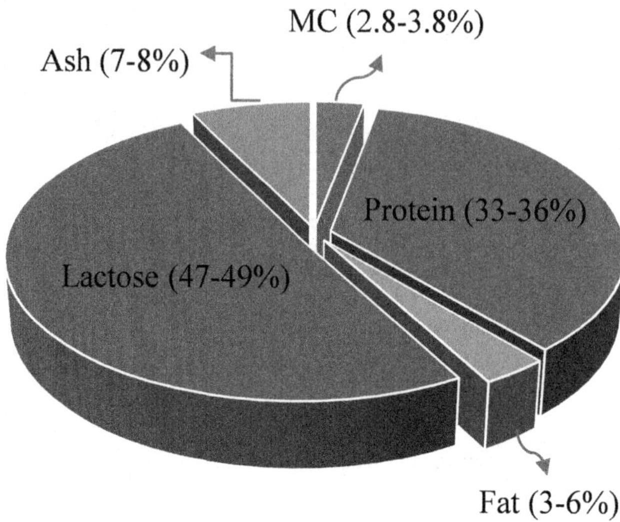

Figure 1.6 The composition of buttermilk powder.

production steps of whey-based powders are shown in Table 1.8 (Deeth and Hartanto, 2009) and Figure 1.3.

1.3.5.1 *Whey powders*

Whey-based powders are produced from liquid whey, a by-product of either cheese manufacturing or casein/caseinate production. Liquid whey contains approximately 93% water. The main components in liquid whey are lactose and proteins which make up about 75 and 11.5% of total whey solids, respectively (Bansal and Bhandari, 2016; Qi and Onwulata, 2011). Depending on the type of agent (acids or rennet) used to precipitate caseins in cheese

Table 1.8 Composition of different types of whey-based powders

Whey-based powders	Composition (%, w/w)				
	MC	Proteins	Fat	Lactose	Ash
		Whey powders			
Acid whey powder	≤3.5	9.0–12.0	0.8	65.0–69.0	11.0–12.0
Sweet whey powder	3.0–6.0	12.0–13.0	0.8–1.5	70.0–73.0	7.5–8.5
Demineralized whey powder	≤3.0	≥11.0	≤1.5	78.0–82.0	≤4.0
Delactosed whey powder	2.0–3.0	18.0–25.0	1.0–4.0	40.0–60.0	11.0–27.0
		Whey protein powders			
WPC 35	4.6	34.0–36.0	2.0–4.0	44.0–53.0	7.0–8.0
WPC 50	4.3	53.0	5.0	35.0	7.0
WPC 65	3.0–4.0	59.0–65.0	4.0–6.0	21.0–22.0	3.5–4.0
WPC 85	4.0–5.0	72.0–81.0	0.3–0.7	2.0–13.0	2.5–6.5
WPI	2.5–6.0	89.0–93.0	0.1–0.7	0.1–0.8	1.4–3.8

and casein/caseinate production, two main types of liquid whey with similar compositions are produced, namely sweet whey and acid whey. Sweet whey is obtained from rennet coagulation of milk at pH 6.0–6.5 while acid whey is a result of acid coagulation of milk at pH 4.5–5.0 (Tunick, 2008). The presence of casein fines and fat in whey has adverse effects on the production of whey-based powders; therefore, they should be immediately separated from whey via cyclones, centrifugal separators, vibrating screens or rotating filters after whey is produced. For short-term storage (less than 8 h), whey must be quickly cooled to 5°C to minimize microorganism growth, while for long-term storage, whey is pasteurized. Depending on the type of whey-based powders (whey powders or whey protein powders), pre-treated whey is concentrated with different membrane-separating processes (Tetra Pak, 2018).

For the production of whey powders, pre-treated whey is often concentrated by reverse osmosis. However, the maximum concentration of whey obtained by reverse osmosis was reported to be about 20% because of the high osmotic pressure created by the concentrated lactose (Pearce, 1992). Thus, reverse osmosis is commonly used in combination with nanofiltration and/or vacuum evaporation to further increase the concentration of whey up to 50–60%. Due to the high content of lactose, which could be the main cause for the stickiness of whey powders during drying, storage and handling, the lactose in whey should be crystallized via rapid cooling and by keeping the concentrated whey at a low temperature before spray drying. Depending on the source of whey (e.g. sweet or acid whey), whey powders can be classified as sweet whey powders or acid whey powders. Regarding delactosed whey powder, a part of the lactose is removed from whey prior to spray drying by either lactose crystallization followed by decanter centrifugation, ultrafiltration, diafiltration or hydrolysis. The high content of minerals in whey can impart a salty taste to whey powders, limiting their applications in food processing. These minerals can be removed from whey using a combination of various unit operations such as nanofiltration, ion exchange and electrodialysis, by which demineralized whey powder is produced.

1.3.5.2 Whey protein concentrate and whey protein isolate powders

The production process of WPC and WPI is quite similar to that of MPC and MPI, respectively. The pre-treated whey is subjected to a combination of various membrane separation processes such as microfiltration, ultrafiltration, nanofiltration, diafiltration and ion exchange to produce whey protein powders with different protein concentrations. Details of the production process for these types of whey-based powders have been well reported by various authors (Bansal and Bhandari, 2016; Boland, 2011; Chegini and Taheri, 2013; Morr and Ha, 1993; Onwulata and Huth, 2009; Zadow, 2012). Due to the nutritional values and functional properties of whey-based powders, they are used in a wide range of food products such as confectionery, bakery, dairy, meat/fish, dietetic foods, infant formula, nutraceuticals and pharmaceuticals (Ramos et al., 2016).

1.3.6 Microencapsulated dairy powders

Microencapsulation is a physical process in which active compounds (e.g. core materials) are covered by a continuous thin coating of wall materials to protect them from the adverse effects of the surrounding environment, to mask their undesired properties and to control their release rate under favored conditions (Barbosa-Cánovas et al., 2005). Among the reported common techniques for the production of microencapsulated food powders, spray drying is emerging as the most common and cheapest technique due to the exclusive characteristics of the spray-drying process. The application of spray drying in the production of microencapsulated food ingredients has been well described by Gharsallaoui et al. (2007). Figure 1.7 illustrates the main steps, together with factors affecting the production of microencapsulated dairy powder via spray drying, in order to control the encapsulation process and consequently to obtain the highest encapsulation efficiency and microcapsule stability (Bhandari et al., 2008). Depending on the properties of the core materials and the final products, the wall materials can be selected from either a single polymer (e.g. maltodextrin, gum arabic, alginate, carrageenan, lactose, sucrose, sodium caseinate, micellar casein, whey protein, soy protein isolate, gelatin and modified starch) or their mixture. Due to the nature of spray drying in which a homogeneous aqueous feed solution is required, an expected wall material should have high solubility in water and it should form a good emulsion with core materials. In many cases, in order to obtain a fine and stable emulsion of core material in a wall solution, an emulsifier is added to the solution of wall and core materials, and a homogenization step is performed. During the spray drying of the obtained emulsion, direct contact of the atomized droplets of emulsion with the high temperature drying air results in rapid water evaporation and fast solidification of the solute. Differences in the drying characteristics of wall materials and core dispersing solutions result in the formation of a coating around the core containing droplets during the falling rate drying period (Anandharamakrishnan, 2017).

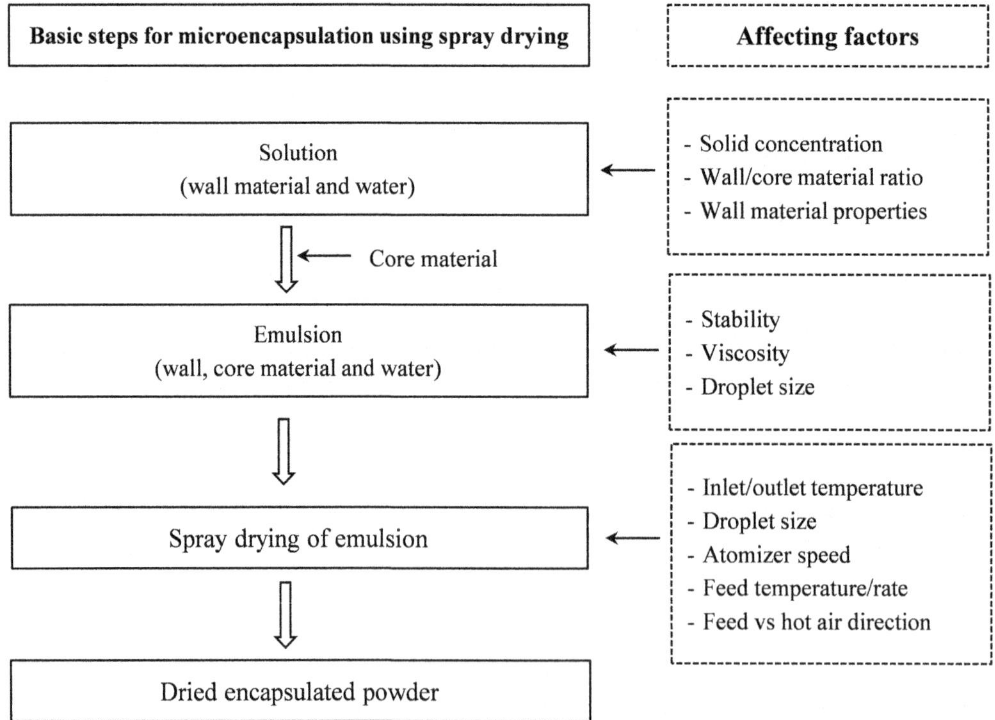

Figure 1.7 A schematic diagram illustrating the basic steps in the production of microencapsulated powder using spray drying and important factors affecting the efficiency of encapsulation. Redrawn with permission from Bhandari et al. (2008).

Many microencapsulated dairy powders in which either the wall materials or the core materials are dairy-based components have been produced by spray drying. In these products, different core materials such as oils, milk fat, volatile compounds and probiotics are encapsulated using milk protein and whey protein powders as coating materials aiming to prevent fat oxidation, to retain and control the release rate of volatile compounds and to improve the viability of microorganisms and their functionality during storage and food processing (Anandharamakrishnan, 2017; Burgain et al., 2015; Gharsallaoui et al., 2007). Some of these are illustrated in Table 1.9.

1.3.7 *Infant milk formula*

Powdered infant milk formula is produced as a substitute for breast milk when it is not available. Thus, its composition, nutritional properties and functions must be similar to

Table 1.9 Some typical applications of spray drying in the production of microencapsulated dairy powders

Core materials	Wall materials	Main results	References
Soybean oil	Sodium caseinate with/without lactose	Encapsulated powders containing up to 85% of oil	Fäldt and Bergenståhl (1995)
Coconut oil		Coconut oil and butterfat encapsulated powders containing up to 75% of oil	
Butterfat			
Rapeseed oil		Almost completely encapsulated	
Anhydrous milk fat (AMF)	Whey protein (WPC & WPI) with lactose	Encapsulated powders containing up to 95% of AMF	Young et al. (1993)
Ethyl butyrate	WPI with/without lactose	Retention levels of 76%	Rosenberg and Sheu (1996)
Ethyl caprylate		Retention levels of 92%	
Soya oil	WPC 75	Encapsulated powders containing up to 75% of oil	Hogan et al. (2001)
Avocado oil	WPI with maltodextrin	Encapsulated powders containing up to 66% of oil	Bae and Lee (2008)
Milk fat	Carbohydrate matrices	Encapsulated powders containing up to 60% of milk fat	Onwulata et al. (1996)
Saccharomyces boulardii	WPI (pH = 4)	High survival rate with ~1.0 log reduction	Diep et al. (2014)
Lactobacillus rhamnosus GG	Micellar casein powder and whey protein powder	Excellent survival rate with <0.5 log reduction	Guerin et al. (2017)
Lactobacillus acidophilus La-5	SMP and sweet whey powder	High survival rate with ~2.0 log reduction and viability almost unchanged during 90-day storage at 4–25°C	Maciel et al. (2014)
Saccharomyces cerevisiae var. *boulardii*	WPC	Survival rate of 91.81%	Arslan et al. (2015)

those of breast milk to provide the same growth and development factors of an exclusively breastfed infant (Blanchard et al., 2013). Generally, powdered infant milk formula is produced via either a "dry mix" or a "wet mix" process, or a combination of both (Schuck, 2013). In the dry mix method, all the ingredients received from suppliers are in powder form and they are simply mixed together in large batches to obtain a uniform mixture. In some cases, a mixture of various ingredients is prepared using the wet mix process to produce a base powder to which the remaining dry ingredients are added (Blanchard et al., 2013). Oversized particles can be removed from the mixture by sifting before inner gas packaging (Jiang and Guo, 2014). Regarding the wet mix processing method, although more complicated and costly than the dry mix method, it allows control over all aspects of the product quality. Thus, it is currently the most widely used method to produce powdered infant milk formula. A typical production line of powdered infant milk formula using the wet mix method is shown in Figure 1.8, which includes three main steps, namely preparation of the mix, evaporation and spray drying. Details of these steps have been well described by Montagne et al. (2009), Guo and Ahmad (2014), Blanchard et al. (2013) and Jiang and Guo (2014). The main advantages and disadvantages of the dry and wet mix process are presented in Table 1.10 (Montagne et al., 2009; Blanchard et al., 2013).

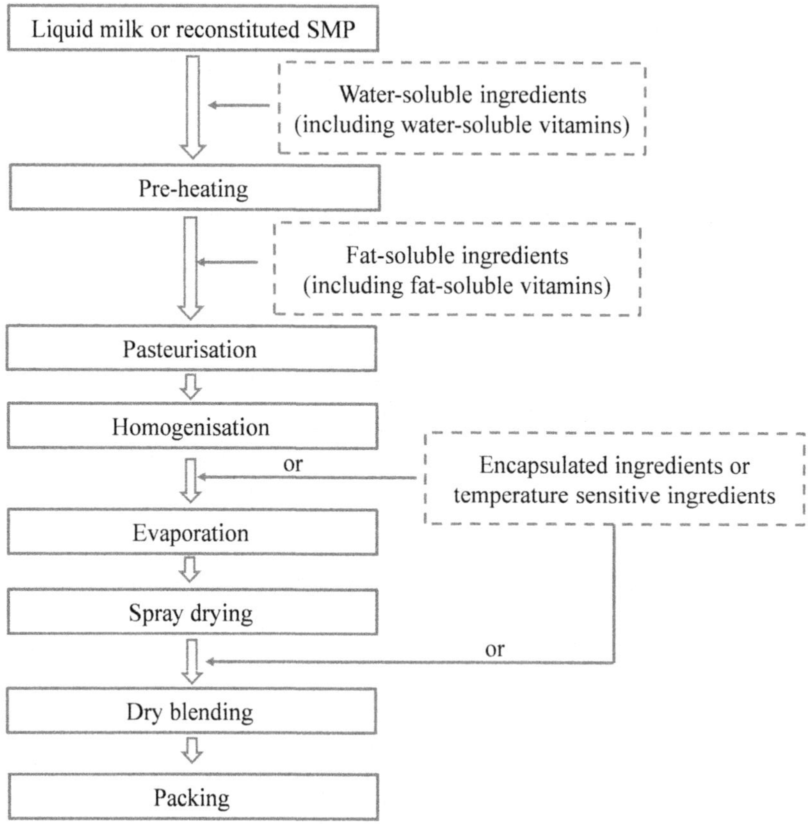

Figure 1.8 A schematic diagram illustrating the basic steps in the production of powdered infant milk formula using the wet mix method. Redrawn with permission from Jiang and Guo (2014).

Table 1.10 Main advantages and disadvantages of two methods to produce infant milk formula powder

	Dry mix method	Wet mix method
Advantages	• Small investment cost • Low energy consumption • Safer from microbiological hazards due to no water involvement	• Ease to incorporate hydrophobic compounds (e.g. oils) • Ease to control all aspects of product quality (microbiology, physical and chemical properties)
Disadvantages	• Impossible to incorporate oils • High dependence of powder properties (physical properties – wettability and solubility, and microbiological quality) on the quality of the raw materials • Possibility of post-process contamination • Inhomogeneity in appearance and composition of the product due to different density ingredients	• High investment and production cost

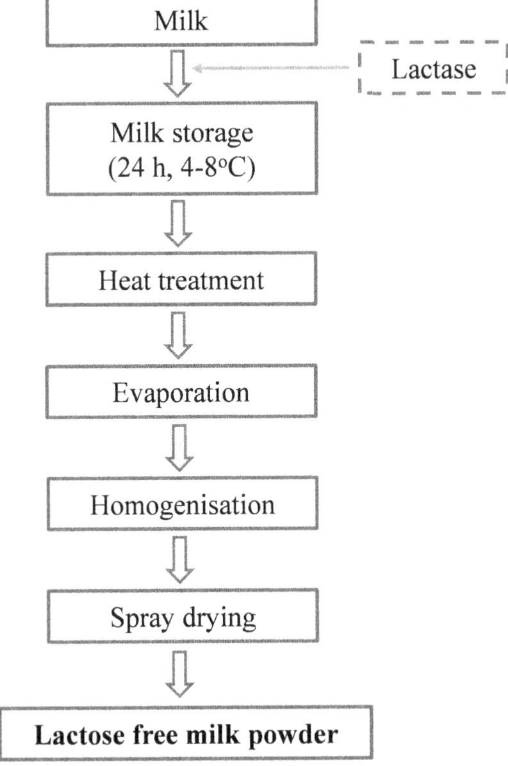

Figure 1.9 A schematic diagram illustrating the basic steps in the production of lactose-free milk powder. Redrawn with permission from Dekker et al. (2019).

1.3.8 Lactose hydrolyzed milk powder

Lactose hydrolyzed milk powder, also known as lactose-free milk powder, is produced using enzyme β-galactosidase (lactase) to hydrolyze the lactose in milk into monosaccharides (glucose and galactose) before the milk is subjected to pasteurization, homogenization, evaporation and spray drying (Dekker et al., 2019). In comparison to regular milk powders, the current market for lactose-free milk powder is still very small due to technical problems regarding its production and storage (Torres et al., 2017). The production of lactose-free milk powder is quite similar to that of whole milk powder (Figure 1.9). However, the formation of low glass transition temperature (T_g) compounds (e.g. glucose with a T_g of ≈31°C and galactose with a T_g of ≈30°C) during the hydrolysis of lactose (T_g of ≈101°C) leads to a drop in the T_g of lactose hydrolyzed milk, resulting in a series of problems during the spray drying and storage of lactose-free milk powder such as agglomeration, stickiness, caking, elevated hygroscopicity and browning (Fialho et al., 2018).

1.4 Conclusion

In this chapter, dried dairy products produced by spray drying are classified and briefly introduced with an emphasis on their composition and manufacturing process. With recent advances in spray-drying technologies, especially a better understanding of the stickiness mechanisms of dairy powders (a major problem in the spray drying of dairy powders), the application of spray drying in the production of dairy powders has greatly expanded. In terms of powder productions, spray drying has been utilized in the manufacture of conventional milk powders such as non-fat milk powder, SMP and WMP; high protein powders such as caseinate powder, MPCs and MPI, whey powders, WPCs and WPI; high fat powders such as cream powder, cheese powder and buttermilk powder; and encapsulated dairy powders. However, spray drying still has several challenges inspiring further research work. Most of them are related to energy consumption and product quality. Currently, the high drying temperature of spray drying prevents its application in the dehydration of other materials which contain high levels of heat-sensitive bioactive compounds, such as camel milk.

References

Anandharamakrishnan, C. (2015). *"Spray drying techniques for food ingredient encapsulation"*. John Wiley & Sons, Chichester, UK.

Anandharamakrishnan, C. (2017). *"Handbook of drying for dairy products"*. John Wiley & Sons, Chichester, UK.

Anandharamakrishnan, C., and Ishwarya, S. P. (2015). Introduction to spray drying. *In "Spray drying techniques for food ingredient encapsulation"* (C. Anandharamakrishnan and S. P. Ishwarya, eds.). John Wiley & Sons, Ltd., Chichester, UK.

Andersson, K., and Lingnert, H. (1997). Influence of oxygen concentration on the storage stability of cream powder. *LWT-Food Science and Technology* **30**(2), 147–154.

Andersson, K., and Lingnert, H. (1998a). Influence of oxygen concentration and light on the oxidative stability of cream powder. *LWT-Food Science and Technology* **31**(2), 169–176.

Andersson, K., and Lingnert, H. (1998b). Influence of oxygen concentration on the flavour and chemical stability of cream powder. *LWT-Food Science and Technology* **31**(3), 245–251.

Arslan, S., Erbas, M., Tontul, I., and Topuz, A. (2015). Microencapsulation of probiotic *Saccharomyces cerevisiae* var. *boulardii* with different wall materials by spray drying. *LWT-Food Science and Technology* **63**(1), 685–690.

Badem, A., and Uçar, G. (2017). Production of caseins and their usages. *International Journal of Food Science and Nutrition* **2**, 4–9.

Bae, E., and Lee, S. (2008). Microencapsulation of avocado oil by spray drying using whey protein and maltodextrin. *Journal of Microencapsulation* **25**(8), 549–560.

Bansal, N., and Bhandari, B. (2016). Functional milk proteins: Production and utilization – Whey-based ingredients. *In "Advanced dairy chemistry"*, pp. 67–98. Springer, New York.

Barbosa-Cánovas, G. V., Ortega-Rivas, E., Juliano, P., and Yan, H. (2005). *"Encapsulation processes"*. Springer, New York.

Bhandari, B., and Howes, T. (2005). Relating the stickiness property of foods undergoing drying and dried products to their surface energetics. *Drying Technology* **23**(4), 781–797.

Bhandari, B. R., Datta, N., and Howes, T. (1997). Problems associated with spray drying of sugar-rich foods. *Drying Technology* **15**(2), 671–684.

Bhandari, B. R., Patel, K. C., and Chen, X. D. (2008). Spray drying of food materials – process and product characteristics. *Drying Technologies in Food Processing* **4**, 113–157.

Blanchard, E., Zhu, P., and Schuck, P. (2013). Infant formula powders. *In "Handbook of food powders"* (B. Bhandari, N. Bansal, M. Zhang and P. Schuck, eds.), pp. 465–483. Woodhead Publishing, Cambridge.

Broadhead, J., Edmond Rouan, S. K., and Rhodes, C. T. (1992). The spray drying of pharmaceuticals. *Drug Development and Industrial Pharmacy* **18**(11–12), 1169–1206.

Boland, M. (2011). 3 – Whey proteins. *In "Handbook of food proteins"* (G. O. Phillips and P. A. Williams, eds.), pp. 30–55. Woodhead Publishing, Cambridge.

Boonyai, P., Bhandari, B., and Howes, T. (2004). Stickiness measurement techniques for food powders: A review. *Powder Technology* **145**(1), 34–46.

Burgain, J., Corgneau, M., Scher, J., and Gaiani, C. (2015). Encapsulation of probiotics in milk protein microcapsules. *In "Microencapsulation and microspheres for food applications"* (M. C. S. Leonard, ed.), pp. 391–406. Elsevier, San Diego.

Cao, X. Q., Vassen, R., Schwartz, S., Jungen, W., Tietz, F., and Stöever, D. (2000). Spray-drying of ceramics for plasma-spray coating. *Journal of the European Ceramic Society* **20**(14–15), 2433–2439.

Carić, M. (1994). *"Concentrated and dried dairy products"*. VCH Publishers Inc., New York.

Chegini, G., and Taheri, M. (2013). Whey powder: Process technology and physical properties: A review. *Middle-East Journal of Scientific Research* **13**, 1377–1387.

Christie, W., Noble, R., and Davies, G. (1987). Phospholipids in milk and dairy products. *International Journal of Dairy Technology* **40**(1), 10–12.

Codex, Alimentarius. (2011). *"Milk and milk products"*. 2nd edition. Joint FAO/WHO Food Standards Programme, Rome.

de Souza, V. B., Thomazini, M., de Carvalho Balieiro, J. C., and Fávaro-Trindade, C. S. (2015). Effect of spray drying on the physicochemical properties and color stability of the powdered pigment obtained from vinification byproducts of the Bordo grape (*Vitis labrusca*). *Food and Bioproducts Processing* **93**, 39–50.

Deeth, H., and Hartanto, J. (2009). Chemistry of milk-role of constituents in evaporation and drying. *In "Dairy powders and concentrated products"* (A. Y. Tamime, ed.), pp. 1–27. Blackwell Publishing, Chichester, UK.

Dekker, P. J., Koenders, D., and Bruins, M. J. (2019). Lactose-free dairy products: Market developments, production, nutrition and health benefits. *Nutrients* **11**(3), 551.

Diep, D. T. N., George, P., Gorczyca, E., and Kasapis, S. (2014). Studies on the viability of *Saccharomyces boulardii* within microcapsules in relation to the thermomechanical properties of whey protein. *Food Hydrocolloids* **42**, 232–238.

DMV (2012). Caseinate – The incredible milk protein. Vol. 2018. *FrieslandCampina DMV* https://www.foodingredientsfirst.com/technical-papers/caseinates-the-incredible-milk-protein.html.

Downton, G. E., Flores-Luna, J. L., and King, C. J. (1982). Mechanism of stickiness in hygroscopic, amorphous powders. *Industrial and Engineering Chemistry Fundamentals* **21**(4), 447–451.

Early, R. (1990). The use of high-fat and specialized milk powders. *International Journal of Dairy Technology* **43**(2), 53–56.

Elling, J., Duncan, S., Keenan, T., Eigel, W., and Boling, J. (1996). Composition and microscopy of reformulated creams from reduced-cholesterol butteroil. *Journal of Food Science* **61**(1), 48–53.

Erbay, Z., Koca, N., Kaymak-Ertekin, F., and Ucuncu, M. (2015). Optimization of spray drying process in cheese powder production. *Food and Bioproducts Processing* **93**, 156–165.

Fäldt, P., and Bergenståhl, B. (1995). Fat encapsulation in spray-dried food powders. *Journal of the American Oil Chemists' Society* **72**(2), 171–176.

Fäldt, P., and Bergenståhl, B. (1996). Changes in surface composition of spray-dried food powders due to lactose crystallization. *LWT-Food Science and Technology* **29**(5–6), 438–446.

Farkye, N. (2006). Significance of milk fat in milk powder. *In "Advanced dairy chemistry Volume 2: Lipids"* (P. F. Fox and P. L. H. McSweeney, eds.), pp. 451–465. Springer, New York.

Fialho, T. L., Martins, E., Silva, C. R. D. J., Stephani, R., Tavares, G. M., Silveira, A. C. P., Perrone, Í. T., Schuck, P., de Oliveira, L. F. C., and de Carvalho, A. F. (2018). Lactose-hydrolyzed milk powder: Physicochemical and technofunctional characterization. *Drying Technology* **36**(14), 1688–1695.

Filkova, I., and Mujumdar, A. S. (1995). Industrial spray drying systems. *In "Handbook of industrial drying"*, Vol. 1, 3rd Edition (A. S. Mujumdar, ed.), pp. 263–308. CRC Press, Boca Raton, FL.

Foster, K. (2002). The prediction of sticking in dairy powders: A thesis for the degree of Doctor of Philosophy in Bioprocess Engineering at Massey University. Massey University.

Fox, P. F., Guinee, T. P., Cogan, T. M., and McSweeney, P. L. H. (2017). Cheese as an ingredient. *In "Fundamentals of cheese science"* (P. F. Fox, T. P. Guinee, T. M. Cogan and P. L. H. McSweeney, eds.), pp. 629–679. Springer US, Boston, MA.

Gharsallaoui, A., Roudaut, G., Chambin, O., Voilley, A., and Saurel, R. (2007). Applications of spray-drying in microencapsulation of food ingredients: An overview. *Food Research International* **40**(9), 1107–1121.

Guerin, J., Petit, J., Burgain, J., Borges, F., Bhandari, B., Perroud, C., Desobry, S., Scher, J., and Gaiani, C. (2017). *Lactobacillus rhamnosus* GG encapsulation by spray-drying: Milk proteins clotting control to produce innovative matrices. *Journal of Food Engineering* **193**, 10–19.

Guo, M., and Ahmad, S. (2014). Formulation guidelines for infant formula. *In "Human milk biochemistry and infant formula manufacturing technology"* (M. Guo, ed.), pp. 141–171. Woodhead Publishing, Cambridge, UK.

Havea, P., Baldwin, A. J., and Carr, A. J. (2009). Specialised and novel powders. *In "Dairy powders and concentrated products"* (A. Y. Tamime, ed.), pp. 268–293. Blackwell Publishing Ltd., Chichester, UK.

Himmetagaoglu, A. B., and Erbay, Z. (2019). Effects of spray drying process conditions on the quality properties of microencapsulated cream powder. *International Dairy Journal* **88**, 60–70.

Ho, T. M., Truong, T., and Bhandari, B. (2017). Spray-drying and non-equilibrium states/glass transition. *In "Non-equilibrium states and glass transitions in foods"* (B. Bhandari and Y. H. Roos, eds.), pp. 111–136. Woodhead Publishing, Duxford, UK.

Hogan, S., McNamee, B., O'Riordan, E., and O'Sullivan, M. (2001). Microencapsulating properties of whey protein concentrate 75. *Journal of Food Science* **66**(5), 675–680.

Jiang, Y. J., and Guo, M. (2014). Processing technology for infant formula. *In "Human milk biochemistry and infant formula manufacturing technology"* (M. Guo, ed.), pp. 211–229. Woodhead Publishing, Cambridge, UK.

Kelly, P. (2011). Milk protein products. *In "Encyclopedia of dairy sciences"*, 2nd Edition (J. W. Fuquay, ed.), pp. 848–854. Academic Press, San Diego, CA.

Kim, E. H. J., Chen, X. D., and Pearce, D. (2002). Surface characterization of four industrial spray-dried dairy powders in relation to chemical composition, structure and wetting property. *Colloids and Surfaces, Part B: Biointerfaces* **26**(3), 197–212.

Koca, N., Erbay, Z., and Kaymak-Ertekin, F. (2015). Effects of spray-drying conditions on the chemical, physical, and sensory properties of cheese powder. *Journal of Dairy Science* **98**(5), 2934–2943.

Maciel, G., Chaves, K., Grosso, C., and Gigante, M. (2014). Microencapsulation of *Lactobacillus acidophilus* La-5 by spray-drying using sweet whey and skim milk as encapsulating materials. *Journal of Dairy Science* **97**(4), 1991–1998.

Masters, K. (1985). *"Spray drying handbook"*, 4th Edition. John Wiley & Sons, New York.

Meena, G. S., Singh, A. K., Panjagari, N. R., and Arora, S. (2017). Milk protein concentrates: Opportunities and challenges. *Journal of Food Science and Technology* **54**(10), 3010–3024.

Montagne, D. H., van Dael, P., Skanderby, M., and Hugelshofer, W. (2009). Infant formulae-powders and liquids. *In "Dairy powders and concentrated products"* (A. Y. Tamine, ed.), pp. 294–331. Wiley, Chichester, UK.

Morr, C., and Ha, E. (1993). Whey protein concentrates and isolates: Processing and functional properties. *Critical Reviews in Food Science and Nutrition* **33**(6), 431–476.

O'Callaghan, D., and Hogan, S. (2013). The physical nature of stickiness in the spray drying of dairy products – a review. *Dairy Science and Technology* **93**(4–5), 331–346.

Onwulata, C., Smith, P., Cooke, P., and Holsinger, V. (1996). Particle structures of encapsulated milkfat powders. *Lebensmittel-Wissenschaft & Technologie* **29**(1–2), 163–172.

Onwulata, C. I., and Huth, P. J. (2009). *"Whey processing, functionality and health benefits"*. John Wiley & Sons, Ames, Iowa.

Papadakis, S. E., and Bahu, R. E. (1992). The sticky issues of drying. *Drying Technology* **10**(4), 817–837.

Patel, B., Patel, J. K., and Chakraborty, S. (2014a). Review of patents and application of spray drying in pharmaceutical, food and flavor industry. *Recent Patents on Drug Delivery and Formulation* **8**(1), 63–78.

Patel, H., Patel, S., and Agarwal, S. (2014b). *"Milk protein concentrates: Manufacturing and applications – Technical report"*, Vol. 2018. US Dairy Export Council, https://www.usdairy.com/~/media/usd/public/mpc-tech-report-final.pdf.

Paterson, A. H., Zuo, J. Y., Bronlund, J. E., and Chatterjee, R. (2007). Stickiness curves of high fat dairy powders using the particle gun. *International Dairy Journal* **17**(8), 998–1005.

Pearce, R. J. (1992). Whey processing. *In "Whey and lactose processing"* (J. G. Zadow, ed.), pp. 73–89. Springer Netherlands, Dordrecht.

Pisecky, J. (1985). Technological advances in the production of spray dried milk. *International Journal of Dairy Technology* **38**(2), 60–64.

Pısecký, J. (1997). *"Handbook of milk powder manufacture"*. Niro A/S, Copenhagen.

Písecký, J. (2005). Spray drying in the cheese industry. *International Dairy Journal* **15**(6–9), 531–536.

Qi, P. X., and Onwulata, C. I. (2011). Physical properties, molecular structures, and protein quality of texturized whey protein isolate: Effect of extrusion moisture content. *Journal of Dairy Science* **94**(5), 2231–2244.

Ramos, O. L., Pereira, R. N., Rodrigues, R. M., Teixeira, J. A., Vicente, A. A., and Malcata, F. X. (2016). Whey and whey powders: Production and uses. *In "Encyclopedia of food and health"* (B. Caballero, P. M. Finglas and F. Toldrá, eds.), pp. 498–505. Academic Press, Oxford.

Ranken, M. D., Kill, R. C., and Baker, C. (1997). Dairy products. *In "Food industries manual"* (M. D. Ranken, R. C. Kill and C. Baker, eds.), pp. 75–138. Springer US, Boston, MA.

Refstrup, E. (1995). Advances in spray drying of food products. *International Journal of Dairy Technology* **48**(2), 50–54.

Roos, Y. H. (2002). Importance of glass transition and water activity to spray drying and stability of dairy powders. *Le Lait* **82**(4), 475–484.

Roos, Y. R., and Karel, M. A. (1992). Crystallization of amorphous lactose. *Journal of Food Science* **57**(3), 775–777.

Rosenberg, M., and Sheu, T. (1996). Microencapsulation of volatiles by spray-drying in whey protein-based wall systems. *International Dairy Journal* **6**(3), 273–284.

Roy, I., Bhushani, A., and Anandharamakrishnan, C. (2017). Techniques for the preconcentration of milk. *In "Handbook of drying for dairy products"* (C. Anandharamakrishnan, ed.), pp. 23–42. John Wiley & Sons, Ltd., Chichester, UK.

Sarode, A., Sawale, P., Khedkar, C., Kalyankar, S., and Pawshe, R. (2016). Casein and caseinate: Methods of manufacture. *In "Encyclopedia of food and health"*, Vol. 1 (B. Caballero, P. Finglas and F. Toldra, eds.), pp. 676–682. Academic Press Elsevier Ltd., Oxford.

Schuck, P. (2002). Spray drying of dairy products: State of the art. *Le Lait* **82**(4), 375–382.

Schuck, P. (2013). Dairy powders. *In "Handbook of food powders"* (B. Bhandari, N. Bansal, M. Zhang and P. Schuck, eds.), pp. 437–464. Woodhead Publishing, Cambridge, UK.

Tetra Pak (2018). *"Dairy processing handbook"*, Vol. 2018. Tetra Pak, https://dairyprocessinghandbook.com.

Torres, J. K. F., Stephani, R., Tavares, G. M., De Carvalho, A. F., Costa, R. G. B., de Almeida, C. E. R., Almeida, M. R., de Oliveira, L. F. C., Schuck, P., and Perrone, Í. T. (2017). Technological aspects of lactose-hydrolyzed milk powder. *Food Research International* **101**, 45–53.

Tunick, M. H. (2008). Whey protein production and utilization: A brief history. *In "Whey processing, functionality and health benefits"* (C. I. Onwulata and P. J. Huth, eds.), pp. 1–13. John Wiley & Sons, Ames, Iowa.

Uluko, H., Liu, L., Lv, J.-P., and Zhang, S.-W. (2016). Functional characteristics of milk protein concentrates and their modification. *Critical Reviews in Food Science and Nutrition* **56**(7), 1193–1208.

Varnam, A., and Sutherland, J. P. (2001). *"Milk and milk products: Technology, chemistry and microbiology"*. Springer Science & Business Media, Surrey.

Wang, S., and Langrish, T. (2007). Measurements of the crystallization rates of amorphous sucrose and lactose powders from spray drying. *International Journal of Food Engineering* **3**(4), 1–17.

Woo, M. (2017). Recent advances in the drying of dairy products. *In "Handbook of drying for dairy products"* (C. Anandharamakrishnan, ed.), pp. 249–268. John Wiley & Sons, Chichester, UK.

Woo, M., and Bhandari, B. (2013). Spray drying for food powder production. *In "Handbook of food powders"* (B. Bhandari, N. Bansal, M. Zang and P. Shuck, eds.), pp. 29–56. Elsevier, Cambridge.

Young, S. L., Sarda, X., and Rosenberg, M. (1993). Microencapsulating properties of whey proteins. 1. Microencapsulation of anhydrous milk fat. *Journal of Dairy Science* **76**(10), 2868–2877.

Zadow, J. (2012). *"Whey and lactose processing"*. Springer Science & Business Media, New York.

chapter 2

Technology, modeling and control of the processing steps

Mathieu Persico, Laurent Bazinet, Gaëlle Tanguy, Pierre Schuck, Christelle Lopez, Meng Wai Woo, Cordelia Selomulya, Xiao Dong Chen, Mathieu Person, Romain Jeantet, Cécile Le Floch-Fouéré, Bernard Cuq and Nima Yazdanpanah

Contents

2.1 *Improvement of whey products spray-drying by use of membranes processes*

Mathieu Persico and Laurent Bazinet

2.1.1 *Introduction*

Whey is the serum that remains after milk has coagulated and casein has been removed. About 200 million tons of whey is produced worldwide each year and production is constantly increasing[1]. Whey represents around 90% of milk volume, contains about 20% of its proteins and almost all of its lactose[2,3]. Hence, it is a by-product rich in proteins, lactose and minerals; however, depending on its manufacturing process, the composition of whey can vary between sweet, salty and acid (Figure 2.1.1 and Table 2.1.1)[4,5]. Sweet whey is usually generated during the manufacture of cheddar and other dry-salted cheeses and is further processed into whey powder, protein concentrate and lactose using a drying step,[6] typically realized by spray drying. Salty whey accounts for 2–5% of the total whey generated from dry-salted cheeses. Acid whey is generated during the manufacture of acid-coagulated dairy products such as Greek yogurts. The past few years have seen a tremendous increase in the popularity of Greek-style yogurts giving rise to a constant increase in the volume of acid whey produced[7]

Spray-drying of sweet whey produces good quality powders, but spray-drying of acid and salty whey produces a powder that is significantly sticky with major operational problems caused by large amounts of calcium and/or lactic acid. The presence of these

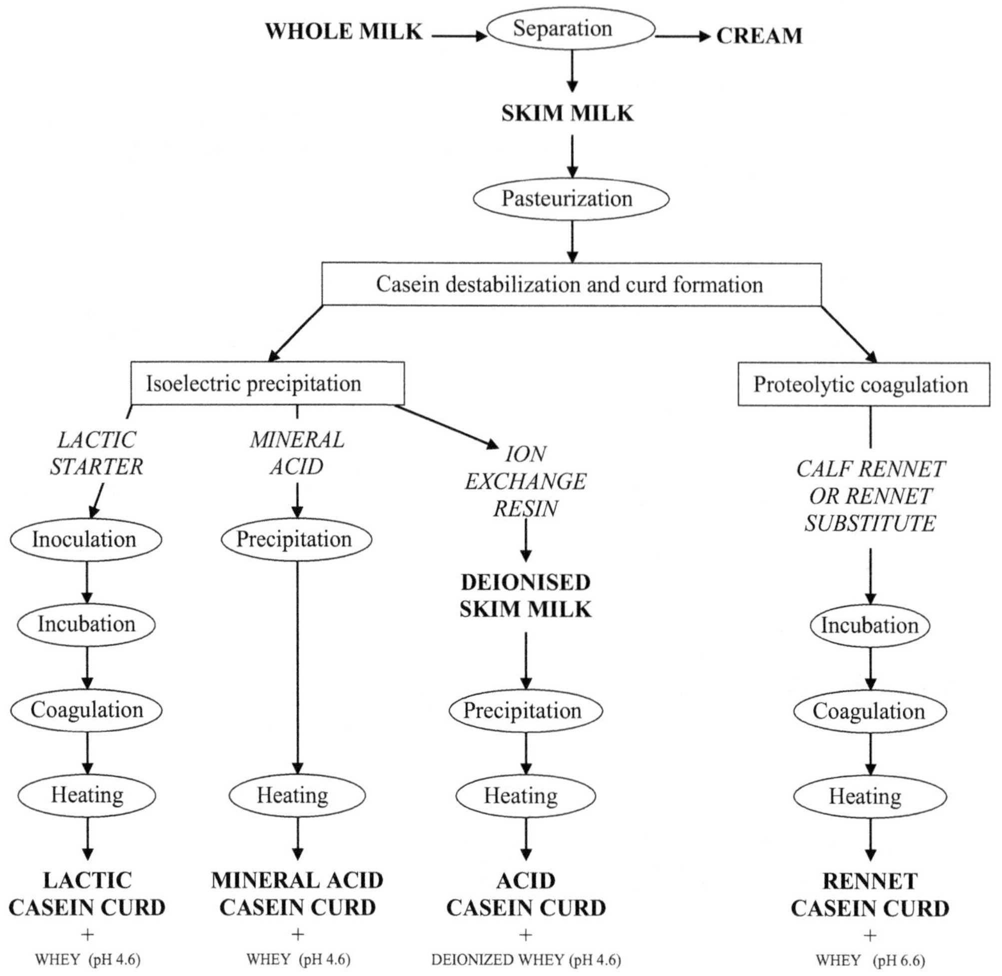

Figure 2.1.1 Processes leading to whey production. (Adapted from [11])

Table 2.1.1 Averaged compositional analysis of four wheys[5]

Component	Native whey	Sweet whey	Acid whey	Salty whey
Total solid (%)	6.95	6.24	5.55	21.9
Ash (%)	0.77	0.65	0.73	14.8
Protein (%)	0.86	1.04	0.24	0.81
Lactose (%)	3.06	3.06	3.16	2.82
Lactic acid (%)	0.2	0.25	0.55	0.28
pH	4.61	5.38	4.52	5.31
Na^+ (%)	0.014	0.31	0.021	5.5
K^+ (%)	0.046	0.12	0.133	0.13
Ca^{2+} (%)	0.115	0.059	0.132	0.072
Mg^{2+} (%)	0.015	0.017	0.017	0.016

components has a strong influence on the hygroscopic character of the powder, lowering the lactose glass transition temperature[8,9]. Indeed, it has been shown that high amounts of calcium and/or lactic acid affect the phase transition of lactose which appears in its amorphous form. This amorphous form prevents lactose crystallization leading to stickiness and caking during spray-drying with greater energy required to remove water from its hydration shell[9]. In these technological and sustainable contexts, pretreatments such as demineralization and deacidification before spray-drying whey or whey products are necessary to improve their quality and extend their storage. In the dairy industry, whey is demineralized using electrodialysis (ED), nanofiltration (NF) or ion-exchange resins. Ion-exchange resins or ED are technologies that allow salt and acid removal. For over 30 years, both techniques have been investigated for their economic and environmental purposes[10]. This chapter aims to review the main membrane techniques and studies related to the treatment of whey and whey products prior to spray-drying. The principles of membrane processes will be presented, followed by techniques for demineralization alone, demineralization and delactosation as well as demineralization and deacidification of whey and whey products.

2.1.2 Membrane processes and their principles

Well-established in the processing of dairy products, membrane technologies tend to reduce the cost of production and are an economical alternative to many important dairy processes such as centrifugation, evaporation or demineralization of whey[12]. There are two main types of membrane processes: pressure-driven processes and electrically driven processes.

2.1.2.1 Pressure-driven membrane processes

Pressure-driven processes are mainly used to fractionate and concentrate dairy liquids such as milk and whey. In pressure-driven membrane processes, the pressure gradient is the driving force for the passage of a liquid through a membrane. These separation techniques are based on the size exclusion of the soluble components in order to obtain two liquid fractions: the retentate and the permeate. There are four types of pressure-driven membrane processes classified according to their cut-off: microfiltration (MF), ultrafiltration (UF), nanofiltration and reverse osmosis (RO) (Figure 2.1.2).

MF is mainly used for its ability to retain microorganisms, casein micelles and fat globules. UF has been proven to be a suitable technology for protein and lactose removal in some cases. For demineralization and deionization, NF has a promising larger field of applications due to its intermediate selectivity (300–1,000 Da) between UF and RO[14]. Moreover, the selectivity of NF seems to be mainly attributed to a combination of size exclusion, Donnan exclusion, dielectric and transport effects occurring at the membrane surface and within the nanopores of its matrix[15]. The size exclusion mechanism influences the transport of neutral solutes. The Donnan exclusion describes the electrostatic attraction or repulsion between charged species and the interface of the charged membrane. Indeed, NF membranes contain ionizable groups on their surface and within their matrix[16,17] which can be dissociated according to the pH of the contacting solution.

2.1.2.2 Electrically driven membrane processes

There are two main industrial electrically driven membrane processes: conventional electrodialysis and bipolar membrane electrodialysis.

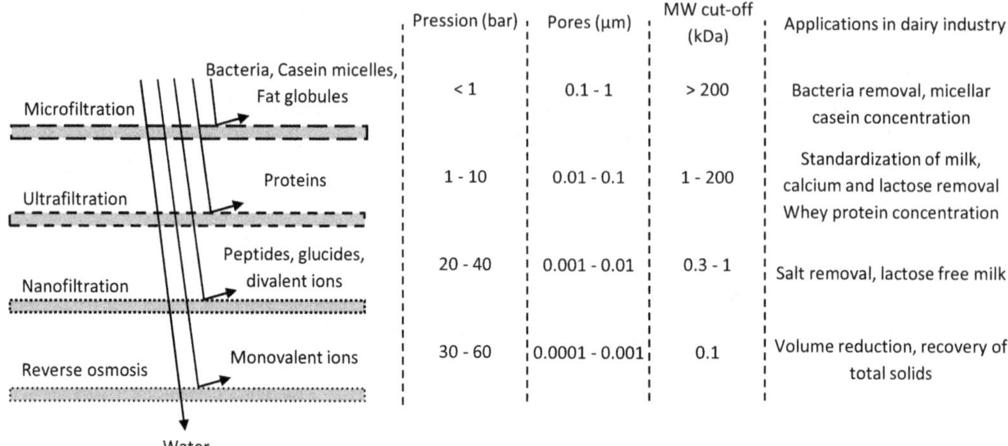

Figure 2.1.2 Filtration spectrum used for milk fractionation. (Adapted from [12,13])

Electrodialysis is an electrochemical process based on the migration of charged molecules through ion-exchange membranes (IEM). The driving force is generated by the application of a potential difference between an anode and a cathode. In conventional ED, anion-exchange membranes (AEM) and cation-exchange membranes (CEM) are arranged in an alternating pattern of up to 300 repeated units (Figure 2.1.3). The main commercial AEM and CEM membranes hold quaternary ammonium and sulfonic exchange groups, respectively, as functional groups[18], allowing the transport of counterions and blocking co-ions based on the Donnan exclusion. Therefore, cations are able to pass through CEM but not through AEM, whereas anions can migrate through AEM but not CEM.

Bipolar membranes electrodialysis (BMED) uses bipolar and conventional ion-exchange membranes. A bipolar membrane is composed of three parts: an anion-exchange layer, a cation-exchange layer and a hydrophilic transition layer at their junction[19]. The diffusion of water molecules from both sides of the bipolar membrane leads to their dissociation into protons and hydroxyl ions when a minimum potential is applied (around 1.8 V), which further migrate from the junction layer through the cation-exchange and anion-exchange layers, respectively (Figure 2.1.4). For a solution circulated in an electrodialysis cell on the cationic side of the bipolar membrane, where the H^+ are generated (Figure 2.1.4), the pH of the solution will decrease. Similarly, a solution circulated on the anionic side of the bipolar membrane, where OH^- are generated (Figure 2.1.4), will experience an increase in pH. However, when there is a depletion of ions in the diffusion boundary layers (DBL) of conventional AEMs and CEMs, the electric current transport is ensured by protons and hydroxyl ions at the membrane interfaces generated by water splitting (Figure 2.1.5).

Both ED and BMED are applied on an industrial scale with the application of a continuous or constant current (CC current) all along the processes. However, a new and promising current mode, called pulsed electric field (PEF), is under study and has been applied at the laboratory scale to ED and very recently to BMED. To generate a pulsed mode, a hashed current is applied: a constant current during a fixed time (pulse) followed by a pause lapse (pause) (Figure 2.1.6). The use of a PEF during ED was first proposed by Karlin et al.[20] on a model salt solution and is based on controlling the concentration polarization phenomenon at the interface of the IEMs. It was proven that desalination can be

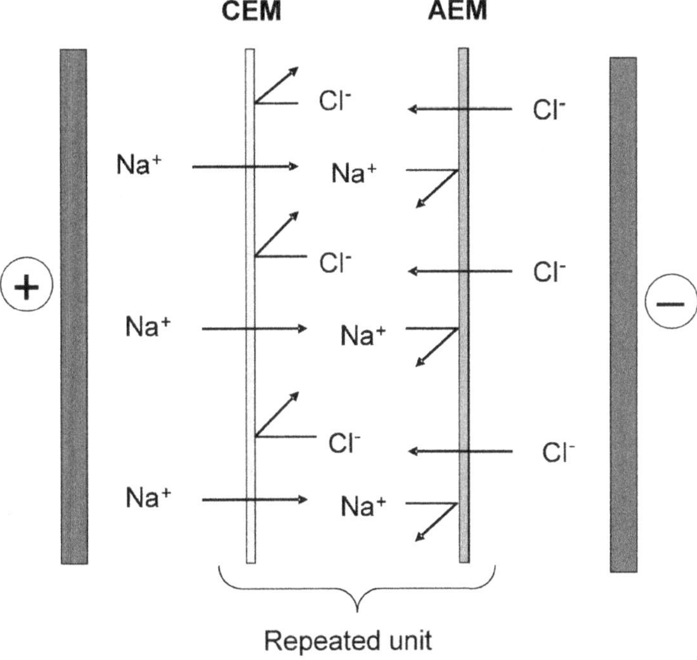

Figure 2.1.3 Principle of electrodialysis and migration of salt ions.

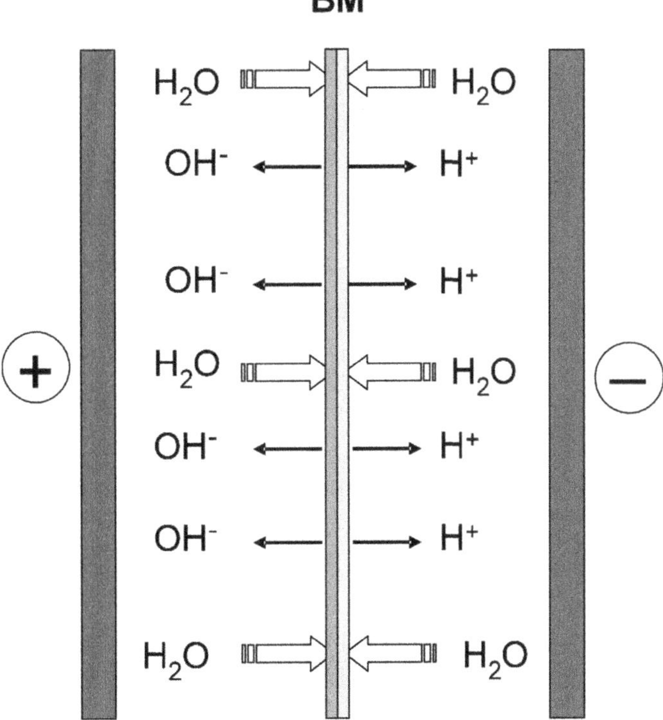

Figure 2.1.4 Principle of a bipolar membrane (BM) and water splitting.

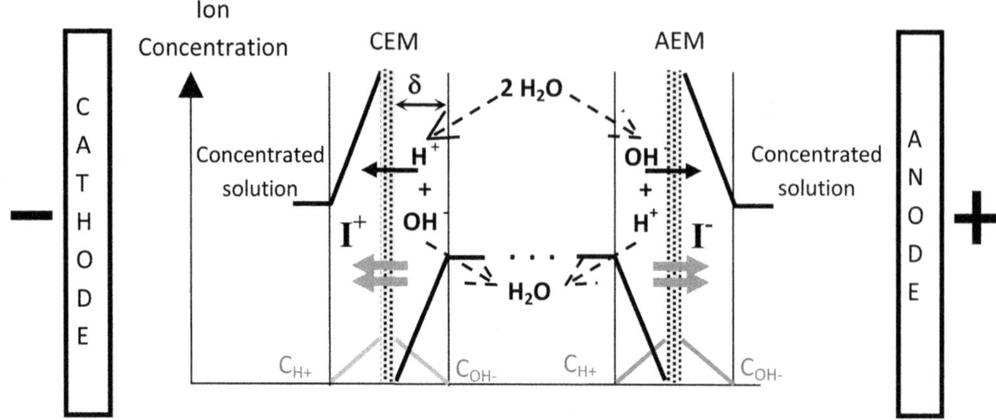

Figure 2.1.5 Concentration polarization and water splitting phenomena in electrodialysis when overpassing the limiting current density. CEM: cation-exchange membrane; AEM: anion-exchange membrane; C_{H+}: concentration in H^+ ions; C_{OH-}: concentration in OH^- ions; δ: diffusion boundary layer. (Adapted from [11,13])

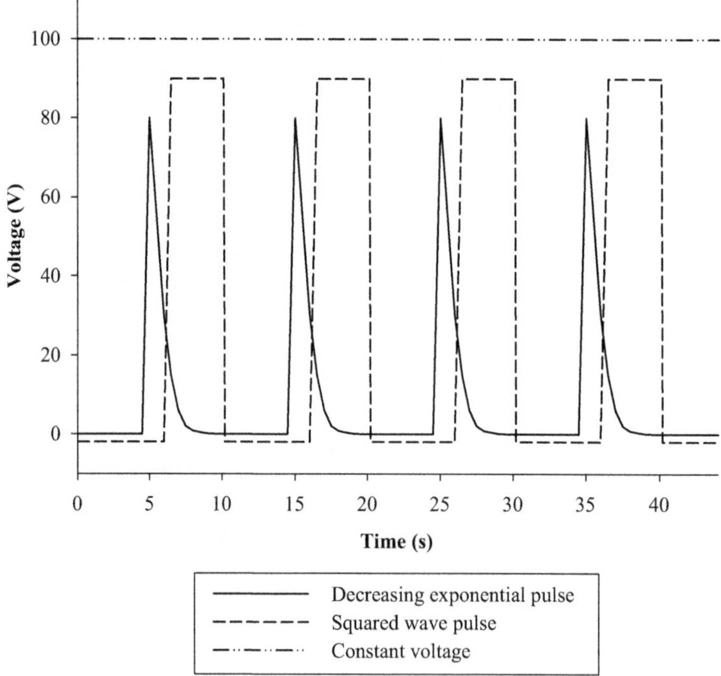

Figure 2.1.6 Principles of pulsed electric field and different types of pulse/pause in comparison with constant voltage (DC current).

intensified several times, depending on the pulse-pause characteristics[21,22]. The improved migration of ions was likely caused by the decrease in the concentration of polarization (CP) with the application of a PEF, and the subsequent reduction in the diffusion layer thickness at the surface of the membrane. The CP is created when the ionic concentration on one side of the membrane is very low while the ionic concentration on the other side of the same membrane is very high[22–24]. PEF perturbations may scale down the CP because, during pause lapse, the transport of ions from the bulk solution to the membrane layer continues by means of diffusion and convection; therefore, the concentration gradient decreases before the application of the pulse lapse[25,26]. Furthermore, the inertial properties of liquid movement can exist after pulse lapse, which can intensify mass transfer as demonstrated by Mishchuk et al.[21] and Mikhaylin et al.[22] on model salt solutions. Such a decrease or the disappearance of the diffusion boundary layer would facilitate ionic mass transfer by electromigration from the solution to the interface of the membrane. If the DBL is small, then the ions would be closer to the membrane when the pulse restarts, promoting improved migration.

2.1.3 *Demineralization*

Mineral salts affect the flavor, functionality and value of whey products. Whey demineralization is therefore an important step in the production of whey powder and has been studied for a long time.

2.1.3.1 *Demineralization by pressure-driven membrane technologies*

In the dairy industry, NF is already extensively used for demineralization and concentration of sweet whey prior to spray drying[27,28] but recently it has also been considered in the demineralization of acid whey[29,30] (Table 2.1.2).

Rasanen et al. (2002) compared four NF membranes in concentration mode in the demineralization of sweet whey: Desal-5 DK, Koch SR1 and two NF45 from different generations[31]. When using Desal-5 DK and the new NF45, loss of lactose was only 0.1% and the demineralization rate (DR) in the retentate was 53%. By comparison, Koch SR1 and the old NF45 led to a total loss of lactose of 1% and a DR of 58%.

Similarly to Rasanen et al., Cuartas-Uribe et al. (2007)[32] compared two other NF membranes in concentration mode in the demineralization of sweet whey: NF200 and DS-5 DL. The retention of lactose was slightly higher for NF200 than for DS-5 DL (98 and 95%, respectively). However, increasing the transmembrane pressure increased the ion retention until equilibrium above 1.5 MPa was reached. The DR of monovalent ions for NF200 was lower than for DS-5 DL with respective values of 20% and >70%. Surprisingly, the Cl^- ions could pass through the DS-5 DL. Indeed, based on the Donnan effect, the Cl^- ions must pass through the membrane in higher proportions to ensure the electroneutrality of the permeate stream. However, the NF200 did not behave in this way because of both a higher zeta potential value causing more electrostatic repulsion of co-ions than DS-5 DL and a smaller pore radii size, which made it difficult for Cl^- ions to pass through the membrane. On the contrary, the DR of polyvalent ions was <10% for both membranes. It was concluded that the DS-5 DL was more appropriate than the NF200 for the demineralization of sweet whey.

Later on, Cuartas-Uribe et al. (2009)[33] worked on the separation of lactose from a sweet whey UF permeate by using NF for demineralization with a DS-5 DL membrane in two main modes: (1) concentration mode and (2) continuous diafiltration mode. Concerning the concentration mode at 1 and 2 MPa, the retention rates were 84 and 89% for lactose while the DRs were 25 and 20% for polyvalent cations and over 70% for monovalent cations at both pressures, respectively. Concerning the continuous diafiltration mode at 1 and 2 MPa, the retention rates of lactose were about 82 and 90% for a volume dilution factor (VDF) of ≈2 and decreased until reaching equilibrium at 65% in both pressure conditions for a VDF of ≈5. Meanwhile, for both pressures, 80 and 99%

Table 2.1.2 Pressure-driven technologies and conditions used for (a) demineralization alone, (b) demineralization and delactosation and (c) demineralization and deacidification

Objective	Authors	Process	Membranes or conditions	% Removal
(a) Demineralization alone	Rasasen et al. (2002)	NF Concentration VCF = 4	Desal-5 DK & New NF45	Demin: 53% Loss of lactose: 0.1%
			Koch SR1 & Old NF45	Demin: 58% Loss of lactose: 2–5%
	Cuartas-Uribe et al. (2007)	NF	NF200	Demin: 20%
			DS-5 DL	Demin: >70%
	Cuartas-Uribe et al. (2007)	NF Concentration VCF = 2	DS-5 DL	Demin: 70% for monovalent ions 20–25% for polyvalent ions Loss of lactose: 11–16%
		NF Diafiltration VDF = 8	DS-5 DL	Demin: 50–99% for monovalent ions 10–50% for polyvalent ions Loss of lactose: 10–18%
	Hinkova et al. (2012)	NF Recycling mode (Concentration)	NTR-7450-S	Demin: 62–95% for monovalent ions 45–55% for polyvalent ions Loss of lactose: 5–15%
			NF-270	Demin: 86–90% for monovalent ions 50% for polyvalent ions Loss of lactose: 13–19%
	Kelly et al. (1994)	NF Concentration VCF = 4	HC-50	Demin: 41–56% for monovalent ions 0% for polyvalent ions
		NF Concentration VCF = 4 +Diafiltration VDF = 4	HC-50	Demin: 62% for monovalent ions 0% for polyvalent ions
	Pan et al. (2011)	NF + Diafiltration		Demin: 72% for total minerals 60% for monovalent ions 10–19% for polyvalent ions
	Suarez et al. (2006)	NF Concentration VCF = 4	Membrane module DK2540C	Demin: 30% for total minerals 30–40% for monovalent ions <10% for polyvalent ions Loss of proteins and lactose: negligible
	Wigers et al. (1998)	NF Dead-end mode	MWI5, MG 17, BQOl, B006, B007 and MX07	Demin: for total minerals: 86–89% in dead-end mode 28–31% in cross-flow mode
(b) Demineralization + Delactosation	Baldasso et al. (2011)	UF (VCF = 5) +5 discontinuous diafiltration	UF-6001	Delactosation after UF: 11% Delact. after UF + diafiltration: 49%
		UF (VCF = 6) +4 discontinuous diafiltration	UF-6001	Delactosation after UF: 21% Delact. after UF + diafiltration: 60%
(c) Demineralization + Deacidification	Chandrapala et al., (2015)	NF	HL, XN45 and DK	Demin: 30–50% for total minerals Deacidification: 45–55% at pH3 <30% over pH4.5 Loss of lactose: <10%
	Bedas et al. (2017)	NF (VCF = 3) +Concentration +Spray-drying		Demin: 37% for total minerals Deacidification: 32% Reduction in energy consumption: 43%

of chloride ions were removed, whereas 50 and 70% of other monovalent anions and only 10 and 50% of polyvalent ions were removed for a VDF of ≈2 and 5, respectively. Therefore, the best-operating conditions were achieved using the continuous diafiltration mode with a VDF of 2 at 2 MPa since lactose losses were low and the monovalent ion DRs were around 80%.

Pan et al. (2011) optimized the DR of sweet whey by adjusting its pH and using NF followed by diafiltration[34]. The best DR (72%) was obtained at pH 4.6 which is the isoelectric point (pI) of whey. Indeed at the pI of whey proteins, the complexation between proteins and ions is weak. Thus, more minerals can be released from proteins and removed by NF more easily. Considering the individual minerals, 60% of monovalent Na^+ and K^+ ions were removed while the removal of divalent Ca^{2+} and Mg^{2+} was only 19 and 10%, respectively. No lactose was detected in the permeate water, meaning no loss.

More recently, Hinkova et al. (2012) compared the efficiency of two NF membranes: NTR-7450-S and NF-270 in recycling mode at 1.5 MPa[35]. Concerning the NTR-7450-S membrane, the retention of lactose was between 85 and 95% whatever the condition tested. Meanwhile, the DRs were very high (62–95%) for monovalent K^+ and Na^+ and about 45–55% for Ca^{2+} ions. Concerning the NF-270 membrane, the retention rate of lactose was slightly lower in comparison with NTR-7450-S (81–87%). Lactose retention was lower in comparison with the NTR-7450 membrane with the same removal of ions, meaning that the NF-270 membrane is less suitable for whey demineralization.

Concerning acid whey, the only studies reported in the literature concerning demineralization alone using pressure-driven process are those of Kelly et al. (1994)[36] who investigated the demineralization of acid casein whey by NF followed by diafiltration[36]. After NF, about 33% of minerals in acid casein whey were removed corresponding to a decrease from 12 to 8% in dry matter. After diafiltration, this removal increased up to 41% with a final concentration of 7% in dry matter. Considering the individual minerals in the retentate after NF, the monovalent Na^+, K^+ and Cl^- decreased from 0.83 to 0.37%, from 2.18 to 0.98% and from 4.25 to 2.5% in dry matter, respectively, corresponding to global DRs of 56, 55 and 41%. On the contrary, the divalent Ca^{2+} and Mg^{2+} could not cross over the NF membrane. Moreover, the permeation of Cl^- ions was enhanced to 70% with the addition of 0.02 M of citrate. After diafiltration (without citrate), the removal of Na^+ and K^+ in dry matter increased up to 60 and 62%, respectively, while the Cl^- ions were significantly removed (from 41% with NF) to 53%. Recently, Roman et al. (2010)[29] and Bédas et al. (2017)[37] tested nanofiltration for acid whey demineralization. (Note that the results will be presented in Section 2.1.4 as they combined the simultaneous effect of demineralization and deacidification.)

Demineralization by NF was also tested on other whey product such as milk permeate and whey protein hydrolysate. Indeed, milk ultrafiltrate could be partially demineralized by at least 30% using an NF membrane module (DK2540C). Again, monovalent Na^+ and K^+ showed higher DRs (30–40%) than divalent Ca^{2+} and Mg^{2+} (<10%), and protein and lactose losses from permeation were negligible[38,39]. The maximum DR was obtained at pH 4.6, close to the whey protein isoelectric point[40]. Demineralization by NF is also possible for a whey protein hydrolysate. Indeed, Wijers et al. (1998) used different NF membranes (negatively charged: MWI5, MG 17, BQOl, B006, B007 and a polyamide one: MX07) and removed monovalent ions from peptides with a high selectivity in dead-end and cross-flow modes[41]. Salt removal was about 86–89% for dead-end and 28–31% for cross-flow filtration. Finally, the authors pointed out the fouling issue which occurred in higher proportions in the dead-end mode. Concerning a special case with milk, Rice et al.[27] described the influence of temperature (from 10 to 50°C) and pH

(5, 6.9 and 9) on membrane fouling during the separation of lactose and calcium on a polyamide NF membrane TFC-SR3 (Koch membranes). A pH above 8.3 reduced the permeate flux by 40% and similarly a pH under 5.5 cut the flux by 20%. For both pHs, the effect of temperature was not significant. On the contrary, at the natural whey pH (6.9) the output was mostly affected by the temperature and fouling increased with heating. This was due to the precipitation of calcium phosphate onto the membrane, forming a mineral cake layer.

2.1.3.2 *Demineralization by electrically driven membrane technologies*

During the ED demineralization process, pasteurized whey is generally concentrated to 20 or 30% of total solids before being electrodialyzed at 20 or 38°C[42,43]. A high conductivity is desirable for the operation of an ED system because it leads to low electrical resistance resulting in the large-scale migration of ions[44]. Continuous ED allows a 60–75% decrease in ash[43,45,46] (Table 2.1.3). Furthermore, a 90–95% DR can be obtained by recirculating the whey in the ED cell (batch recirculation process)[43,47–49]. Monovalent K^+ and Na^+ can be removed faster with concentration reductions of 83–100%, whereas divalent Ca^{2+} and Mg^{2+} contents decrease by 61–96%[50]. ED treatments enhance the stability of frozen skim milk and concentrated skim milk proteins: the removal of more than 40% of calcium by ED allows an increase in the stability of protein stored at −8°C from 1 week to 17 weeks. The demineralization of calcium by 70% increases the protein stability to 53 weeks[51]. The application of ED demineralization to UF whey permeates and retentate has been studied[52]. ED is also applied to the demineralization of skim milk, deproteinated whey, delactosed whey and UF whey permeate and retentate used in the production of whey protein concentrate (WPC). The WPC (35% protein) is demineralized by ED in both Japan and the United States. Desalted WPC is mixed with lactose and non-fat milk solids to produce infant formula[43].

Bipolar membrane electrodialysis has also been tested for whey demineralization, but demineralization is not the first concern. In such an application, the first attempt is acidification of the dairy products, while demineralization is an important side effect. Bazinet et al. (1999) were the first to propose the use of BMED to precipitate bovine milk protein, without adding acids, and allowing the production of an acid whey partially demineralized[53]: According to the membrane used (monovalent-ion permselective membrane or conventional ion-exchange membrane)[54], ionic strength[55] and milk DR prior to BMED treatment[55], the range of DRs for acid whey was between 25 and 85%. During BMED, potassium is the predominant cation to migrate across the CEM, with a decrease of 76.8% in its concentration. The sodium, magnesium and calcium concentrations decreased by 60.5, 33.7 and 31.7%, respectively[56].

Balster et al. (2007) developed a specific process configuration to electroacidify milk for demineralization and acid/base production[57]. The milk was acidified by the protons generated with the bipolar membrane, causing precipitation of caseins. Afterward, these precipitated caseins were removed by solid/liquid separation while the supernatant whey was fed to a second bipolar membrane stack for desalination. By combining electroacidification with desalination, the acid whey could be neutralized by the produced base (NaOH) while the acid (HCl) was recovered for further applications. BMED has also been used for the electroacidification of whole and skimmed wheys with a 38% demineralization rate[58].

Very recently, ED with PEF was tested for sweet whey demineralization[59]. It was demonstrated that PEF modes with higher frequencies and a pulse/pause ratio of 1 were an effective approach to sweet whey demineralization. Indeed, the PEF 0.1s-0.1s condition, corresponding to a frequency of 5 Hz, enhanced the DR by more than 81% as well as reducing the energy consumption by about 16% on the same DR basis in comparison with the conventional continuous current condition. This improvement was explained by a gain in voltage followed by the possible formation of electroconvection at the beginning of each pulse lapse. In addition, the PEF mode with a pulse/pause ratio of 1 could hamper the pH-sensitive scaling and organic fouling formation on the membrane by reducing the pH variations in ED compartments[60]. Furthermore, when ED-PEF was applied to model solutions containing whey proteins and salts, a DR of 80% was achieved versus 55% for conventional ED[61].

2.1.4 *Demineralization and delactosation*

UF is usually used in the separation of proteins from lactose and minerals, whereas NF and diafiltration are used to separate lactose and minerals (Table 2.1.2). Baldasso et al. (2011) used UF in combination with discontinuous diafiltration to concentrate and purify whey proteins[64]. Two experiments with different strategies were conducted to optimize the yield of protein purification by demineralizing and recovering lactose. The initial whey contained 15% (dry basis) proteins, 72% lactose and 12% minerals. After UF, the retentates contained 35 and 40% proteins, 64 and 57% lactose and almost no ash for Experiments 1 and 2, respectively. After diafiltration, its retentates contained 60 and 70% proteins and 37 and 29% lactose for Experiments 1 and 2, respectively. A higher protein content was obtained during Experiment 2 with fewer diafiltration steps and with a smaller volume of water required to obtain protein concentrates.

Concerning a particular case with milk, but interesting in terms of lactose separation, Atra et al. (2005)[65] ultrafiltered fresh milk containing 3.5% proteins and 5% lactose and obtained a retentate phase of 12–14% proteins and a permeate phase made of 0.5% proteins and 5% lactose. Then, using a 400 Da cut-off membrane pressure at 20 bars, this permeate was nanofiltered into a retentate of 20–25% lactose and a permeate of 0.1–0.2% lactose, corresponding to a lactose loss lower than 2%. An operating temperature of 30°C was found to be optimum. Unfortunately, the levels of minerals were not measured in this study, although the minerals are likely to migrate into the NF permeate. Therefore, NF could be used to concentrate the lactose for a further spray-drying step.

2.1.5 *Demineralization and deacidification*

Pretreatments for removing lactic acid and/or calcium from lactose, in order to favor its crystallization, are necessary prior to spray drying.

2.1.5.1 *Demineralization and deacidification by pressure-driven processes*

Studies reported in the literature on the demineralization and deacidification of whey or whey products are mainly focused on NF (Table 2.1.2). Hence, Barrantes et al. (1997)[67] deacidified and demineralized acid whey from cottage cheese using a sequential fractionation process divided into three steps: (1) NF to concentrate the acid whey; (2) diananofiltration (DNF) of the retentate; and (3) NF to concentrate the DNF retentate[67]. About 85% of the total solids of the initial acid whey were recovered in the first NF retentate and 67% in the DNF retentate. About 94% of the initial lactose was recovered in the retentate during the first NF step and about 71% remained in the retentate after DNF. Meanwhile, lactic acid remained at 69% in the first NF retentate, and at 53% in the final NF retentate. The initial minerals were recovered at 56% in the final retentate. Lactic acid was most effectively removed at pH 6.5 whereas minerals were most effectively removed at pH 5.5.

Later on, Chandrapala and Vasiljevic published numerous articles with different collaborators related to the removal of lactic acid and minerals from acid whey and the mechanism of lactose crystallization[9,30,68–70]. Chandrapala et al. (2016) extracted lactic acid from acid whey using three commercial NF membranes: HL, XN45 and DK[69]. The removal of lactic acid through NF membranes increased when the pH of whey decreased. Indeed, lactic acid is a weak acid that can be dissociated into a lactate anion with a pKa of 3.86 at 25°C. It was shown that the permeation rate was closely correlated with the dissociation rate of the lactic acid described by the Henderson–Hasselbalch equation, suggesting that the charge repulsion of lactate hindered the separation[69]. Meanwhile, the NaCl retention was about 50–70% at any pH value. In contrast to lactate, lactose retention was over 90% and pH independent since it is an uncharged molecule.

Very recently, Bedas et al. (2017) attempted to assess the ability of NF in the production of partially demineralized acid whey powder, regarding the dryability of its concentrate and the quality of the resulting powder at a semi-industrial scale[37]. Initial acid whey (5.9% dry matter) was processed in two ways: (1) pre-concentrated by vacuum evaporation (33% dry matter), concentrated by vacuum evaporation (60% dry matter) and finally spray-dried into a non-NF powder (95% dry matter); and (2) nanofiltered by a volume concentration factor (VCF) of 3 leading to a retentate (33% dry matter) which was concentrated by vacuum evaporation (62% dry matter) and spray-dried into an NF powder (97% dry matter). The reduction in lactic acid, ash, potassium, sodium and chloride contents were 30, 27, 53, 60 and 53% in the NF acid whey retentate. Similarly for the NF powder, lactic acid, ash, potassium, sodium and chloride contents were reduced by 32, 37, 46, 60 and 47%, respectively. In terms of powder quality, the non-NF powder showed a higher hygroscopicity than the NF powder (22.6 and 14.1% for relative humidity at 80°C, respectively) lowering its quality. This was explained by the hygroscopic behavior of monovalent ions and lactic acid which were removed by NF in the NF powder. Therefore, using NF reduced the energy costs of the overall process by 43% because the water removal required a lower specific energy by migration under a pressure gradient with NF membrane filtration than heating by spray drying (10–40 kJ/kg water vs. enthalpy of water vaporization: 2258 kJ/kg water at 100°C).

Furthermore, NF has been used for the recovery of lactic acid from fermentation broths in which the concentration of lactic acid is about 10–20 times that in acid whey[71]. For this product, the authors noted that maximum impurities rejections and lactate recovery were obtained at maximum transmembrane pressures. Interestingly, divalent Mg^{2+} and Ca^{2+} ions permeation were 64 and 72%, respectively. Meanwhile, about 40% of SO_4^{2-} and PO_4^{3-} ions were also partially removed from the broth. Similar studies dealing with broth are available in the literature[72,73].

2.1.5.2 *Demineralization and deacidification by electrically driven processes*

The removal of salts and acid from whey can be achieved by ED (Table 2.1.3). Kusavskii (1992) was the first to use ED to demineralize and neutralize acid whey[74]. The process was conducted at a DR of up to 90% and a pH of the obtained demineralized whey equal to 6.0–6.5. This method is claimed to simplify the technology of demineralization and neutralization processes, to reduce the number of stages and to eliminate the use of expensive reagents.

More recently, Chen et al. (2016) tested ED demineralization and deacidification in batch mode at three temperatures and two pHs for acid whey and model solutions made of lactic acid, lactose and NaCl[75]. Since salt anions have a smaller ionic radius and a higher mobility than lactate ions, they were preferentially removed to lactate ions at the beginning of the process. The required energy varied from 4 to 9 kWh/kg of lactic acid to remove the first 40% of lactic acid, because energy was also consumed to remove other anions. Nonetheless, the removal of these anions was faster at high temperatures. Interestingly, when 80% of the lactate ions were removed to achieve a similar ratio of lactic acid to lactose as found in sweet whey, 90% of the minerals were simultaneously removed. About 0.014 kWh/kg acid whey was required to achieve a DR of 90%.

Very recently, a comparison between conventional ED and ED with bipolar membranes was investigated by Dufton et al. (2018)[76] for whey deacidification and demineralization. Indeed, a bipolar membrane was recently successfully used for juice and wine deacidification, with up to 80% acid removal[77,78]. In this study, two different ED configurations (Figure 2.1.7) were tested.

Both configurations reached around 44% deacidification and 67% demineralization rates after three hours of treatment. Interestingly, 31–41% of lactate removed from the initial whey was recovered in the concentrate compartment. However, even though the appearance of fouling or scaling has never been reported by Chen et al. (2016)[75], different types of scaling were observed on membranes using both configurations. These scaling formations were linked to the migration of divalent ions and the water splitting phenomenon caused by a high demineralization rate or by an already formed significant scaling[76]. Following the appearance of scaling, Dufton et al. (2019) tested the use of PEF for demineralization and deacidification in the conventional CACAC configuration[79]. The application of a PEF 50s/10s pulse/pause condition allowed the mitigation of scaling visually but was not statistically different from results obtained for the CC current condition, whereas a 25s/25s pulse/pause combination allowed the amount of scaling to diminish drastically and to decrease the final system electrical resistance by 32%. The use of a pulsed current also improved the removal of lactic acid by 10 and 16% for PEF 50s/10s and 25s/25s, respectively, in comparison with the CC current condition. Moreover, the migration enhancement also applied to mineral ions such as calcium, magnesium and potassium with a DR of 53, 32 and 85%, respectively, which are 5 to 10% higher than the rates obtained with the CC current. According to the authors, the results obtained for PEF 25s/25s are due to two mechanisms: (1) the mitigation of the concentration polarization phenomenon and (2) the rinsing of the membranes during the pause periods[79].

2.1.6 *Conclusion*

In this chapter, techniques or operating conditions for enhancing the spray drying of whey and whey products were described. Pressure-driven and electrically driven processes are able to separate whey compounds such as lactose and proteins from lactic acid and minerals for purification and further spray drying in an efficient way. However, in the case of pressure-driven membrane processes, a deacidification rate over 50% cannot be reached while for electrically driven membrane processes, scaling is still the main concern. The best way to demineralize, deacidify and/or remove lactose is coupling different operation units, but this significantly increases the cost of the whole process and decreases its eco-efficiency. Hence, NF could be used as a pretreatment step since whey is simultaneously concentrated and partly demineralized; thereafter, for further demineralization, ED is more appropriate. Up to 80–95% of salt and 30–80% of lactic acid can be removed from the whey, enabling better lactose crystallization.

In addition, there are currently promising areas of research which could benefit from the development of membrane-based technologies for use in the dairy industry. This is the

Table 2.1.3 Electrically driven technologies and conditions used for (a) demineralization alone and (b) demineralization and deacidification

Objective	Authors	Process	Membranes or conditions	% Removal
(a) Demineralization alone	Ahlgren et al. (1972) Marshall et al. (1982) Batchelder et al. (1986)	ED		Demin: 60–75% for total minerals
	Delbeke et al. (1975) Diblikova et al. (2010) Baldasso et al. (2014)	ED		Demin: 90–95% for total minerals
	Diblikova et al. (2013)	ED		Demin: 90–95% for total minerals 83–100% for monovalent ions 61–96% for polyvalent ions
	Lin Teng Shee et al. (2005)	ED with bipolar membranes	BP-1 and CMX-SB	Demin: 38% for total minerals
	Lemay et al. (2019)	ED with PEF	CMX and PEF 0.1s-0.1s	Demin: 81% for total minerals
	Casademont et al. (2009)	ED	CMX-S	Demin: 55% for total minerals
		ED with PEF	CMX-S	Demin: 80% for total minerals
	Greiter et al. (2002)	ED		Demin: 90% for total minerals Energy demand: 12.8 kWh
		Ion-exchange resins		Demin: 99% for total minerals Energy demand: 35.2 kWh (2.75 times higher than using ED) Waste emission: Four times higher than ED
(b) Demineralization + Deacidification	Kusavskii (1992)	ED	(NA)	Demin: 90%

(*Continued*)

Table 2.1.3 (Continued) Electrically driven technologies and conditions used for (a) demineralization alone and (b) demineralization and deacidification

Objective	Authors	Process	Membranes or conditions	% Removal
	Chen et al. (2016)	ED	CMB and AHA	For model solutions: Demin: 90% for total minerals Deacidification: 80% For acid whey: Demin: 55–75% for total minerals Deacidification: ~40%
	Dufton et al. (2018)	ED with bipolar membranes ED	BP-1, CMX and AMX CMX and AMX	Demin: 67% for total minerals Deacidification: 44%
	Dufton et al. (2019)	ED with bipolar membranes and PEF	CMX and AMX in CACAC configuration	Demin: 64% in conventional ED 67% for 25s/25s Deacidification: 48% for 50s/10s 51% for 25s/25s

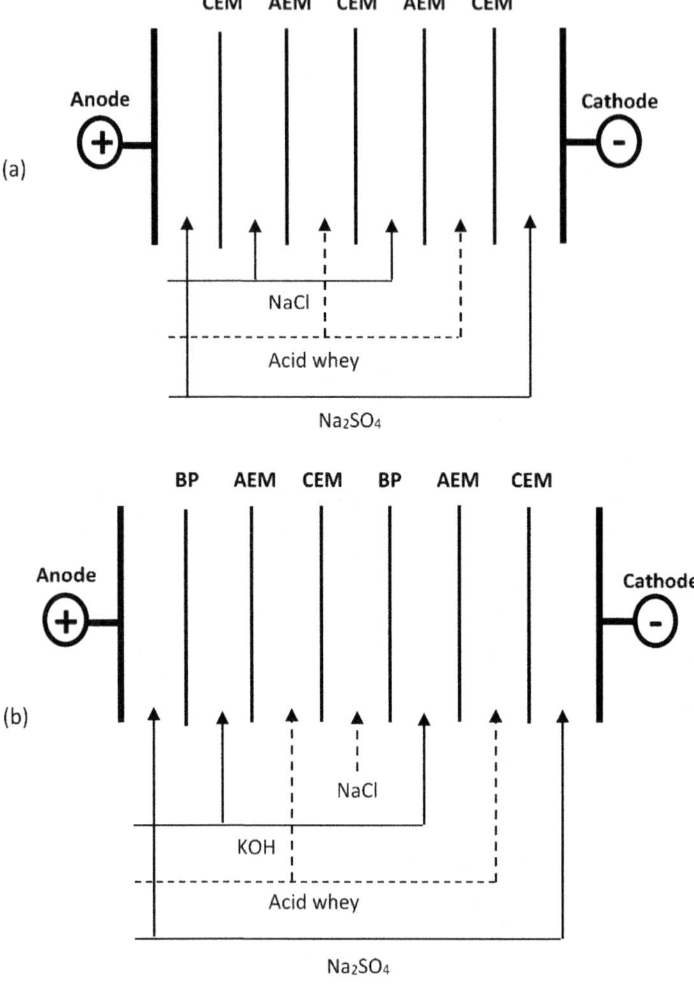

Figure 2.1.7 Electrodialysis configurations (a) CACAC and (b) BACBAC used for acid whey deacidification.

case for example with the use of back-pulse pressure and pulsed electric field which could be used specifically as an unavoidable operational unit before spray drying.

Such a new type of pressure/current application mode would decrease or eliminate the risks of fouling/scaling in a more efficient way than the conventional cleaning procedure, to decrease the concentration polarization phenomenon and to increase mass transfer.

Nomenclature

AEM: Anion-exchange membrane
BMED: Bipolar membrane electrodialysis
CC: Continuous current
CEM: Cation-exchange membrane
CP: Concentration of polarization
Da: Dalton

DBL: Diffusion boundary layer
DNF: Diananofiltration
DR: Demineralization rate
ED: Electrodialysis
IEM: Ion-exchange membrane
MF: Microfiltration
NF: Nanofiltration
PEF: Pulsed electric field
RO: Reverse osmosis
UF: Ultrafiltration
VCF: Volume concentration factor
VDF: Volume dilution factor

Acknowledgments

This project was supported by the NSERC Industrial Research Chair on Electromembrane Processes aiming toward the ecoefficiency improvement of biofood production lines (Grant IRCPJ 492889-15 to Laurent Bazinet) and the NSERC Discovery Grants Program (Grant SD RGPIN-201804128 to Laurent Bazinet).

References

1. 3A Business Consulting – Dairy and whey ingredients. – Market and industry trends 2010. Chicago, April 27, 2010. Tage Affertsholt, 3A Business Consulting.
2. Siso, M. I. G. The biotechnological utilization of cheese whey: A review. *Bioresource Technology* 1996, *57*(1), 1–11.
3. Ryan, M. P.; Walsh, G. The biotechnological potential of whey. *Reviews in Environmental Science and Bio/Technology* 2016, *15*(3), 479–498.
4. Blaschek, K. M.; Wendorff, W. L.; Rankin, S. A. Survey of salty and sweet whey composition from various cheese plants in Wisconsin. *Journal of Dairy Science* 2007, *90*(4), 2029–2034.
5. Nishanthi, M.; Chandrapala, J.; Vasiljevic, T. Properties of whey protein concentrate powders obtained by spray drying of sweet, salty and acid whey under varying storage conditions. *Journal of Food Engineering* 2017, *214*, 137–146.
6. Tsakali, E.; Petrotos, K.; Allessandro, A. D. A review on whey composition and the methods used for its utilization for food and pharmaceutical products. 6th Int. Conf. Simul. Model. Food Bio-Industry. FOODSIM 8. 2010.
7. Schultz, E. J.; Parekh, R. Strong consumer demand pushes Greek yogurt into a dairy-aisle battlefield *Advertising Age* 2011, *82*, 8.
8. Saffari, M.; Langrish, T. Effect of lactic acid in-process crystallization of lactose/protein powders during spray drying. *Journal of Food Engineering* 2014, *137*, 88–94.
9. Chandrapala, J.; Vasiljevic, T. Properties of spray dried lactose powders influenced by presence of lactic acid and calcium. *Journal of Food Engineering* 2017, *198*, 63–71.
10. Greiter, M.; Novalin, S.; Wendland, M.; Kulbe, K.-D.; Fischer, J. Desalination of whey by electrodialysis and ion exchange resins: Analysis of both processes with regard to sustainability by calculating their cumulative energy demand. *Journal of Membrane Science* 2002, *210*(1), 91–102.
11. Pourcelly, G.; Bazinet, L. Developments of bipolar membrane technology in food and BIO-industries. In *Handbook of membrane separations: Chemical, pharmaceutical, food, and biotechnological applications.* Pabby, A. K., Rizvi, S. S. H., and Sastre A. M., Eds. CRC Press, Boca Raton, FL, 2008, 581–657.
12. Pouliot, Y. Membrane processes in dairy technology—From a simple idea to worldwide panacea. *International Dairy Journal* 2008, *18*(7), 735–740.

13. Bazinet, L.; Castaigne, F. *Concepts de génie alimentaire : Procédés associés, applications à la conservation et transformation des aliments.* 2nd edition. Ed. Presses Internationales Polytechniques, 2019, 597–628.
14. Kumar, P.; Sharma, N.; Ranjan, R.; Kumar, S.; Bhat, Z. F.; Jeong, D. K. Perspective of membrane technology in dairy industry: A review. *Asian-Australasian Journal of Animal Sciences* 2013, *26*(9), 1347–1358.
15. Mohammad, A. W.; Teow, Y. H.; Ang, W. L.; Chung, Y. T.; Oatley-Radcliffe, D. L.; Hilal, N. Nanofiltration membranes review: Recent advances and future prospects. *Desalination* 2015, *356*, 226–254.
16. Ernst, M.; Bismarck, A.; Springer, J.; Jekel, M. Zeta-potential and rejection rates of a polyethersulfone nanofiltration membrane in single salt solutions. *Journal of Membrane Science* 2000, *165*(2), 251–259.
17. Hagmeyer, G.; Gimbel, R. Modelling the salt rejection of nanofiltration membranes for ternary ion mixtures and for single salts at different pH values. *Desalination* 1998, *117*(1), 247–256.
18. Dammak, L.; Larchet, C.; Grande, D. Ageing of ion-exchange membranes in oxidant solutions. *Separation and Purification Technology* 2009, *69*(1), 43–47.
19. Mani, K. N. Electrodialysis water splitting technology. *Journal of Membrane Science* 1991, *58*(2), 117–138.
20. Karlin, Y. V.; Kropotov, V. N. Electrodialysis separation of Na+ and Ca2+ in a pulsed current mode. *Russian Journal of Electrochemistry* 1995, *31*(5), 517–521.
21. Mishchuk, N. A.; Koopal, L. K.; Gonzalez-Caballero, F. Intensification of electrodialysis by applying a non-stationary electric field. *Colloids and Surfaces A: Physicochemical and Engineering Aspects* 2001, *176*(2–3), 195–212.
22. Mikhaylin, S.; Nikonenko, V.; Pourcelly, G.; Bazinet, L. Intensification of demineralization process and decrease in scaling by application of pulsed electric field with short pulse/pause conditions. *Journal of Membrane Science* 2014, *468*, 389–399.
23. Mishchuk, N. A.; Verbich, S. V.; Gonzales-Caballero, F. Concentration polarization and specific selectivity of membranes in pulse mode. *Colloid Journal* 2001, *63*(5), 586–595.
24. Uzdenova, A. M.; Kovalenko, A. V.; Urtenov, M. K.; Nikonenko, V. V. Effect of electroconvection during pulsed electric field electrodialysis. Numerical experiments. *Electrochemistry Communications* 2015, *51*, 1–5.
25. Nikonenko, V. V.; Pismenskaya, N. D.; Belova, E. I.; Sistat, P.; Huguet, P.; Pourcelly, G.; Larchet, C. Intensive current transfer in membrane systems: Modelling, mechanisms and application in electrodialysis. *Advances in Colloid and Interface Science* 2010, *160*(1–2), 101–123.
26. Malek, P.; Ortiz, J. M.; Richards, B. S.; Schäfer, A. I. Electrodialytic removal of NaCl from water: Impacts of using pulsed electric potential on ion transport and water dissociation phenomena. *Journal of Membrane Science* 2013, *435*, 99–109.
27. Rice, G.; Kentish, S.; O'Connor, A.; Stevens, G.; Lawrence, N.; Barber, A. Fouling behaviour during the nanofiltration of dairy ultrafiltration permeate. *Desalination* 2006, *199*(1), 239–241.
28. Okawa, T.; Shimada, M.; Ushida, Y.; Seki, N.; Watai, N.; Ohnishi, M.; Tamura, Y.; Ito, A. Demineralisation of whey by a combination of nanofiltration and anion-exchange treatment: A preliminary study. *International Journal of Dairy Technology* 2015, *68*(4), 478–485.
29. Román, A.; Popović, S.; Vatai, G.; Djurić, M.; Tekić, M. N. Process duration and water consumption in a variable volume diafiltration for partial demineralization and concentration of acid whey. *Separation Science and Technology* 2010, *45*(10), 1347–1353.
30. Chandrapala, J.; Duke, M. C.; Gray, S. R.; Zisu, B.; Weeks, M.; Palmer, M.; Vasiljevic, T. Properties of acid whey as a function of pH and temperature. *Journal of Dairy Science* 2015, *98*(7), 4352–4363.
31. Räsänen, E.; Nyström, M.; Sahlstein, J.; Tossavainen, O. Comparison of commercial membranes in nanofiltration of sweet whey. *Lait* 2002, *82*(3), 343–356.
32. Cuartas-Uribe, B.; Alcaina-Miranda, M. I.; Soriano-Costa, E.; Bes-Piá, A. Comparison of the behavior of two nanofiltration membranes for sweet whey demineralization. *Journal of Dairy Science* 2007, *90*(3), 1094–1101.
33. Cuartas-Uribe, B.; Alcaina-Miranda, M. I.; Soriano-Costa, E.; Mendoza-Roca, J. A.; Iborra-Clar, M. I.; Lora-García, J. A study of the separation of lactose from whey ultrafiltration permeate using nanofiltration. *Desalination* 2009, *241*(1), 244–255.

34. Pan, K.; Song, Q.; Wang, L.; Cao, B. A study of demineralization of whey by nanofiltration membrane. *Desalination* 2011, *267*(2), 217–221.
35. Hinkova, A.; Zidova, P.; Pour, V.; Bubnik, Z.; Henke, S.; Salova, A.; Kadlec, P. Potential of membrane separation processes in cheese whey fractionation and separation. *Procedia Engineering* 2012, *42*, 1425–1436.
36. Kelly, J.; Kelly, P. Desalination of acid casein whey by nanofiltration. *International Dairy Journal* 1995, *5*(3), 291–303.
37. Bédas, M.; Tanguy, G.; Dolivet, A.; Méjean, S.; Gaucheron, F.; Garric, G.; Senard, G.; Jeantet, R.; Schuck, P. Nanofiltration of lactic acid whey prior to spray drying: Scaling up to a semi-industrial scale. *LWT - Food Science and Technology* 2017, *79*, 355–360.
38. Suárez, E.; Lobo, A.; Álvarez, S.; Riera, F. A.; Álvarez, R. Partial demineralization of whey and milk ultrafiltration permeate by nanofiltration at pilot-plant scale. *Desalination* 2006, *198*(1), 274–281.
39. Suárez, E.; Lobo, A.; Alvarez, S.; Riera, F. A.; Álvarez, R. Demineralization of whey and milk ultrafiltration permeate by means of nanofiltration. *Desalination* 2009, *241*(1), 272–280.
40. Boer, R.; Robbertsen, T. Electrodialysis and ion exchange processes: The case of milk whey. In *Progress in food engineering: Solid extraction, isolation and purification*. Cantarelli, C. and Peri, C., Eds. Proceedings of the European Symposium of the Food Working Party of the E.F.C.E. Forster Publishing, Kusnacht, Switzerland, 1981, 393–403.
41. Wijers, M. C.; Pouliot, Y.; Gauthier, S.; F.; Pouliot, M.; Nadeau, L. Use of nanofiltration membranes for the desalting of peptide fractions from whey protein enzymatic hydrolysates. *Lait* 1998, *78*(6), 621–632.
42. Higgins, J. J.; Short, J. L. Demineralization by electrodialysis of permeates derived from ultrafiltration of wheys and skim milk. *New Zealand Journal of Dairy Science and Technology* 1980, *15*(3), 277–288.
43. Batchelder, B. T. Electrodialysis applications in whey processing. International Whey Conference, Chicago, October 28, Paper no TP 343 ST 1986.
44. Johnston, K. T.; Hill, C. G.; Amundson, C. H. Electrodialysis of raw whey and whey fractionated by reverse osmosis and ultrafiltration. *Journal of Food Science* 1976, *41*, 770–777.
45. Ahlgren, R. M. Electromembrane processing of cheese whey. In *Industrial processing with membranes*. Lacey, R.E. and Loeb, S., Eds. Wiley-Interscience, New York, 1972, 71–82.
46. Marshall, K. R. Industrial isolation of milk proteins : Whey proteins. *Development in Dairy Chemistry* 1982, 339–373.
47. Delbeke, R. La déminéralisation par électrodialyse du lactosérum doux de fromagerie. *Le Lair* 1975, *641–542*, 76–94.
48. Diblíková, L.; Čurda, L.; Homolová, K. Electrodialysis in whey desalting process. *Desalination and Water Treatment* 2010, *14*(1–3), 208–213.
49. Baldasso, C.; Marczak, L. D. F.; Tessaro, I. C. A comparison of different electrodes solutions on demineralization of permeate whey. *Separation Science and Technology* 2014, *49*(2), 179–185.
50. Diblíková, L.; Čurda, L.; Kinčl, J. The effect of dry matter and salt addition on cheese whey demineralisation. *International Dairy Journal* 2013, *31*(1), 29–33.
51. Lonergan, D. A.; Fennema, O.; Amundson, C. H. Use of electrodialysis to improve the protein stability of frozen skim milks and milk concentrates. *Journal of Food Science* 1982, *47*(5), 1429–1434.
52. Pérez, A.; Andrés, L. J.; Álvarez, R.; Coca, J.; Hill, C. G. Electrodialysis of whey permeates and retentates obtained by ultrafiltration. *Journal of Food Process Engineering* 1994, *17*(2), 177–190.
53. Bazinet, L.; Lamarche, F.; Ippersiel, D.; Amiot, J. Bipolar membrane electroacidification to produce bovine milk casein isolate. *Journal of Agricultural and Food Chemistry* 1999, *47*(12), 5291–5296.
54. Bazinet, L.; Ippersiel, D.; Montpetit, D.; Mahdavi, B.; Amiot, J.; Lamarche, F. Effect of membrane permselectivity on the fouling of cationic membranes during skim milk electroacidification. *Journal of Membrane Science* 2000, *174*(1), 97–110.
55. Bazinet, L.; Ippersiel, D.; Gendron, C.; Mahdavi, B.; Amiot, J.; Lamarche, F. Effect of added salt and increase in ionic strength on skim milk electroacidification performances. *Journal of Dairy Research* 2001, *68*(2), 237–250.

56. Bazinet, L.; Ippersiel, D.; Gendron, C.; Beaudry, J.; Mahdavi, B.; Amiot, J.; Lamarche, F. Cationic balance in skim milk during bipolar membrane electroacidification. *Journal of Membrane Science* 2000, *173*(2), 201–209.
57. Balster, J.; Pünt, I.; Stamatialis, D. F.; Lammers, H.; Verver, A. B.; Wessling, M. Electrochemical acidification of milk by whey desalination. *Journal of Membrane Science* 2007, *303*(1), 213–220.
58. Lin Teng Shee F.; Angers, P.; Bazinet, L. Relationship between electrical conductivity and demineralization rate during electroacidification of Cheddar cheese whey. *Journal of Membrane Science* 2005, *262*(1), 100–106.
59. Lemay, N.; Mykhaylin, S.; Bazinet, L. Voltage spike and electroconvective vortices generation during electrodialysis under pulsed electric field : Impact on demineralisation process efficiency and energy consumption. *Innovative Food Science and Emerging Technology* 2019, *52*, 221–231.
60. Ayala-Bribiesca, E.; Pourcelly, G.; Bazinet, L. Nature identification and morphology characterization of cation-exchange membrane fouling during conventional electrodialysis. *Journal of Colloid and Interface Science* 2006, *300*(2), 663–672.
61. Casademont, C.; Sistat, P.; Ruiz, B.; Pourcelly, G.; Bazinet, L. Electrodialysis of model salt solution containing whey proteins: Enhancement by pulsed electric field and modified cell configuration. *Journal of Membrane Science* 2009, *328*(1–2), 238–245.
64. Baldasso, C.; Barros, T. C.; Tessaro, I. C. Concentration and purification of whey proteins by ultrafiltration. *Desalination* 2011, *278*(1), 381–386.
65. Atra, R.; Vatai, G.; Bekassy-Molnar, E.; Balint, A. Investigation of ultra- and nanofiltration for utilization of whey protein and lactose. *Journal of Food Engineering* 2005, *67*(3), 325–332.
67. Barrantes, L. D.; Morr, C. V. Partial deacidification and demineralization of cottage cheese whey by nanofiltration. *Journal of Food Science* 1997, *62*(2), 338–341.
68. Chandrapala, J.; Wijayasinghe, R.; Vasiljevic, T. Lactose crystallization as affected by presence of lactic acid and calcium in model lactose systems. *Journal of Food Engineering* 2016, *178*, 181–189.
69. Chandrapala, J.; Chen, G. Q.; Kezia, K.; Bowman, E. G.; Vasiljevic, T.; Kentish, S. E. Removal of lactate from acid whey using nanofiltration. *Journal of Food Engineering* 2016, *177*, 59–64.
70. Chandrapala, J.; Duke, M. C.; Gray, S. R.; Weeks, M.; Palmer, M.; Vasiljevic, T. Strategies for maximizing removal of lactic acid from acid whey – Addressing the un-processability issue. *Separation and Purification Technology* 2017, *172*, 489–497.
71. Bouchoux, A.; Roux-de Balmann, H.; Lutin, F. Investigation of nanofiltration as a purification step for lactic acid production processes based on conventional and bipolar electrodialysis operations. *Separation and Purification Technology* 2006, *52*(2), 266–273.
72. Sikder, J.; Chakraborty, S.; Pal, P.; Drioli, E.; Bhattacharjee, C. Purification of lactic acid from microfiltrate fermentation broth by cross-flow nanofiltration. *Biochemical Engineering Journal* 2012, *69*, 130–137.
73. Wang, C.; Li, Q.; Tang, H.; Yan, D.; Zhou, W.; Xing, J.; Wan, Y. Membrane fouling mechanism in ultrafiltration of succinic acid fermentation broth. *Bioresource Technology* 2012, *116*, 366–371.
74. Kusavskij, A. M. Method for processing milk whey. Sweden Patent SU 1729378-A1. 1992.
75. Chen, G. Q.; Eschbach, F. I. I.; Weeks, M.; Gras, S. L.; Kentish, S. E. Removal of lactic acid from acid whey using electrodialysis. *Separation and Purification Technology* 2016, *158*, 230–237.
76. Dufton, G.; Mikhaylin, S.; Gaaloul, S.; Bazinet, L. How electrodialysis configuration influences acid whey deacidification and membrane scaling. *Journal of Dairy Science* 2018, *101*(9), 7833–7850.
77. Calle, E. V.; Ruales, J.; Dornier, M.; Sandeaux, J.; Sandeaux, R.; Pourcelly, G. Deacidification of the clarified passion fruit juice (*P. edulis f. flavicarpa*). *Desalination* 2002, *149*(1), 357–361.
78. Serre, E.; Rozoy, E.; Pedneault, K.; Lacour, S.; Bazinet, L. Deacidification of cranberry juice by electrodialysis: Impact of membrane types and configurations on acid migration and juice physicochemical characteristics. *Separation and Purification Technology* 2016, *163*, 228–237.
79. Dufton, G.; Mikhaylin, S.; Gaaloul, S.; Bazinet, L. Positive impact of pulsed electric field on lactic acid removal, demineralization and membrane scaling during acid whey electrodialysis. Special issue on "Ion and Molecule Transport in Membrane Systems". *International Journal of Molecular Sciences* 2019, *20*(4), 797, Open access.

2.2 *Concentration by vacuum evaporation*

Gaëlle Tanguy

2.2.1 *Introduction*

During the last decade, the growing demand for the constituents of whey (protein, lactose, etc.) that represent the main components of infant milk powders, and the development of membrane technologies have led to the diversification of dairy powders, with some becoming high value-added products. The manufacture of dairy powders includes several unit operations. Two such unit operations remove water and convert a liquid product into a powder: concentration using vacuum evaporation or membrane filtration, and spray drying.

Many efforts have been made to control and improve the efficiency of the spray-drying step, but few studies have dealt with the vacuum evaporation process. This process is carried out predominantly in falling-film evaporators that can be considered a combination of shell-and-tube heat exchangers. Improvements in the design and working of the equipment aimed to save energy. Nowadays, falling-film evaporators are energy-efficient devices compared to the energy required for water evaporation: their energy consumption is only 100–300 kWh.t^{-1} of removed water (or even less), whereas it is about 1000–2000 kWh.t^{-1} for spray-drying towers (Jeantet et al., 2008). In the manufacture of dairy powders, it makes sense to use vacuum evaporation before spray drying in order to concentrate the dry matter content of the product as high as possible, with the objective to have a minimum quantity of water to remove during the drying step and to reduce the overall energy cost of the whole production process. However, the concentrate must remain sprayable at the inlet of the drying tower. The apparent viscosity of the concentrate thus determines the maximum achievable concentration for the concentration step. In fact, the maximal concentration at the outlet of the falling-film evaporator is expressed in terms of dry matter content (DM in % w/w). The maximum DM is about 50–55% for skim milk, 55–60% for whey and 20–25% for protein isolates (Schuck, 2011), mostly depending on the protein content.

The complex configuration of the falling-film evaporation unit arises from the main objective of energy savings. For the skim milk concentration, different configurations of evaporation units and operating parameters exist, which generate variable temperature conditions and residence times in the equipment. Furthermore, the evaporation process and the resulting increasing concentration of the product along its passage in the evaporator induce several changes in the product regarding its physicochemical and structural properties. Control of the vacuum concentration step is of primary importance in the manufacture of dairy powders as the properties of the resulting concentrate affect those of the final powder. As an example, the viscosity of the concentrate influences the size of the droplets at the inlet of the drying tower; it affects the time-temperature history of the droplets, the size distribution of the granules and, as a consequence, the end-use properties of the powder (flowability, rehydration properties, etc.) (Baldwin et al., 1980; Bienvenue et al., 2003a; Ferguson, 1989; Fernandez-Martin, 1972).

This chapter consists of four main sections providing an overview of the operation of concentration by vacuum evaporation. Section 2.2.2 recalls the working principle and the design of the evaporators optimized for energy savings. Section 2.2.3 focuses on the physicochemical and structural changes induced by the evaporation process. Section 2.2.4 deals with the flow and heat transfer phenomena involved in the operation. Lastly, Section 2.2.5 focuses on the fouling and cleaning of evaporators. As with any heat exchanger, fouling problems occur after several hours of production runs that require appropriate cleaning procedures to recover their optimal performances.

2.2.2 Concentration in falling-film evaporators

2.2.2.1 Working principle of falling-film evaporators

Concentration by evaporation consists of exposing a product to temperature and pressure conditions allowing the vaporization of the solvent (generally water in food materials) and the concentration of non-volatile components (Schuck et al., 2012). Moreover, evaporation carried out under reduced pressure allows water boiling at a lower temperature and prevents heat damage to the product.

Different types of industrial evaporators are currently available, which differ in terms of the circulation of the products inside the equipment and the geometry of the heating surfaces; however, falling-film evaporators are predominantly used in the dairy industry. They are composed of three main parts: an evaporation housing filled with numerous high vertical tubes, a separator and a condenser (Figure 2.2.1). On the tube side, a liquid product flows down the inside surface in the form of a thin falling film of a few millimeters, whereas on the shell side, steam condenses and provides its latent heat for water evaporation within the product. A multi-hole distribution plate or a spray feed achieves a uniform and continuous distribution of the product at the top of the evaporation tubes. For early development of the film conditions, the inlet product must be at its boiling temperature

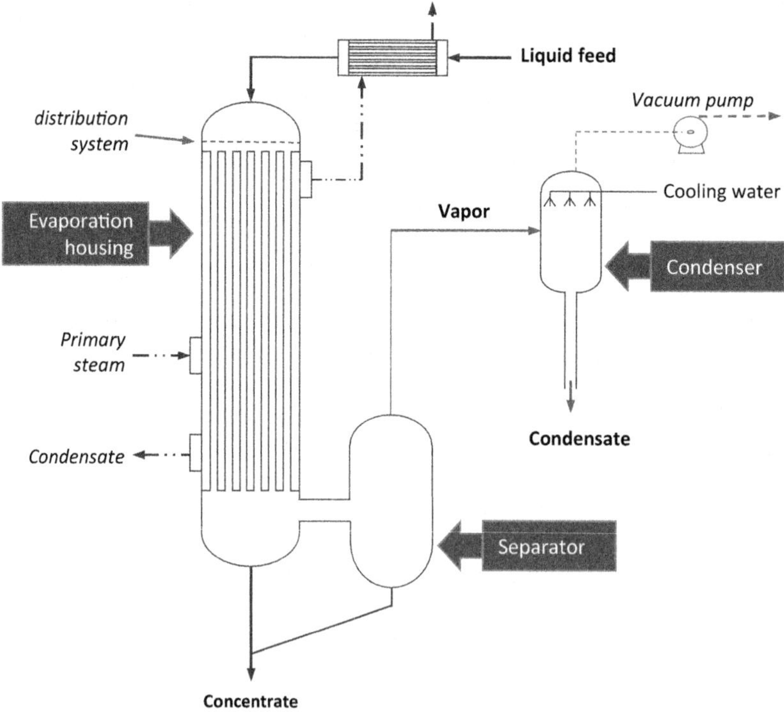

Figure 2.2.1 Single-effect falling-film evaporator.

and some vapor must be present within the distribution system. Consequently, the inlet temperature must be slightly higher than the water boiling temperature to achieve flash evaporation conditions (Gray, 1981). The concentrated product and the generated vapors flow down co-currently in the evaporation tubes and are separated in the so-called separator. The latent heat of the vapor is then used either to preheat the inlet feed stream, or as a "secondary vapor" in a subsequent evaporation tube in the case of a multiple-effect evaporator. A vacuum is generated through the condensation of the vapor in the condenser in contact with a cold medium (water). The saturation vapor pressure at the boiling temperature of the product sets the value of the vacuum.

The distribution system at the top of the evaporation tubes is a key component of the falling-film evaporators. It must achieve uniform distribution of the feed stream to all tubes and provide sufficient flow into each tube to realize a complete film covering the inner tube surface at any time. If this is not the case, incomplete wetting and overheating of the surface result, and fouling is likely to occur. An example of a distribution plate is presented in Figure 2.2.2. The distribution plate is perforated with small holes (6–8 mm diameter) and each inlet evaporation tube is surrounded by groups of three or six small holes (Figure 2.2.2). Morison (2015) mentioned that the liquid level over the plate must be equal to at least 20 mm to ensure the uniform distribution of the feed stream.

At the top of each evaporation effect, a part of vapor coming from flash evaporation and different evaporation rates in the passes of a same effect may flow upward (Figure 2.2.3). This horizontal vapor stream whose velocity is up to 25 m.s^{-1} may affect the liquid flow coming from the distribution plate (Morison and Broome, 2014). The use of a distribution system equipped with vapor tubes is thus recommended to minimize interference between the liquid and the vapor flows (Morison, 2015).

Falling-film evaporators are well adapted to heat-sensitive materials as they work under reduced pressure. Moreover, they are characterized by (i) high heat transfer coefficients related to the thin layer of the product and short residence times (between 0.5 and 2 minutes for a single pass in an evaporation tube) (Decloux and Rémond, 2009); (ii) low temperature differences between the condensing steam and the concentrate (about 2–3°C) (Ferguson, 1989; Mackereth, 1995; Winchester, 2000); and (iii) high heat transfer surfaces – an industrial falling-film evaporator can be composed of 4000 tubes from 30 to 50

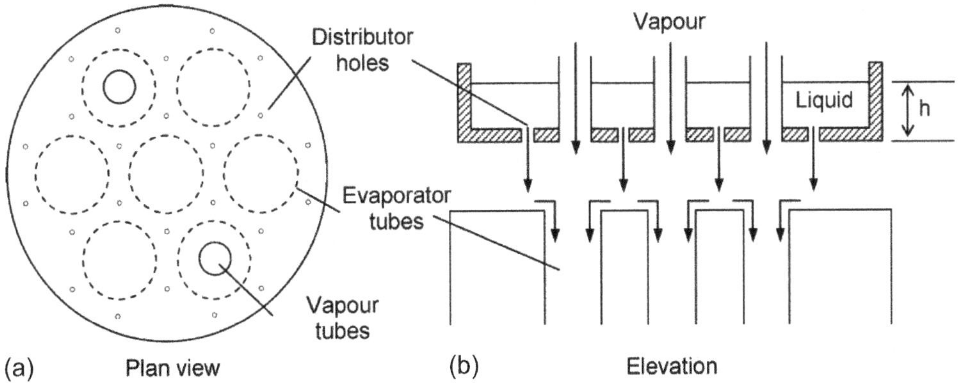

Figure 2.2.2 Distribution system at the inlet of a falling-film evaporator: (a) plane view and (b) vertical cross section.

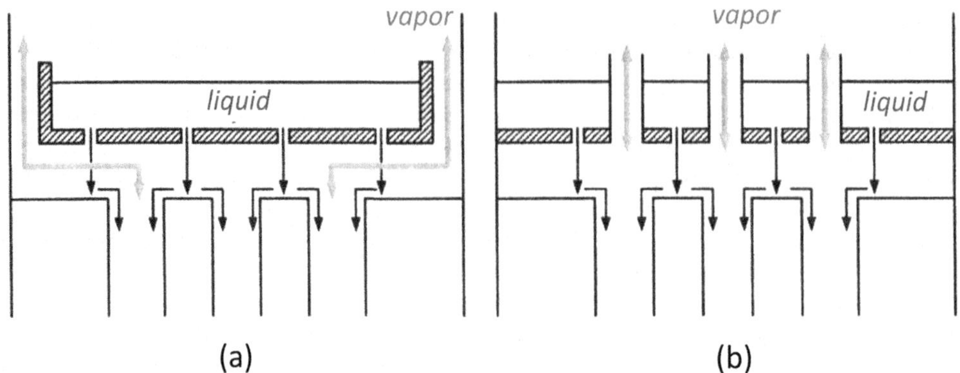

Figure 2.2.3 Vertical cross section of two distributor designs showing the liquid product and possible product vapor flows: (a) without vapor tubes and (b) with vapor tubes. Morison (2015).

mm in diameter and up to 22 meters high (Decloux and Rémond, 2009). However, falling-film evaporators are quite susceptible to fouling because of a falling-film breakdown that leads to the incomplete wetting of the heating surfaces, and low falling-film shear stresses. As the film flows down by gravity, particle deposition on the heating surfaces is favored (Winchester, 2000).

2.2.2.2 *Design of multistage falling-film evaporators to increase energy efficiency*

Whatever the process (concentration, drying, etc.), water removal using evaporation is an energy-intensive operation because an energy supply equivalent to the latent heat of water vaporization is needed. The evaporation of 1 kg of water in an evaporator is assumed to require 1.1 kg of condensing steam which means 2300 kJ.kg^{-1} of evaporated water. Currently, energy savings are made through the arrangement of evaporators and the use of vapor recompression systems.

A multiple-effect evaporator is composed of several evaporation stages in series where the vapor generated in an effect is reused as the heating medium to perform the evaporation in a subsequent effect (Figure 2.2.4). The energy cost for an evaporator with n effects is then approximately equal to 2300/n kJ.kg^{-1} water removed. However, the evaporation temperature decreases all along the evaporator due to the heat transfer from the condensing steam to the boiling product, the pressure drop and the boiling point elevation of the concentrates (Decloux and Rémond, 2009). The temperature decrease from one effect to another is greater than or equal to 5°C (Schuck et al., 2012).

The difference between the boiling temperature in the first effect and that in the latter effect determines the maximal number of effects for an evaporator. The heat sensitivity of the product limits the boiling temperature in the first effect, whereas the minimal achievable pressure sets the boiling temperature in the last effect. It depends on the temperature and the flow rate of the cooling water in the condenser as well as the concentrate viscosity. Nowadays, most evaporators for dairy products operate between 70°C in the first effect and 45°C in the last effect (Bouman et al., 1993; Gray, 1981; Kjaergaard Jensen and Oxlund, 1988). A high number of effects come with greater energy savings but imply larger heating surfaces, which means higher investment costs, higher cleaning costs and a lower final stage temperature, which means that a higher viscosity concentrate and lower pressure are ensured in the system. Consequently, the number of effects for an industrial evaporator is between three and six.

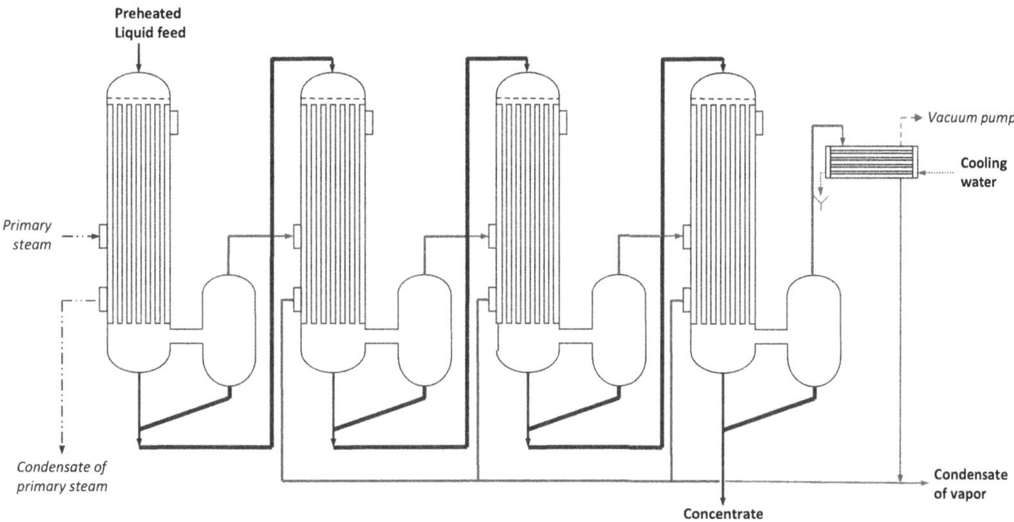

Figure 2.2.4 Multiple-effect falling-film evaporator.

Another way to save energy in falling-film evaporators is to use vapor recompression. Some or all of the vapor from one effect is compressed, resulting in an increase in temperature and it is then used to heat the same effect again. Compression can be carried out either thermally (thermal vapor recompression [TVR]) or mechanically (mechanical vapor recompression [MVR]). TVR allows an increase in the enthalpy of the vapor from an evaporation unit by mixing it with live steam (8–12 bars) previously accelerated using a venturi, mostly at a 1:1 w/w ratio. MVR uses volumetric or centrifugal compressors to recompress the vapor from an evaporation unit. In general, MVR is associated with effects characterized by low product concentrations and high evaporation rates, whereas TVR is associated with effects at high concentrations and low evaporation rates (Decloux and Rémond, 2009).

Evaporation plants are composed of a combination of these different technologies including preheaters, coolers and regenerating systems. The arrangement of the plants depends on the type of product to be concentrated and the final DM content expected. An example of an arrangement for an industrial multi-effect evaporator is presented in Morison (2015). It is a four-effect evaporator where the two first effects are associated with one MVR and the two last effects with one TVR. The first effect is subdivided into two passes (i.e. sections) and the second effect is subdivided into five passes, whereas there is only one pass in the third and fourth effects The subdivision of an effect into several passes is used to ensure good wettability of the inner surface of the evaporation tubes.

During the concentration of whey and its derivatives at a DM content of over 55%, spontaneous lactose crystallization can occur in evaporators. It is then necessary to maintain the evaporation temperature over the lactose solubility in the last effect, where there is a higher concentration of lactose. A solution is that the concentrate may pass through some effects in reverse order. Kjaergaard Jensen and Oxlund (1988) provide the case for whey concentration in a seven-effect falling-film evaporator (numbered from 1 to 7). The feed stream passes through the effects 1–2-3–4-7–6-5 where their evaporation temperatures are 68–65-61–57-39–45-50°C, respectively. The temperature of the last effect is then higher than the previous two effects.

2.2.3 Changes caused by vacuum concentration

The evaporation process leads to the concentration of all components within the product, which can induce major physicochemical and structural modifications. First, the pH of the concentrates decreases due to both the concentration and heating. Indeed, the increasing ionic strength related to the increasing ion content leads to an increase in the ion solubility constants, a decrease in the pKa of acid functions and the conversion of species into less protonated forms with a concomitant release of protons. Likewise, the increasing concentration and decreasing temperature along the evaporator lead to the precipitation of inverse soluble salts such as calcium phosphate with the release of protons. During skim milk concentration, the pH decreases from 6.66 at 9% DM (w/w) to 6.64, 6.47 and 6.17 at 12, 20 and 48% DM, respectively (Oldfield et al., 2005). Bienvenue et al. (2003b) also reported a decrease of 0.5–0.6 units between a liquid skim milk at 9% DM and the resulting concentrate at 45% DM, whereas the ionic strength of the latter would be almost five times higher compared to liquid skim milk.

Another consequence of calcium phosphate precipitation in milk is the transfer of calcium and phosphate from the soluble to the colloidal phase (Le Graët and Brulé, 1982; Liu et al., 2012; Vujicic and DeMan, 1966). Le Graët and Brulé (1982) noticed that the soluble calcium and phosphate contents increase by 2.2 and 2.5, respectively, for a concentration factor of 5.

The Ca^{2+} activity increases slightly but not proportional to the concentration factor due to the decreasing activity coefficients and the transfer of a part of the soluble calcium to the colloidal phase in the case of milk (Anema, 2009; Le Graët and Brulé, 1982). In the meantime, the ratio of monovalent to divalent cations increases markedly (Singh and Newstead, 1992).

During concentration, protein association reactions are favored due to: (i) the concentration itself that leads to closer packing of casein micelles and serum proteins; (ii) the decrease in pH and the increase in calcium ion activity that promote less electrostatic repulsions; and (iii) the increase in the ionic strength leading to a thinner electrical double layer. Moreover, in comparison to heat exchangers and due to the decreasing evaporation temperature from one effect to another, several combinations of temperature conditions and residence times are applied to a concentrate all along its passage in the evaporator. Singh and Newstead (1992) provided some values in the case of a four-effect evaporator used for milk concentration from 9.5 to 48% DM. The residence time in each effect is about 60 seconds and the evaporation temperatures are about 72, 64, 56 and 50°C in the first, second, third and fourth effect, respectively. These different heat treatments could favor whey protein denaturation leading to some modifications in the protein state and the functional properties of the concentrate. However, some authors suggested that the level of denaturation would be minimal in falling-film evaporators (Gray, 1981; Kjaergaard Jensen and Oxlund, 1988; Oldfield et al., 2005). Singh and Creamer (1991) followed the denaturation of whey proteins during the manufacture of skim milk powder. They showed that the denaturation of the two major whey proteins, β-lactoglobulin and α-lactalbumin, mainly occurs during preheating and that the denaturation rate is all the higher as the heat treatment is intense . They added that further evaporation increased the extent of whey protein denaturation, in contrast with subsequent drying. As an example, for a preheat treatment at (72°C, 15 seconds), the percentage of β-lactoglobulin undenatured after preheating, vacuum evaporation and drying was 94.2, 92.7 and 92.5%, respectively, whereas for a preheat treatment at (120°C, 2 min), this percentage decreased to 14.2, 9.7 and 9.0%. Likewise, Oldfield et al. (2005) showed that preheating had a great impact on the level of whey proteins denaturation, whereas it was minimal in the evaporator. The combined effect of a decreasing temperature and an increasing concentration of milk components along the evaporation process may explain this result. However, the extent of whey proteins denaturation is not only a function of temperature and holding time, but it is also considerably influenced by pH, ionic strength, protein and calcium concentrations as well as lactose content, all parameters evolving with concentration. Moreover, the susceptibility of whey proteins to heating depends on each type of protein. In milk and whey concentrates, a high lactose concentration stabilizes the conformation of β-lactoglobulin and delays denaturation for heating temperatures between 75 and 90°C, whereas α-lactalbumin denaturation is unaffected (Anema, 2000; Plock et al., 1998a, 1998b).

Concerning the effects of the evaporation process on casein micelles, Liu et al. (2012) noticed that during milk concentration, water is removed preferentially from the serum, which means that water within the micelles is only removed after the removal of most of the bulk water. Bienvenue et al. (2003b) suggested that the progressive decrease in pH and the increase in ionic strength would reduce the net negative charge on the micelles. This would lead to a decrease in electrostatic repulsions between the micelles. In the meantime, the quantity of colloidal calcium phosphate increases due to the transfer of a part of the soluble calcium and phosphate to the casein micelles, which contributes to increasing their size, leading to their destabilization. As noticed by Liu et al. (2012), there is some uncertainty around where the additional calcium is deposited on the micelles. It can be on the micelles themselves or on the existing colloidal calcium phosphate nanoclusters. If calcium ions deposit on micelles, a shielding of charge may occur and the hairy k-casein layer may collapse.

The evaporation process modifies the physical properties of a product such as its density, surface tension, thermal conductivity, boiling point elevation and flow properties. All these parameters are useful to understand the flow patterns and the heat and mass transfer phenomena and ultimately to design evaporators. However, data relative to concentrated systems are scarce in the literature and much of the data are related to milk concentrates. Some recent works (Madoumier et al., 2015; Munir et al., 2016; Zhang et al., 2014) deal with the modeling of milk properties for further simulation of the operation of vacuum concentration using commercial process simulators. Milk is considered a mixture of water and four components (fat, proteins, sugar and minerals) modeled as "pseudo-components" in the process simulator. These authors compared the sensitivity of empirical models relative to milk properties with experimental data and selected the most relevant models; some selected by Madoumier et al. (2015) to model skim milk concentration in an industrial falling-film evaporator are presented in Table 2.2.1.

Among the physical properties previously mentioned, the concentrate viscosity is the critical parameter for a falling-film evaporator as it determines the maximum achievable DM content. Both an increase in DM content and a decrease in temperature along the evaporator contribute to increasing the viscosity (Jebson and Iyer, 1991), leading to film thickening and lower heat transfer performances and requiring higher liquid flow rates to maintain a continuous and uniform film (Gray, 1981). A high concentrate viscosity also affects the drying step and the final powder quality (Baldwin et al., 1980; Bienvenue et al., 2003a), influencing the size of the droplets at the time of spraying, which modifies the subsequent drying kinetics in the dryer and the size distribution of the powder grains. The drying of a viscous concentrate results in a lower bulk density and wetting ability of the powder and a higher solubility index.

In general, the viscosity of a fluid depends on the inherent viscosity of the continuous phase and the volume fraction occupied by the dispersed elements. In dairy concentrates, the viscosity of the continuous phase and the volume fraction are determined,

Table 2.2.1 Selected property models for milk and food components

Property	Model	Reference
Heat capacity (Cp)	$Cp = \sum Cp_i \times w_i$ with: $Cp_{fat} = 1.9842 + 1.4733 \times 10^{-3} \times T - 4.8008 \times 10^{-6} \times T^2$ $Cp_{nitrogen\ content} = 2.0082 + 1.2089 \times 10^{-3} \times T - 1.3129 \times 10^{-6} \times T^2$ $Cp_{carbohydrate} = 1.5488 + 1.9625 \times 10^{-3} \times T - 5.9399 \times 10^{-6} \times T^2$ $Cp_{ash} = 1.0926 + 1.8896 \times 10^{-3} \times T - 3.6817 \times 10^{-6} \times T^2$	(Choi and Okos, 1986)
Boiling point elevation (BPE)	$\text{BPE} = \left[\dfrac{1}{\dfrac{1}{T_{SAT}} + \dfrac{R \times \ln(1 - x_{SOL})}{H_{VAP}}} \right] - T_{SAT}$	(Winchester, 2000)
Density (ρ)	$\rho = \dfrac{1}{\sum w_i / \rho_i}$ with: $\rho_{fat} = 925.59 - 4.1757 \times 10^{-1} \times T$ $\rho_{nitrogen\ content} = 1329.9 - 5.1840 \times 10^{-1} \times T$ $\rho_{carbohydrate} = 1599.1 - 3.1046 \times 10^{-1} \times T$ $\rho_{ash} = 2423.8 - 2.8063 \times 10^{-1} \times T$ $\rho_{water} = 997.18 + 0.0031439 \times T - 0.0037574 \times T^2$	(Choi and Okos, 1986)
Thermal conductivity (λ)	$\lambda = \left(326.58 + 1.0412 \times T - 3.37 \times 10^{-3} \times T^2\right) \times \left(0.46 + 0.54 \times w_{water}\right) \times 1.73 \times 10^{-3}$	(Riedel, 1949)
Viscosity (μ)	$\eta = 1000 \times \eta_{water} \times \exp\left(\dfrac{\sum A_i \times w_i}{w_{water}}\right)$ with: $A_{fat} = 3.46 - 0.025 \times T + 1.6 \times 10^{-4} \times T^2$ $A_{carbohydrate} = 3.35 - 2.38 \times 10^{-2} \times T + 1.25 \times 10^{-4} \times T^2$ $A_{native\ WP} = 8.24 - 0.0367 \times T + 0.000528 \times T^2$ $A_{denaturated\ WP} = 9.66 - 0.0434 \times T + 0.000438 \times T^2$ $A_{casein} = 22.578 - 0.592 \times T + 0.058 \times T^2$	(Morison et al., 2013)
Surface tension (σ)	General model (whole and skim milk): $\sigma = 55.6 - 0.163 \times T + 1.8 \times 10^{-4} \times T^2$	(Bertsch, 1983)

Source: Adapted from Madoumier et al. (2015).

respectively, by the lactose concentration and the proteins (caseins and native and denatured whey proteins). According to Snoeren et al. (1982), the viscosity of milk concentrates is better described as a function of the volume fraction than the DM content. Moreover, any treatment and factor that increases the volume fraction leads to an increase in viscosity (Anema et al., 2014). The volume fraction of a dairy concentrate depends on the protein composition, the concentration and the hydration state of each type of protein. The pH and the ionic strength also influence the protein hydration, which explains the higher viscosity of nanofiltered concentrated acid wheys compared to standard concentrated acid wheys (Jeantet et al., 1996). Heat treatment is a technological treatment that modifies protein hydration due to the heat denaturation of whey proteins for temperatures higher than 70°C. It promotes protein association through denatured whey proteins, which results in an increase in protein voluminosity and, consequently, an increase in viscosity (Snoeren et al., 1982).

The rheological behavior of concentrates may change during concentration. As an example, skim milk exhibits a Newtonian behavior but becomes non-Newtonian (shear-thinning) from about 35% DM. At higher concentrations (above 40% DM), a yield stress factor appears and a small change in concentration leads to a sharp increase in viscosity. Therefore, viscosity measurements depend on the operating conditions, mainly the temperature and shear rate. Typical shear rates for process operations are 10 to 10^3 for mixing and stirring, 1 to 10^3 for pipe flowing and 10^3 to 10^4 for spraying (Steffe, 1992). The shear rate of a falling film is estimated at 100 s^{-1} (Ang, 2011).

Some viscosity data on dairy concentrates, mainly skim milk, have been published, including the works of Velez-Ruiz and Barbosa-Canovas (1997), Snoeren et al. (1982) and Bloore and Boag (1981). Morison et al. (2013) have developed a model from these experimental data that allows predicting milk and cream viscosities as a function of their composition and their temperature at a shear rate of 2000 s^{-1}.

In the manufacturing process of dairy powders, Westergaard (2004) advises a concentrate viscosity of less than 100 mPa.s^{-1} at the spraying step. Likewise, Winchester (2000) mentioned that industrial drying towers operate at viscosities of between 30 and 70 mPa.s^{-1}. In practice, however, the operation limits for falling-film evaporators are expressed in terms of DM content, as mentioned in the introduction to this chapter.

Preheating before evaporation is an efficient tool to ensure food safety and to control the extent of protein denaturation and the concentrate viscosity (Gray, 1981; Kjaergaard Jensen and Oxlund, 1988). This goal is achieved by means of an appropriate combination of preheat temperature and holding time. Viscosity increases when the concentration and the exposure time to heat increases. Consequently, for a constant DM content, heat treatment should be as short as possible to reduce the concentrate viscosity. It implies a low evaporation temperature in a falling-film evaporator and a high preheat temperature before evaporation (Ferguson, 1989). In the same way, Bloore and Boag (1981) showed that heat treatment at a high temperature and a short holding time produced a skim milk concentrate at a lower viscosity compared to a treatment at a low temperature and a long holding time and this, for the same extent of protein denaturation.

The storage time between the falling-film evaporator and the spray dryer affects the concentrate viscosity, with viscosity increasing rapidly with time at a high temperature and a high DM content. This phenomenon is called age-thickening. Bienvenue et al. (2003b) followed the evolution of the apparent viscosity of a 45% DM skim milk concentrate stored at 50°C from 0 to 19 hours. The apparent viscosity of the concentrate increased considerably after storage for 4 hours at low shear rates, and after 6 hours at all shear rates. After storage for 19 hours, the apparent viscosity was five times higher at 1000 s^{-1}.

2.2.4 *Heat transfer and film flow*

As for heating equipment, the efficiency of the concentration depends on the heat transfer through the heating surface and the flow conditions of the concentrate.

The heat flux through the tube wall is a function of the global heat transfer coefficient h (expressed in $kW.m^{-2}.K^{-1}$), the heat exchange surface area A and the temperature difference between the steam condensing film and the concentrate ΔT (Figure 2.2.5). In industrial multiple-effect evaporators, h ranges between 0.3 and 3.0 $kW.m^{-2}.K^{-1}$ (Jebson and Chen, 1997) with a mean value close to 1.65 $kW.m^{-2}.K^{-1}$, but it mainly depends on the nature of the product and the operating conditions. Some values are given in Walstra et al. (2006).

The global heat transfer coefficient h is a combination and a function of four specific heat transfer coefficients:

- Heat transfer coefficient on the heating steam side, in the range 5–10 $kW.m^{-2}.K^{-1}$ without incondensable gas and superheated vapor (Bimbenet et al., 2002).
- Heat transfer coefficient in the tube wall thickness, around 15 $kW.m^{-2}.K^{-1}$ for a stainless steel wall of 1–2 mm thickness (Bimbenet et al., 2002).
- Heat transfer coefficient through the eventual fouling layer. The fouling layer induces a sharp decrease in the global heat transfer coefficient h but it is difficult to measure.
- Heat transfer coefficient on the product side at the interface with the boiling liquid. This latter is a key parameter for the design of evaporators, but it requires knowledge of numerous data related to the product (viscosity, surface tension, conductivity, etc.), the process (heat flux, wetting rate, etc.) and the heating surface (nature, geometry, roughness, etc.). Thus, it is difficult to predict the values of this coefficient from the existing formula available in the literature (Adib and Vasseur, 2008).

The boiling regime of the film strongly influences the heat transfer coefficient on the product side (Adib and Vasseur, 2008; Bouman et al., 1993; Broome, 2005). In the case

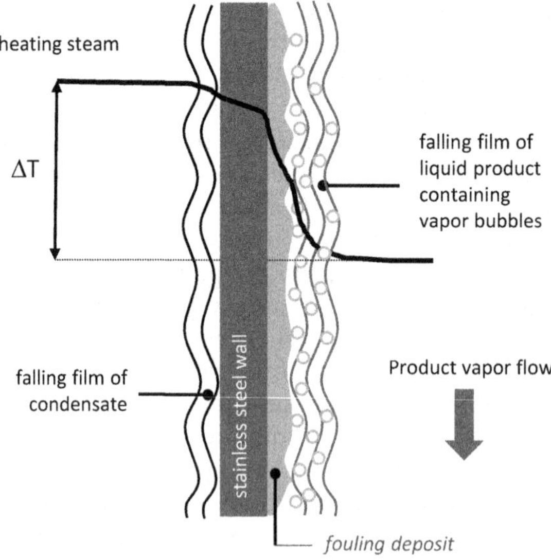

Figure 2.2.5 Heat transfer through a wall between the condensate from heating steam and the boiling liquid.

of non-nucleate boiling (convective boiling), i.e. at low temperature difference between the steam condensing film and the concentrate, the heat flux passes through the liquid film and evaporation takes place at the film–vapor interface, without bubble formation at the metal surface. The heat transfer coefficient on the product side is then a function of the flow pattern. In the case of nucleate boiling (i.e. at larger temperature differences), some bubbles are formed at the metal surface. They grow and travel to the film–vapor interface. The transition from one regime to the other is not well defined. Bouman et al. (1993) stated that nucleate boiling starts at a temperature difference of 0.5°C for milk and about 5°C for water, the difference being due to the lower surface tension of milk.

The nucleate boiling regime improves heat transfer compared to the convective regime (Bouman et al., 1993; Gourdon et al., 2015), but fouling is more susceptible to occur due to the drying of the surface below the bubble (Mackereth, 1995). Bubble formation also depends on the surface material properties (roughness) and the presence of particles (fat globules, impurities, etc.) that may act as nuclei (Adib et al., 2009; Chen and Jebson, 1997; Jebson and Chen, 1997). Gourdon et al. (2015) demonstrated that the formation of bubbles is caused by vapor entrapment in waves rather than due to nucleate boiling. This means that the flow of concentrate waves at different velocities leads to the formation of a secondary film that may collapse on the main one, trapping underneath a column of bubbles formed at the liquid–vapor interface.

In industrial falling-film evaporators, the vapors flow down co-currently with the liquid film but a part may also flow upwards in the first stages (Morison and Broome, 2014). Their velocities have been estimated to be between 10 and 30 $m.s^{-1}$ (Broome, 2005) and they may influence the performances of evaporators. Their flow exerts a shear stress at the film surface and influences the flow pattern. The co-flowing vapor influences the film thickness, velocity and surface state (steam bubble) (Mura and Gourdon, 2016). Besides, vapors generate a pressure drop all along the tubes which affects the local evaporation temperature (Bouman et al., 1993; Gourdon and Mura, 2017).

The flow in a falling-film evaporator is characterized by the wetting rate Γ. It is defined as the mass flow rate of the product per unit of circumference of a tube and it is expressed in $kg.s^{-1}.m^{-1}$. A minimum flow rate is required to maintain a complete falling film on the surface and it is expressed using the minimum wetting rate Γ_{min}. If the liquid is not sufficient, dry patches will appear on the surface of the evaporation tube which will reduce the heat exchange surface. Some authors such as Paramalingam et al. (2000) have determined Γ_{min} for the concentration of dairy products. They measured liquid-tube advancing contact angles for water and milk (10–40% DM) and used the correlation of Hartley and Murgatroyd (1964) (Equation 2.2.1) to determine Γ_{min}.

$$\Gamma_{min} = 1.69\left(\frac{\eta\rho}{g}\right)^{\frac{1}{5}}\left[\sigma(1-\cos\theta)\right]^{\frac{3}{5}} \tag{2.2.1}$$

where η is the viscosity (Pa.s); ρ is the density ($kg.m^{-3}$); σ is the surface tension ($N.m^{-1}$); and θ is the contact angle (rad). At 55°C, Γ_{min} is equal to 0.189 and 0.145 $kg.m^{-1}.s^{-1}$ for water and concentrated milk at 40% DM, respectively.

Morison et al. (2006) experimentally determined Γ_{min} for aqueous solutions of glycerol, ethanol and calcium chloride, covering a wide range of properties such as viscosity, surface tension, density and contact angle. Γ_{min} ranges between 0.084 and 0.193 kg.m^{-1}.s^{-1} depending on the products and the following correlation was deduced:

$$\Gamma_{min} = 0.13\left((1-\cos\theta)\sigma\right)^{0.764}\rho^{0.255}\mu^{-0.018} \tag{2.2.2}$$

In fact, a wetting rate of 1000 L.h^{-1}.m^{-1} is recommended for industrial falling-film evaporators with a safety margin and in function as the viscosity (Adib, 2008).

The flow of a product in a falling-film evaporator can also be characterized using the film Reynolds number and the classification of Bird et al. (1960) that distinguish three regions for the flow regime: (i) laminar flow if $Re_f < 25$, (ii) wavy laminar flow if $Re_f < 1000$–2000 and (iii) turbulent flow if $Re_f > 1000$–2000. In falling-film evaporators, Re_f would be less than 500 (Mackereth, 1995).

The thickness of the falling film mainly depends on its viscosity and the wetting rate, but some other factors such as air bubbles and the flow of vapor may also have an influence. For Newtonian fluids in laminar conditions, the film thickness is defined as follows:

$$\delta = \sqrt[3]{\frac{3\eta\Gamma}{\rho^2 g}} \tag{2.2.3}$$

For non-Newtonian fluids, the Power Law and Herschel–Bulkley equations must be used.

2.2.5 *Fouling and cleaning of falling-film evaporators*

2.2.5.1 *Fouling of evaporators*

A fouling deposit on the inner tube surface induces a sharp decrease in the heat transfer coefficient due to the low thermal conductivity of the deposit, which in turn decreases the process efficiency and the duration of production runs. Additionally, surface deposits may become a potential site for the growth of thermophilic bacteria that can be further released into the product (Hinton, 2003). The formation of biofilms in industrial falling-film evaporators is a real problem as the cleaning of evaporators is mainly imposed by a bacterial contamination rather than fouling (Broome, 2005). During the manufacture of skim milk powder, the preheating sections and the falling-film evaporator are the main growth sites for thermophilic spores due to their operating temperature range (45–75°C) (Murphy et al., 1999; Scott et al., 2007). Contamination may occur due to insufficient heat treatment (Murphy et al., 1999) or a residual fouling deposit after cleaning the evaporators, where bacteria can be entrapped (Hinton, 2003; Scott et al., 2007).

Few authors have studied the fouling of evaporators but they agree on the mechanisms involved in deposit formation and the composition of the resulting deposit that should be different from those evidenced in heat exchangers (Jeurnink and Brinkman, 1994; Morison, 2015). As an example, proteins are present in the fouling deposit formed during skim milk concentration although they were previously denatured in the preheating sections and the evaporation temperature is less than 70°C (Jeurnink and Brinkman, 1994). The increasing concentration and viscosity of the product modify its biochemical and physical properties (precipitation of oversaturated calcium phosphate, lower protein thermal stability, etc.), which favors the deposition of some components on the heating surfaces. Therefore, deposits in evaporators have a complex composition

Table 2.2.2 Composition of deposit (in grams) obtained in an industrial falling-film evaporator per tonne of processed product

	Whey 5.7% DM	Whey 28% DM	Whole milk
COD	27	806	192
Proteins	12	79	52
Fat	0.6	20	34
Calcium	17	178	9.6
Phosphate	23	32	12.6
Citrate	6	631	n.d.
Lactose	3	77	n.d.

Source: Jeurnink and Brinkman (1994).

and structure made of mineral and organic substances as well as particles from the crystallization of some elements, such as lactose in the case of dairy products (Decloux and Rémond, 2009).

The composition of fouling deposits formed during the concentration of whey at 5.7% DM, whey at 28% DM and whole milk in an industrial falling-film evaporator are given in Table 2.2.2. The deposits from whey concentration are mainly mineral whereas the deposit from milk concentration is mainly proteinaceous. They also contain lactose to a lesser extent, although it is the main component of whey. More deposit is formed during the concentration of whey at 28% DM rather than during the concentration of whey at 5.7% DM. It contains mostly calcium citrate and little phosphate. The chemical oxygen demand (COD) is high due to the contribution of citrate. Moreover, the presence of calcium citrate has recently been evidenced in deposits formed during the concentration of whey permeate (Vavrusova et al., 2017) and the concentration of hydrochloric acid whey (Tanguy et al., 2019).

During the concentration of acid whey, deposit formation is more important at pH 5.9 than at pH 4.5 (Kessler, 1986). Moreover, the mineral proportion (phosphate, calcium, citrate) is higher at pH 5.9 as calcium phosphate is more soluble at a low pH (Kessler, 1986).

Several factors may influence deposit formation:

- *DM content.* After an induction period, the formation rate of the deposit is higher at high DM content.
- *Evaporation temperature.* A high evaporation temperature at low DM content leads to a large increase in the formation rate of the deposit. The effect is less pronounced at higher DM content, at least for low evaporation temperatures (<60°C) (Kessler, 1986). A maximum evaporation temperature of 70°C is recommended for a milk concentration at low DM content and 58–60°C at high DM content (Ferguson, 1989).
- *Heat flux.* High heat flux favors deposit formation (Gray, 1981; Kessler, 1986).
- *Dissolved air.* In the case of incomplete tube wetting and nucleate boiling onto a heating surface, air bubbles may act as potential nuclei for deposit formation (Daufin and Labbé, 1998; Jeurnink et al., 1996).
- *Produced vapor.* A falling-film breakdown may occur if the flow of vapor is too fast. This phenomenon is likely to occur in the first effect of evaporators where the viscosity of the inlet feed stream is low (Jebson and Iyer, 1991).

- *Preheating.* A preheat treatment and the holding time of the feed stream before concentration may reduce the fouling of evaporators during the concentration of wheys and their derivatives (Kessler, 1986; Morison and Tie, 2002).
- *Design of the distribution system.* As mentioned previously, the optimal design of the distribution system is of great importance for limiting fouling. Morison (2015) provides some recommendations concerning the design of a distribution system and the inlet feed stream.

2.2.5.2 Cleaning of evaporators

The duration of production runs depends on the processed fluids: up to 18–20 hours for skim milk concentration and only 10 hours for MPC concentration (Broome, 2005). After a production run, complete cleaning of an industrial evaporator lasts about 3–4 hours. Cleaning the evaporators generates supplementary production costs with the use of chemicals, the water consumption, the wastewater treatment and the shutdown of production. Morison (2015) considered that an increase in the production time from 20 to 21 hours leads to a decrease of about 5% of these costs.

A cleaning sequence is generally composed of several successive steps: (1) water rinsing, (2) alkaline flush (i.e. cleaning solution put in the sewer), (3) alkaline recirculation, (4) water rinsing, (5) acid recirculation and (6) water rinsing. An alkaline cleaning solution removes proteinaceous components from the deposit whereas an acid cleaning solution removes minerals.

Jeurnink and Brinkman (1994) carried out a complete study on the cleaning of industrial falling-film evaporators during skim milk and whey concentrations. They focused on the efficiency of cleaning sequences as a function of the order of the steps, the concentrations and the flow rates of cleaning solutions in evaporators. They proposed some mechanisms of deposit removal during the cleaning steps. The following two sections deal with the results of their study.

2.2.5.2.1 Cleaning after milk processing The experimental results showed that about 96% of proteins and 52% of calcium are removed during an alkaline flush whereas only 2% of proteins are removed during the subsequent recirculation step, which implies that the duration of the alkaline recirculation step (generally about 40 minutes) could be greatly reduced. The acid recirculation step in turn allows mineral deposits and some remaining organic components to be removed.

According to Jeurnink and Brinkman (1994), the same mechanism of deposit removal would occur in heat exchangers and falling-film evaporators, the deposit being a spongy proteinaceous matrix in which minerals and fat are entrapped. The first step would be a swelling of the deposit layer in contact with the alkaline solution. This means a diffusion of the solution into the deposit and the subsequent formation of cracks that favors the diffusion of the cleaning solution inside. This swelling would result in a loosening of large lumps of deposit (containing organic and mineral components) due to shear stresses exerted by the fast flow of the cleaning solution. The final step would be the removal of the remaining minerals attached to the heating surface during acid cleaning.

A lower flow rate or a higher alkali concentration (for example, in the last effects of falling-film evaporators due to water evaporation) leads to less efficient cleaning. As an example, a 1% NaOH concentration is more efficient than a 2% concentration. Likewise, a wetting rate of 0.860 $L.m^{-1}.s^{-1}$ is better than 0.222 $L.m^{-1}.s^{-1}$. Generally speaking, this has been evidenced by Mercadé-Prieto et al. (2007, 2008), who showed that a high alkali concentration, typically >0.2 M, comes with lower swelling and consequently poorer dissolution of the deposit, although higher potential cleavage kinetics are expected; this results in a non-maximum but optimum concentration for cleaning dairy deposits (Bird and Fryer, 1991). Moreover, Jeurnink and Brinkman (1994) suggested that the residual deposit characterized by a brown color and a rubber-like top layer and observed in the cases of a lower flow rate and a higher alkali concentration would result from a gelation or polymerization reaction of proteins in contact with the alkaline solution. The formed gel would prevent the diffusion of the cleaning solution into the deposit and would reduce the efficiency of the cleaning. This underlines the great importance of the deposit swelling in the deposit removal process.

2.2.5.2.2 Cleaning after whey processing Jeurnink and Brinkman (1994) showed that the order of the cleaning steps is a critical parameter for efficient cleaning of evaporators after whey processing. Cleaning was more efficient if the cleaning procedure was acid cleaning followed by alkaline cleaning instead of starting with an alkaline cleaning. This result is related to the deposit composition, mainly mineral (Table 2.2.2). Besides, whey deposit has a more compact structure and is less spongy than the deposit formed during milk concentration. It should be then a complex of proteins associated with calcium salts (calcium phosphate and calcium citrate).

Jeurnink and Brinkman (1994) noticed that some lumps of the deposit are already removed during whey processing and water rinsing, only due to the effect of shear stresses exerted by the high flowrate of cleaning solutions. They suggested that the cohesive forces between elements of the deposit are stronger than the adhesive forces between the deposit and the inner surface of the evaporation tubes. In addition, there is no swelling of the deposit in contact with an acid solution, contrary to the mechanism proposed during alkaline cleaning.

2.2.6 *Conclusion*

In the process scheme for the manufacture of dairy powders, the vacuum concentration step aims at removing as much water as possible from the product before spray drying in order to reduce the overall energy cost related to the production of powders. Consequently, the performance of this operation step is mainly evaluated through the maximum achievable DM content of the concentrate at the outlet of evaporators and the energy consumption of the equipment. Falling-film evaporators are energy-efficient devices due to the use of different energy recovery systems; however, their performances remain closely linked to the behavior of the concentrate during concentration. Indeed, the evaporation process induces strong modifications of the properties of the concentrates that affect their biochemical and physical properties such as viscosity. Moreover, the properties of the concentrate affect the properties of the final powder. As a consequence, better control of the global process scheme for the manufacture of powders requires a better understanding of this step and of the changes induced by vacuum evaporation. Therefore, it would be important to consider both dehydration steps, vacuum concentration and drying, as a whole.

References

Adib, T.A., 2008. *Estimation et lois de variation du coefficient de transfert de chaleur surface/liquide en ébullition pour un liquide alimentaire dans un évaporateur à flot tombant*. Agro Paris Tech, Massy, France.

Adib, T.A., Heyd, B., Vasseur, J., 2009. Experimental results and modeling of boiling heat transfer coefficients in falling film evaporator usable for evaporator design. *Chemical Engineering and Processing: Process Intensification* 48(4), 961–968.

Adib, T.A., Vasseur, J., 2008. Bibliographic analysis of predicting heat transfer coefficients in boiling for applications in designing liquid food evaporators. *Journal of Food Engineering* 87(2), 149–161.

Anema, S.G., 2009. Effect of milk solids concentration on the pH, soluble calcium and soluble phosphate levels of milk during heating. *Dairy Science and Technology* 89(5), 501–510.

Anema, S.G., 2000. Effect of milk concentration on the irreversible thermal denaturation and disulfide aggregation of beta-lactoglobulin. *Journal of Agriculture and Food Chemistry* 48(9), 4168–4175.

Anema, S.G., Lowe, E.K., Lee, S.K., Klostermeyer, H., 2014. Effect of the pH of skim milk at heating on milk concentrate viscosity. *International Dairy Journal* 39(2), 336–343.

Ang, K.L.J., 2011. *Investigation of rheological properties of concentrated milk and the effect of these parameters on flow within falling film evaporators*. University of Canterbury, Christchurch, New Zealand.

Baldwin, A.J., Baucke, A.G., Sanderson, W.B., 1980. The effect of concentrate viscosity on the properties of spray-dried skim milk powder. *New Zealand Journal of Dairy Science and Technology* 15, 289–297.

Bertsch, A.J., 1983. Surface tension of whole and skim-milk between 18°C and 135°C. *Journal of Dairy Research* 50, 259–267.

Bienvenue, A., Jimenez-Flores, R., Singh, H., 2003a. Rheological properties of concentrated skim milk: Influence of heat treatment and genetic variants on the changes in viscosity during storage. *Journal of Agriculture and Food Chemistry* 51(22), 6488–6494.

Bienvenue, A., Jimenez-Flores, R., Singh, H., 2003b. Rheological properties of concentrated skim milk: Importance of soluble minerals in the changes in viscosity during storage. *Journal of Dairy Science* 86(12), 3813–3821.

Bimbenet, J.J., Duquenoy, A., Trystram, G., 2002. *Génie des Procédés Alimentaires, Techniques industrielles et sciences de l'ingénieur*. Dunod, Paris, France.

Bird, R.B., Stewart, W.E., Lightfoot, E.N., 1960. *Transport phenomena*. John Wiley and Sons, New York.

Bird, M.R., Fryer, P.J., 1991. Experimental study of the cleaning of surfaces fouled by whey proteins. *Food and Bioproducts Processing* 69, 13–21.

Bouman, S., Waalewijn, R., Dejong, P., Vanderlinden, H.J.L.J., 1993. Design of falling-film evaporators in the dairy-industry. *Journal of the Society of Dairy Technology* 46, 100–106.

Bloore, C.G., Boag, I.F., 1981. Some factors affecting the viscosity of concentrated skim milk. *New Zealand Journal of Dairy Science and Technology* 16, 143–154.

Broome, S.R., 2005. *Liquid distribution and falling film wetting in dairy evaporators*. University of Canterbury, Christchurch, New Zealand.

Chen, H., Jebson, R.S., 1997. Factors affecting heat transfer in falling film evaporators. *Food and Bioproducts Processing* 75(2), 111–116.

Choi, Y., Okos, M.R., 1986. Effects of temperature and composition on the thermal properties of fooods, in: Le Maguer, M., Jelen P. (Eds.), *Food Engineering and Process Applications, Volume 1: Transport Phenomena*. Elsevier Applied Science Publishers, London, UK, pp. 93–101.

Daufin, G., Labbé, J.-P., 1998. Equipment fouling in the dairy application: Problem and pretreatment, in: Z. Amjad (Ed.), *Calcium phosphates in biological and industrial systems*. Kluwer Academic Publishers, Norwell, MA, p. 437.

Decloux, M., Rémond, B., 2009. Evaporation – Principes généraux, in: *Techniques de l'ingénieur*. Référence F3003 Paris. https://www.techniques-ingenieur.fr/base-documentaire/procedes-chimie-bio-agro-th2/operations-unitaires-du-genie-industriel-alimentaire-42430210/evaporation-f3003/.

Ferguson, P.H., 1989. Developments in the evaporation and drying of dairy products. *Journal of the Society of Dairy Technology* 42(4), 94–101.

Fernandez-Martin, F., 1972. Influence of temperature and composition on some physical properties of milk and milk concentrates. II. Viscosity. *Journal of Dairy Research* 39(1), 75.

Gourdon, M., Innings, F., Jongsma, A., Vamling, L., 2015. Qualitative investigation of the flow behaviour during falling film evaporation of a dairy product. *Experimental Thermal and Fluid Science* 60, 9–19.

Gourdon, M., Mura, E., 2017. Performance evaluation of falling film evaporators in the dairy industry. *Food and Bioproducts Processing* 101, 22–31.

Gray, R.M., 1981. Subject "skim milk" – Technology of skim milk evaporation. *Journal of the Society of Dairy Technology* 34, 53–57.

Hartley, D.E., Murgatroyd, W., 1964. Criteria for the break-up of thin liquid layers flowing isothermally over solid surfaces. *International Journal of Heat and Mass Transfer* 7(9), 1003–1015.

Hinton, A.R., 2003. *Thermophiles and fouling deposits in milk powder plants*. Massey University, Palmerston North, New Zealand.

Jeantet, R., Croguennec, T., Mahaut, M., Schuck, P., Brulé, G., 2008. *Les produits laitiers*, 2e ed. Tec&Doc Lavoisier, Paris, France.

Jeantet, R., Schuck, P., Famelart, M.H., Maubois, J.L., 1996. Nanofiltration benefit for production of spray-dried demineralized whey powder. *Le Lait* 76(3), 283–301.

Jebson, R.S., Chen, H., 1997. Performances of falling film evaporators on whole milk and a comparison with performance on skim milk. *Journal of Dairy Research* 64(1), 57–67.

Jebson, R.S., Iyer, M., 1991. Performances of falling film evaporators. *Journal of Dairy Research* 58(1), 29–38.

Jeurnink, T.J.M., Brinkman, D.W., 1994. The cleaning of heat exchangers and evaporators after processing milk or whey. *International Dairy Journal* 4(4), 347–368.

Jeurnink, T.J.M., Walstra, P., deKruif, C.G., 1996. Mechanisms of fouling in dairy processing. *Netherlands Milk and Dairy Journal* 50, 407–426.

Kessler, H.G., 1986. Multistage evaporation and water vapour recompression with special emphasis on high dry matter content, product losses, cleaning and energy savings, in: *Milk – The vital force*. Proceedings of the 22nd International Dairy Congress. Reidel Publishing Co., Dordrecht, The Hague, pp. 545–558.

Kjaergaard Jensen, G., Oxlund, J.K., 1988. Concentration and drying of whey and permeates. *Bulletin of the International Dairy Federation* 233, 4–20.

Le Graët, Y., Brulé, G., 1982. Effets de la concentration par évaporation et du séchage sur les équilibres minéraux dans le lait et les rétentats. *Le Lait* 62, 113–125.

Liu, D.Z., Dunstan, D.E., Martin, G.J.O., 2012. Evaporative concentration of skimmed milk: Effect on casein micelle hydration, composition, and size. *Food Chemistry* 134(3), 1446–1452.

Mackereth, A.R., 1995. *Thermal and hydraulic aspects of falling film evaporation*, University of Canterbury, Christchurch, New Zealand.

Madoumier, M., Azzaro-Pantel, C., Tanguy, G., Gésan-Guiziou, G., 2015. Modelling the properties of liquid foods for use of process flowsheeting simulators: Application to milk concentration. *Journal of Food Engineering* 164, 70–89.

Mercadé-Prieto, R., Sahoo, P.K., Falconer, R.J., Paterson, W.R., Wilson, D.I., 2007. Polyelectrolyte screening effects on the dissolution of whey protein gels at high pH conditions. *Food Hydrocolloids* 21(8), 1275–1284.

Mercadé-Prieto, R., Wilson, D.I., Paterson, W.R., 2008. Effect of the NaOH concentration and temperature on the dissolution mechanisms of beta-lactoglobulin gels in alkali. *International Journal of Food Engineering* 4(5), (art. 9). http://dx.doi.org/10.2202/1556-3758.1421.

Morison, K.R., 2015. Reduction of fouling in falling-film evaporators by design. *Food and Bioproducts Processing* 93, 211–216.

Morison, K.R., Broome, S.R., 2014. Upward vapour flows in falling film evaporators and implications for distributor design. *Chemical Engineering Science* 114, 1–8.

Morison, K.R., Phelan, J.P., Bloore, C.G., 2013. Viscosity and non-Newtonian behaviour of concentrated milk and cream. *International Journal of Food Properties* 16(4), 882–894.

Morison, K.R., Tie, S.-H., 2002. The development and investigation of a model milk mineral fouling solution. *Food and Bioproducts Processing* 80(4), 326–331.

Morison, K.R., Worth, Q.A.G., O'Dea, N.P., 2006. Minimum wetting and distribution rates in falling film evaporators. *Food and Bioproducts Processing* 84(4), 302–310.
Munir, M.T., Zhang, Y., Yu, W., Wilson, D.I., Young, B.R., 2016. Virtual milk for modelling and simulation of dairy processes. *Journal of Dairy Science* 99(5), 3380–3395.
Mura, E., Gourdon, M., 2016. Interfacial shear stress, heat transfer and bubble appearance in falling film evaporation. *Experimental Thermal and Fluid Science* 79, 57–64.
Murphy, P.M., Lynch, D., Kelly, P.M., 1999. Growth of thermophilic spore forming bacilli in milk during the manufacture of low heat powders. *International Journal of Dairy Technology* 52(2), 45–50.
Oldfield, D.J., Taylor, M.W., Singh, H., 2005. Effect of preheating and other process parameters on whey protein reactions during skim milk powder manufacture. *International Dairy Journal* 15(5), 501–511.
Paramalingam, S., Winchester, J., Marsh, C., 2000. On the fouling of falling film evaporators due to film break-up. *Food and Bioproducts Processing* 78(2), 79–84.
Plock, J., Spiegel, T., Kessler, H.G., 1998a. Influence of the lactose concentration on the denaturation kinetics of whey proteins in concentrated sweet whey. *Milchwissenschaft* 53, 389–393.
Plock, J., Spiegel, T., Kessler, H.G., 1998b. Influence of the dry matter on the denaturation kinetics of whey proteins in concentrated sweet whey. *Milchwissenschaft* 53(6), 327–331.
Riedel, L., 1949. Thermal conductivity measurements on sugar solutions, fruit juices and milk. *Chemie Ingenieur Technik* 21, 340–341.
Schuck, P., 2011. Modifications des propriétés fonctionnelles des poudres de protéines laitières: Impact de la concentration et du séchage. *Innovations Agronomiques* 13, 71–99.
Schuck, P., Dolivet, A., Jeantet, R., 2012. *Les poudres laitières et alimentaires. Techniques d'analyse.* Tec&Doc Lavoisier, Paris, France.
Scott, S.A., Brooks, J.D., Rakonjac, J., Walker, K.M.R., Flint, S.H., 2007. The formation of thermophilic spores during the manufacture of whole milk powder. *International Journal of Dairy Technology* 60(2), 109–117.
Singh, H., Creamer, L.K., 1991. Denaturation, aggregation and heat-stability of milk protein during the manufacture of skim milk powder. *Journal of Dairy Research* 58(3), 269–283.
Singh, H., Newstead, D.F., 1992. Aspects of protein in milk powder manufacture, in: Fox, P.F. (Ed.), *Advanced dairy chemistry: 1. Proteins*. Elsevier Science Publishers Ltd, London, UK, p. 781.
Snoeren, T.H.M., Damman, A.J., Klok, H.J., 1982. The viscosity of skim-milk concentrates. *Netherlands Milk and Dairy Journal* 36, 305–316.
Steffe, J.F., 1992. *Rheological methods in food process engineering*, 1st ed. Freeman Press, East Lansing, MI.
Tanguy, G., Tuler-Perrone, I., Dolivet, A., Santellani, A.-C., Leduc, A., Jeantet, R., Schuck, P., Gaucheron, F., 2019. Calcium citrate insolubilization drives the fouling of falling film evaporators during the concentration of hydrochloric acid wheys. *Food Research International* 116, 175–183.
Vavrusova, M., Johansen, N.P., Garcia, A.C., Skibsted, L.H., 2017. Aqueous citric acid as a promising cleaning agent of whey evaporators. *International Dairy Journal* 69, 45–50.
Velez-Ruiz, J.F., Barbosa-Canovas, G.V., 1997. Effects of concentration and temperature on the rheology of concentrated milk. *Transactions of the ASAE (American Society of Agricultural Engineers)* 40, 1113–1118.
Vujicic, I., DeMan, J.M., 1966. Soluble-colloidal equilibria of constituents of concentrated milks. *Milchwissenschaft* 21, 346–349.
Walstra, P., Wouters, J.T.M., Guerts, T.J., 2006. *Dairy science and technology*, 2nd ed. CRC Press, Boca Raton, FL.
Westergaard, V., 2004. *Milk powder technology: Evaporation and spray drying*. Niro A/S, Copenhagen, Denmark.
Winchester, J., 2000. Model based analysis of the operation and control of falling-film evaporators. A thesis presented for the degree of Doctor of Philosophy in Technology and Engineering, Massey University, Palmerston North, New Zealand.
Zhang, Y., Munir, M.T., Yu, W., Young, B.R., 2014. Development of hypothetical components for milk process simulation using a commercial process simulator. *Journal of Food Engineering* 121, 87–93.

2.3 *Lactose crystallization of whey, permeate and lactose*

Pierre Schuck

2.3.1 *Introduction*

Milk sugar or lactose is a disaccharide ($C_{12}H_{22}O_{11}$) that is only found in milk. This carbohydrate exists in two isomeric forms (α and β). Both forms can crystallize and their physicochemical relationships are very complex (Jenness and Koops, 1962). Moreover, lactose crystallization consists of a set of reactions that strongly depend on the experimental conditions used. The three fundamental steps in aqueous solutions are nucleation, crystal growth and mutarotation. Previously, these steps were often studied separately, with certain simplifications regarding the other steps (e.g. non-rate-limiting mutarotation), but considerable improvements in the understanding of these phenomena have been achieved.

In the dairy industry, whey is a liquid product obtained during the manufacture of cheeses, caseins or similar products from the separation of curd after the coagulation of milk and/or the derivative products of milk. Its weight corresponds to nine times the weight of cheeses manufactured and contains not less than 50% of the dry matter of the milk transformed into cheese. For three decades, the whey industry has developed considerably in Europe and now globally due to the growth in the demand for cheeses from which whey results. The increase is also explained by the development of new valorizations of wheys and their derivatives due to their techno-functional (texturing, foaming, emulsifying, etc.) and nutritional properties (content of proteins, rich in essential amino acids). Today, whey powders and their derivatives (whey proteins, lactose, α-lactalbumin, lactoferrin, etc.) are included in the formulations for diet foods, pharmaceuticals and infant milk products (Blanchard et al., 2014).

In the processing schemes of dairy powders, water evaporation during spray drying is so fast that despite saturation, lactose cannot crystallize and it remains in the powder as amorphous lactose, or lactose glass (King, 1965; Nickerson, 1974; Nickerson and Moore, 1974; Pisecky, 1997; Roos, 1997). Amorphous lactose is very hygroscopic and can cause caking problems during and after spray drying, in powders such as whey powders that have a high lactose content. To avoid caking, lactose must be pre-crystallized as α-lactose monohydrate, which is non-hygroscopic (Berlin et al., 1968a, 1986b, 1970; Holsinger, 1997). This is performed through lactose crystallization in tanks after concentration by vacuum evaporation.

The first steps in the production of whey/permeate and lactose powders are quite similar. They include pre-processing steps (heat treatment, membrane filtration, etc.) and concentration by vacuum evaporation. The processing schemes change during and after lactose crystallization. For lactose production, the lactose crystals formed must be large enough (200–500 μm) with a narrow size distribution to be easily separated from the mother liquor and washed afterward. For whey/permeate production, the lactose crystals formed must be small (25–80 μm) with a narrow size distribution to avoid cakiness (Carpin et al., 2016).

The aim of this chapter is to focus on lactose crystallization for whey, permeate and lactose powders regarding process–product interactions.

2.3.2 Lactose

2.3.2.1 The different forms of lactose

Lactose is a disaccharide composed of a D-galactose and a D-glucose unit bonded through a β-1.4-glycosidic linkage. Lactose (4-O-ß-D-galactopyranosyl-D-glucopyranose, $C_{12}H_{22}O_{11}$) can exist in α and β forms. As can be seen in Figure 2.3.1, the two forms are stereoisomers, which differ by the spatial arrangement of the hydroxyl group at carbon number 1 of the hemiacetal group. The α-form has the greater optical rotation in the dextro direction (Holsinger, 1988). In solution, the rate of transformation between the α- and β-anomers, called mutarotation, is temperature and pH dependent. On the other hand, the ratio at equilibrium depends slightly on temperature and is not affected by pH (Holsinger, 1997).

Lactose can be found in a crystalline state, an amorphous state or a mixture of both. By definition, crystalline lactose presents a very ordered structure, with the exact morphology of the crystal depending on the crystallization conditions. In amorphous lactose, the lactose molecules are not organized according to a regular lattice. Moreover, lactose is polymorphic, meaning that it can crystallize into different forms. The six currently known forms of lactose are presented in Table 2.3.1. Even if the crystalline form, α-lactose monohydrate, has a different chemical composition due to the inclusion of water in the crystal structure, it is often presented as a lactose polymorph in the literature (Carpin et al., 2016).

The most common and stable form of lactose under normal temperature and humidity conditions is α-lactose monohydrate. The crystal structure includes one water molecule for every lactose molecule. This water molecule is crucial to the structure and stabilization of the crystal lattice as it links together the oxygen of four lactose molecules. It also explains

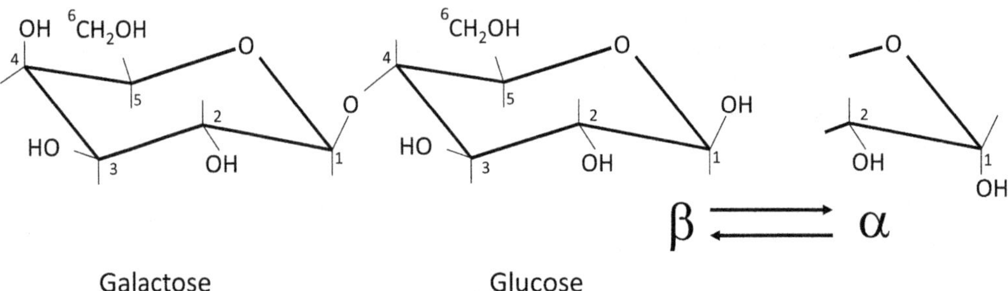

Figure 2.3.1 Molecular formula of α- and β-lactose.

Table 2.3.1 Currently known forms of lactose

Crystalline	**Monohydrate**	α-lactose
	Anhydrous	Unstable α-lactose
		Stable α-lactose
		β-lactose
		Compound β/α lactose[a]
Amorphous		Mixture of α-lactose and β-lactose

Source: From Listiohadi et al. (2005).

Note: [a] Can be obtained at different molar ratios but no consensus to date on the actual existence of this form of lactose (Hourigan et al., 2013).

Figure 2.3.2 Picture of α-lactose monohydrated crystal.

the relative non-hygroscopicity of α-lactose monohydrate. This form of lactose is stable at 25°C below 95% relative humidity (RH) (Salameh et al., 2006). The most common crystal shapes for α-lactose monohydrate are tomahawk and prism (Schuck, 2011a), as seen in Figure 2.3.2. To obtain the tomahawk crystal shape, a commonly used process in the dairy industry is slow cooling of whey permeate concentrate in a crystallizer.

A summary of the inter-relationships between the types of lactose was first published by King (1965), updated several times and more recently by Hourigan et al. (2013) (Figure 2.3.3). In milk, the lactose in solution is in α and β forms according to the temperature, such as $\beta/\alpha = 1.64-(0.0027 \times \text{Temperature})$. During the process, if the lactose is in the supersaturation area, it can be transformed into β anhydrous lactose (crystal form) at a temperature greater than 93.5°C, and into α monohydrate lactose (crystal form) at a temperature less than 93.5°C. In the case of rapid freezing or drying, lactose does not have enough time to crystallize and lactose in the amorphous state is produced. This amorphous lactose can be modified to crystal form in β anhydrous lactose or in α monohydrate lactose by taking up water according to the temperature (> or <93.5°C) and the relative humidity [RH >57% (α-form) or <57% (β-form)].

2.3.2.2 *Lactose solubility*

According to the review of Schuck (2011a), lactose has a final solubility (i.e. at the mutarotation equilibrium of the α and β forms) equal to the sum of the concentration of the α-form, corresponding to its intrinsic solubility, and the concentration of the β-form checking the constant of the mutarotation equilibrium. The final solubility is not the sum of the intrinsic solubilities of each anomeric form. The intrinsic solubility of the β-form is in fact much

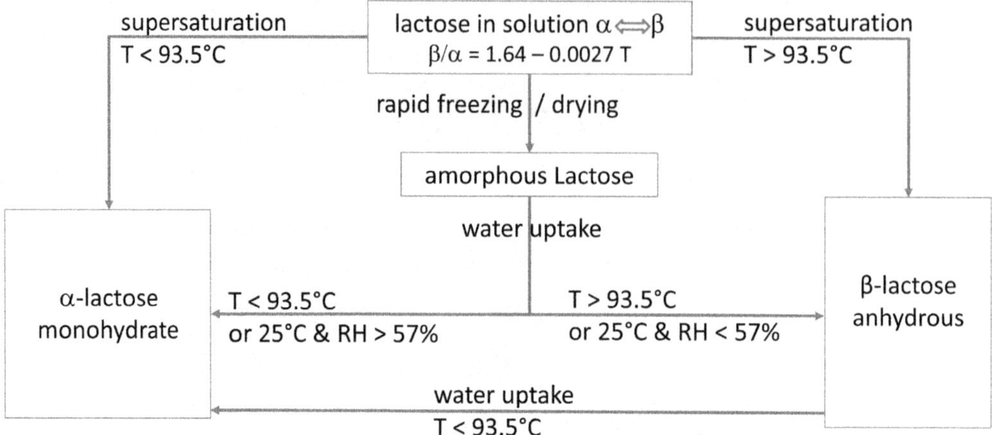

Figure 2.3.3 Inter-relationships between the types of lactose. From King (1965) and updated by Hourigan et al. (2013).

higher (50 g.100 g^{-1} water at 15°C) (Morrissey, 1985). The solubility of the α-form and the final solubility increase with temperature (Jenness and Patton, 1976), whereas the constant of the mutarotation equilibrium changes little with temperature (Holsinger, 1988). As an example, the solubility of the α-form and the final solubility are equal to 9.65 and 24.81 g.100 g^{-1} water at 30°C and 23.36 and 58.40 g.100 g^{-1} water at 60°C, respectively (Figure 2.3.4). The influence of certain "impurities" (riboflavin, urea, ammoniac, etc.) on the final solubility of lactose can be the result of an indirect effect on the value of the constant of mutarotation (Mimouni et al., 2009; Gernigon et al., 2013). A displacement of the mutarotation equilibrium involved (a reduction in the β/α ratio for example) would induce a decrease in the final solubility of lactose without having to modify the solubility of the α-form. Knowledge of the factors influencing the solubility of a substance is very important in the study of solid–liquid balances. Modifications of these factors change the solubility and thus are the origin of the phase change and the appearance of crystals.

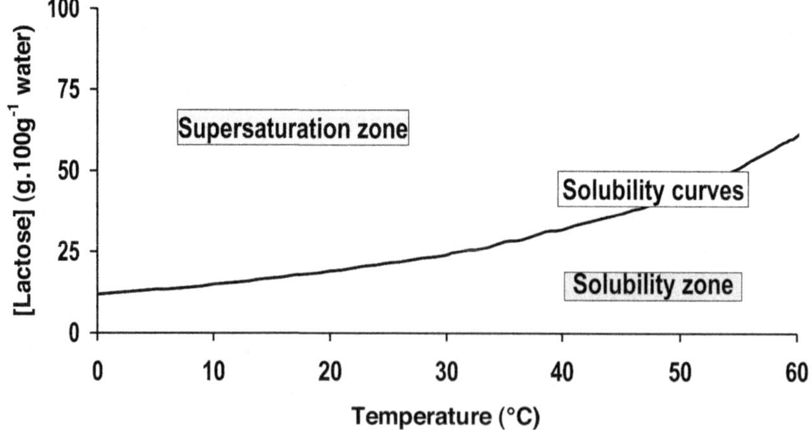

Figure 2.3.4 Solubility curves of lactose. Adapted from Schuck (2011a).

2.3.2.3 Lactose supersaturation

A solution is said to be supersaturated when the concentration of a solute exceeds its solubility value (Figure 2.3.4). From a thermodynamic point of view, the change in phase takes place when a difference exists between the chemical potential of the supersaturated solution and the chemical potential of the solution when the concentration of the solute is equal to its solubility. The potential difference is the "thermodynamic engine" of the phase change, which is called the "driving force" (Myerson and Ginde, 2002; Schuck, 2011a).

2.3.2.4 Metastability, induction time and nucleation

2.3.2.4.1 Stable, metastable and unstable zones According to Schuck (2011a), supersaturation is a requirement for a phase shift in a solution. However, it is not a sufficient condition. It is the case for the cooling of an under-saturated solution (Figure 2.3.5). By cooling the solution from state A to B, the system exceeds the solubility curve but only very slightly. It may be left to stand for one day or one year without the appearance of a crystalline phase. If another sample initially at state A is quickly cooled to state C and then maintained at this state, a few hours is sufficient to notice the appearance of crystals. If the cooling is more intensive (D) and beyond a certain limit, the appearance of crystals will be immediate (or spontaneous). Moreover, in states B and C, if crystals are added to the solution, they become bigger. States B and C characterize supersaturated solutions considered as stable, at least during a given time, whereas they are obviously in a non-thermodynamic equilibrium (growth of the crystals added to the solution). These states are called metastable states. At constant temperature and pressure, the limit of the metastable zone will depend on numerous other conditions of the system. For example, it will be closer to the solubility curve when the system is agitated. In the same way, the chemical nature of the coupled aqueous solution-solvent, the temperature, the mode and the kinetics of supersaturation, the type and quantity of impurities, the thermal and mechanical shocks, etc., influence the width of the metastable zone.

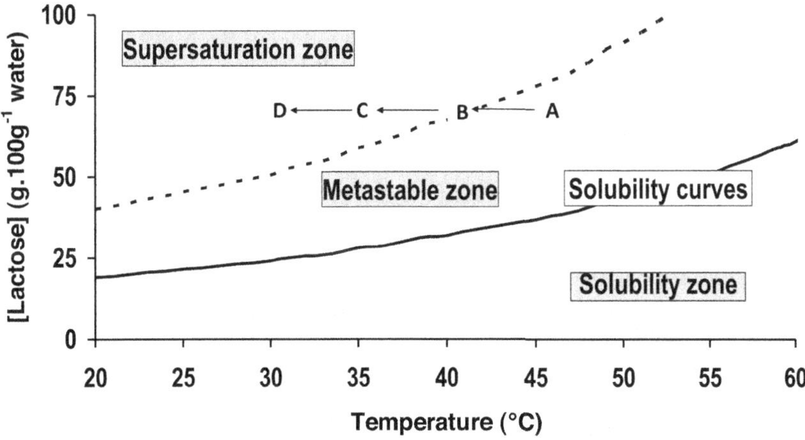

Figure 2.3.5 Solubility curves of lactose and cooling of an undersaturated solution of lactose in water. Adapted from Schuck (2011a).

2.3.2.4.2 Induction time The duration of States B and C (Figure 2.3.5) influences the beginning of the supersaturation and the phase shift. This duration is called time of induction. It decreases with the increase in supersaturation and ends beyond the limit of the metastable zone. The more stable the state of the system (metastable), the longer the induction time. All factors that reduce the width of the metastable zone reduce the induction time in the same way.

2.3.2.4.3 Nucleation and metastability The kinetic expression of the mechanism that controls the first step in the formation of the solid phase is called nucleation (Schuck, 2011a). The mechanisms themselves are complex and still not fully known. The classical theory distinguishes various types of nucleation:

- the primary nucleation, which describes the birth of a species' germs in the absence of this species' crystals in suspension. It is said to be homogeneous in pure solution and heterogeneous in the presence of soluble and solid impurities;
- the secondary nucleation of a species that proceeds in the presence of crystals of the same species.

The numerous factors used in the nucleation mechanisms and the complexity of the phenomena involved make the control of nucleation difficult, especially in an industrial context. However, nucleation is the key step in the control of crystallization and therefore the quality of the resulting products. The number, the intermediate size, the size distribution and the forms of the crystals are controlled by the nucleation kinetics and they have a major role, in particular in the final quality of the food products (texture, conservation).

2.3.2.4.4 Crystal growth When the nucleus reaches its critical size, it grows bigger in its function as the "mother" solution from which it draws the molecules. The crystal growth is the result of several steps (Schuck, 2011a):

- Diffusion of aqueous molecules (lactose, water, etc.) to the solid interface.
- Dissolution of the molecules in the aqueous solution.
- Counter-diffusion of the solvent molecules to the solution.
- Adsorption of molecules on the crystalline surface.
- Migration on the crystalline surface to suitable incorporation sites.
- Incorporation of the growth units into the crystal network.

As for the nucleation, the growth kinetics generally depends on factors related to the type of aqueous solution and on external factors related to the operating conditions (supersaturation, temperature, agitation, presence of impurities).

Figure 2.3.6 summarizes the nucleation and growth of lactose crystals. The lower the temperature, the more favored the nucleation, and the higher the temperature (but always below the solubility curves), the greater the increase in the growth of crystals. Mimouni et al. (2009) and Gernigon et al. (2013) showed that the size of the crystals

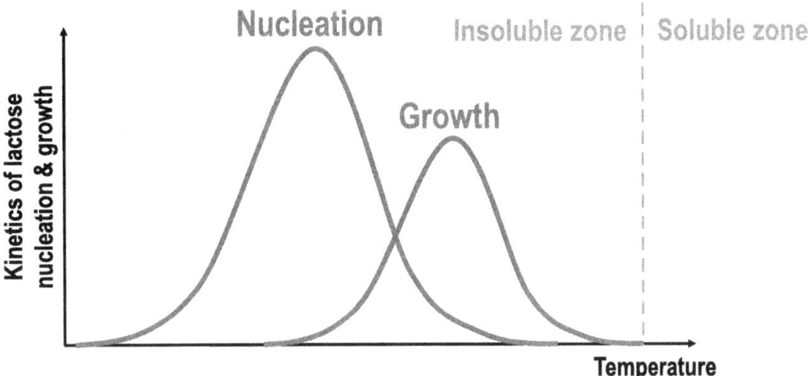

Figure 2.3.6 Kinetics of lactose nucleation and growth.

depends on the number of nuclei. The higher the number of nuclei, the smaller the size of the crystals.

2.3.3 *Process*

2.3.3.1 Principal steps for the production of lactose powder

According to Carpin et al. (2016), lactose powder can be produced directly from whey. However, whey proteins have become extremely valuable and for this reason, whey is nowadays usually ultrafiltered (UF) to separate the protein stream (retentate) from the lactose and soluble mineral stream (permeate). The UF permeate is then further processed to produce edible-grade lactose. The first steps, calcium phosphate removal and nanofiltration, are optional but increase the runtime and efficiency of the process. If not removed upstream, calcium phosphate precipitates in the evaporators due to elevated temperature and fouled heating surfaces. Nanofiltration aims to concentrate and demineralize the whey permeate, which reduces the loss of lactose solids in the separation step and increases the yield (Durham, 2009). Reverse osmosis is also often used to concentrate the permeate before evaporation. Concentration by falling-film evaporators increases the dry matter content to about 60% (Hourigan et al., 2013). The concentrated whey permeate is then pumped into a crystallizer where it is subjected to slow batch cooling for 20–24 hours. The lactose crystals formed should be large enough (200–500 µm) with a narrow size distribution to be easily separated from the mother liquor and washed afterward. For this, the lactose crystallization step is opposite to the one used for whey/permeate powders. It aims to produce large crystals, which have a high impact on the yield of the process (Paterson, 2017). The concentrates must always be in the metastable zone Since, in order to grow large crystals, it is important to avoid having to many nucleation sites. The counterbalance to the large size is the low crystallization ratio (no more than 60%) and the long duration of the crystallization step (1–2 days).

The mother liquor contains minerals, peptides, organic acids and some dissolved lactose. The separation and washing steps will therefore determine the final level of impurities in the product. Too many fines in the lactose crystals can clog the separator leading to poorer separation of the mother liquor and higher levels of impurities and moisture. Wash water can be recirculated to increase the yield of the process. However, depending on the chemical composition (lactose and impurities) of the washing water and the step at which this water returns to the process, recirculation can negatively impact the purity of the final product. After decantation, the crystals usually have an 88–90% dry matter content and must be dried further. Drying takes place in a fluidized bed dryer with dehumidified air. The final total water content of the crystals is 4.5% to –5.5%. The drying temperature and time should be controlled to avoid the formation of amorphous glass and lactose polymorphs other than α-lactose monohydrate (Durham, 2009). Finally, lactose should be cooled prior to packing to avoid caking due to moisture migration from the hot to the cold regions in the bag during storage (Carpin et al., 2017a, 2017b). Additionally, lactose crystals can be milled and sieved to obtain the desired size specifications.

2.3.3.2 *Principal steps for the production of whey and permeate powders*

After the pre-processing steps (heat treatment, demineralization, filtrations, etc.), the unit operations used in the production of whey and permeate powders are concentration by vacuum evaporation, lactose crystallization, spray drying and fluidization. Concentration by vacuum evaporation aims to remove some of the water (85–95%) from whey and to increase the dry matter from 6.5% to 60–65% (w/w). Then, lactose, whose concentration is much higher than the solubility value, changes state during the crystallization step. The "crystallized" whey concentrate is then completely dehydrated using spray drying and further fluidization.

Lactose supersaturation in wheys and permeate is initially performed by the evaporation of water during the concentration step, the final total solid of the concentrate (50–65% w/w) corresponding to a lactose concentration of between 65 and 150 g.100 g^{-1} water according to the type of whey/permeate. Supersaturation is then followed by fast cooling using a "flash cooler" between 25 and 35°C and finally by slow cooling (or not) up to 20–25°C. Under these conditions of pressure (atmospheric pressure) and temperature (lower than 93.5°C), the lactose crystal in α-monohydrated form is the most thermodynamically stable form (i.e. that corresponding to a minimum of free enthalpy). Consequently, although the α and β forms are both supersaturated, only the α-form crystallizes, in the α-monohydrated form.

This change of state during the crystallization step is not instantaneous and depends on a complex kinetics in relation to many factors such as the composition of the whey/permeate, lactose, protein and ash contents, temperature, stirring conditions, seeding, etc. For example, Mimouni et al. (2009) have demonstrated that whey proteins decrease the size of the crystals. Using factorial analysis, Gernigon et al. (2013) have shown that the presence of organic acids (lactate, citrate) leads to faster crystallization and confirmed the fact that whey proteins slow down the crystal growth step. Another example from the studies of Simeão et al. (2017) shows the effect of the crystallizer and the stirrer on the lactose crystallization rate.

A crystallization process is proposed to produce small lactose crystals (50–80 μm) at a high crystallization ratio (close to 80%) in a few hours (Figure 2.3.7).

a. Starting with an empty crystallizer tank and a low temperature in the jacket (4°C).
b. Lowest possible flash cooling before blocking the extraction pump.
c. Seeding with lactose or the previous whey/permeate powder.
d. High-velocity stirring.
e. Stop the cooling and increase the jacket temperature from 4°C to 25°C.
f. Low-velocity stirring.

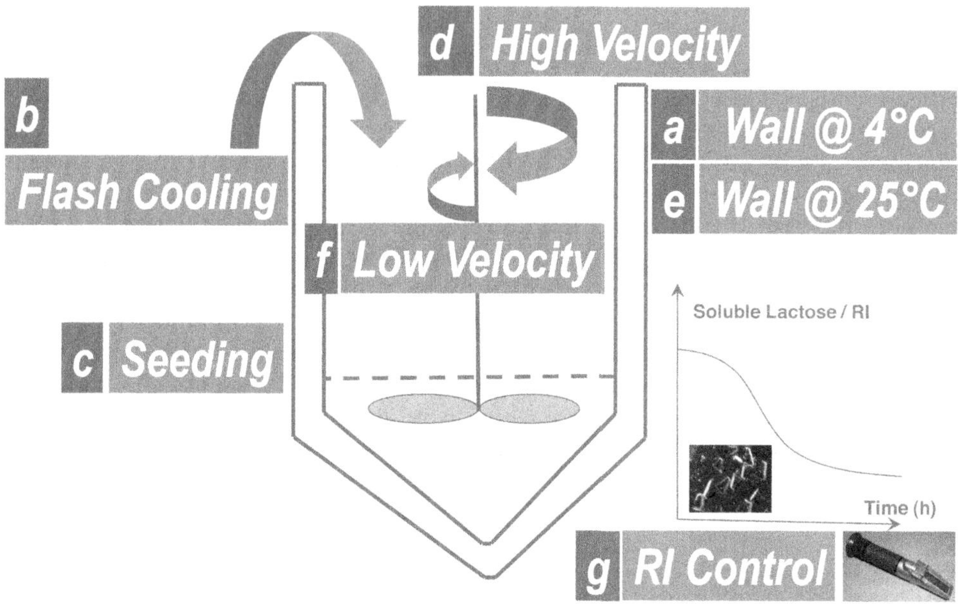

Figure 2.3.7 Optimization of lactose crystallization for whey/permeate powders.

g. Check every 30 min that the refraction index follows the decrease in the lactose solubility versus time to determine the crystallization ratio according to Mimouni et al. (2005). The optimal target in the crystallization rate must be close to 80%.

The most frequently used technique for the dehydration of whey and permeate powders is spray drying. The wide range of whey/permeate concentrates that differ mainly in terms of their composition make it difficult to determine the parameters for spray drying. As a consequence, according to the whey/permeate composition, a higher content of lactic acid, ashes, glucose and galactose means that the spray-drying parameters must be reduced to 50% of the productivity, compared to a sweet whey powder to avoid stickiness in the spray dryer and cakiness in the silo, big bags or bags (Schuck, 2011b,c, 2016).

Regarding the stabilization of the whey/permeate powder during storage, water activity (a_w) is one of the key parameters. The a_w of milk powders composed of non-fat milk solids and milk fat is mainly controlled by the moisture content expressed in non-fat solids, since fat has no influence. Thus, the differences in a_w are mostly due to the state of the proteins and the physical state of the lactose. The a_w should be close to 0.2 at 25°C for optimal preservation (Efstathiou et al., 2002). The ideal moisture content can be determined for the optimal stabilization (at 0.2 a_w and at 25°C) of some dairy powders by using practical or theoretical sorption isotherms (Schuck et al., 2009, 2012). For example, to be at 0.2 a_w, the corresponding free moisture content must be close to 3.4%, 3.1%, 2.3% and 1.6%, for whey powders at 0%, 20%, 50% and 80% of the lactose crystallization rate, respectively. The more the lactose is crystallized, the more the free moisture content must be decreased, to have the same water activity with a constant total moisture content close to 5% at any lactose crystallization rate.

2.3.4 Conclusion

Lactose crystallization is a key stage in the manufacture of whey powders. Controlling this stage at an industrial level should increase the prospects of improving the process and the physicochemical qualities of the powders.

Lactose crystallization occurs in highly supersaturated solutions, indicating that the phenomena of nucleation and crystal growth can take place simultaneously. In addition, in the case of lactose, a preliminary stage of mutarotation between anomeric forms, each with its own kinetics, is present. Although the lactose crystallization mechanisms have been widely described by many authors, and the parameters of the kinetics (orders and constants) calculated for the mutarotation, nucleation and lactose crystal growth stages, few investigations have been undertaken into the interactions between the kinetics using integrated approaches (Mimouni et al., 2009; Schuck, 2011a; Gernigon et al. (2013). To control the nucleation, which is related to the size of a crystal (big for lactose powder, small for whey and permeate powder), and the crystallization ratio, it is very important to follow the decrease in lactose solubility versus time using a refractometer (Mimouni et al., 2005).

References

Berlin E., Anderson A.B., Pallansch M.J. Water vapor sorption properties of various dried milks and wheys. *Journal of Dairy Science* 51(9) (1968a) 1339–1344.

Berlin E., Anderson A.B., Pallansch M.J. Comparison of water vapor sorption by milk powder components. *Journal of Dairy Science* 51(12) (1968b) 1912–1915.

Berlin E., Anderson A.B., Pallansch M.J. Effect of temperature on water vapor sorption by dried milk powders. *Journal of Dairy Science* 53(2) (1970) 146–149.

Blanchard E., Zhu P., Schuck P. Infant formula powders. In *Handbook of Food Powders* Bhandari B, Bansal N, Zhang M and Schuck P (Eds.), Cambridge, UK: Woodhead Publishing Limited, pp. 465–483, 2014.

Carpin M., Bertelsen H., Bech J.K., Jeantet R., Risbo J., Schuck P. Caking of lactose: A critical review. *Trends in Food Science & Technology* 53 (2016) 1–12.

Carpin M., Bertelsen H., Dalberg A., Roiland C., Risbo J., Schuck P., Jeantet R. Impurities enhance caking in lactose powder. *Journal of Food Engineering* 198 (2017a) 91–97.

Carpin M., Bertelsen H., Dalberg A., Bech J.K., Risbo J., Schuck P., Jeantet R. How does particle size influence caking in lactose powder? *Journal of Food Engineering* 209 (2017b) 61–67.

Durham R.J. Modern approaches to lactose production. In *Dairy-Derived Ingredients – Food and Nutraceutical Uses* Corredig M (Ed.), Woodhead Publishing Series in Food Science, Technology and Nutrition, Cambridge, UK: Woodhead Publishing Ltd., 2009, pp. 103–144. https://doi.org/10.1533/9781845697198.1.103.

Efstathiou T., Feuardent C., Méjean S., Schuck P. The use of carbonyl analysis to follow the main reactions involved in the process of deterioration of dehydrated dairy products: Prediction of most favourable degree of dehydration. *Le Lait* 82(4) (2002) 423–439.

Gernigon G., Baillon F., Espitalier F., Le Floch-Fouéré C., Schuck P., Jeantet R. Effects of the addition of various minerals, proteins and salts of organic acids on the principal steps of α-lactose monohydrate crystallization. *International Dairy Journal* 30(2) (2013) 88–95.

Holsinger V.H. Fundamentals of dairy chemistry. In *Lactose* Wong NP (Ed.), New York: Van Nostrand Reinhold, pp. 279–342, 1988.

Holsinger V.H. Physical and chemical properties of lactose. In *Advanced Dairy Chemistry, Vol. 3: Lactose, Water Salts and Vitamins* Fox PF (Ed.), London, UK: Chapman & Hall, pp. 1–38, 1997.

Hourigan J., Lifran E., Vu L., Listiohadi Y., Sleigh R. Lactose: Chemistry, processing, and utilization. In *Advances in Dairy Ingredients* (1st ed.) Smithers GW and Augustin MA (Eds.), Hoboken, NJ: John Wiley & Sons, Inc., pp. 31–69, 2013.

Jenness R., Koops J. Preparation and properties of a salt solution which simulates milk ultrafiltrate. *Netherlands Milk & Dairy Journal* 16 (1962) 153–164.

Jenness R., Patton S. *Principles of Dairy Chemistry*, Wiley & Sons, London, 1976.

King N. The physical structure of dried milk. *Dairy Science Abstract* 27 (1965) 91–104.

Listiohadi Y., Hourigan J., Sleigh R., Steele R. Properties of lactose and its caking behaviour. *Australian Journal of Dairy Technology* 60(1) (2005) 33–52.

Mimouni A., Schuck P., Bouhallab S. Kinetics of lactose crystallization and crystal size as monitored by refractometry and laser light scattering: Effect of proteins. *Lait* 85(4–5) (2005) 253–260.

Mimouni A., Schuck P., Bouhallab S. Isothermal batch crystallization of alpha-lactose: A kinetic model combining mutarotation, nucleation and growth steps. *International Dairy Journal* 19(3) (2009) 129–136.

Morrissey P.A. Lactose: Chemical and physicochemical properties. In *Developments in Dairy Chemistry: Lactose and Minor Constituents* Fox PF (Ed.), New York: Elsevier, pp. 1–67, 1985.

Myerson A.S., Ginde R. Crystals, crystal growth, and nucleation. In *Handbook of Industrial Crystallization* Myerson AS (Ed.), London: Butterworth-Heinemann, pp. 33–63, 2002.

Nickerson T.A. Lactose. In *Fundamentals of Dairy Chemistry* Weeb BH and Johnson AH (Eds.), Westport, UK: Avi Publishing Co., pp. 273–324, 1974.

Nickerson T.A., Moore E.E. Factors influencing lactose crystallization. *Journal of Dairy Science* 5(11) (1974) 1315–1319.

Paterson A.H.J. Lactose processing: From fundamental understanding to industrial application. *International Dairy Journal* 67 (2017) 80–90. https://doi.org/10.1016/j.idairyj.2016.07.018.

Pisecky J. *Handbook of Milk Powder Manufacture*, Copenhagen, Denmark: Niro A/S, 1997.

Roos Y.H. Water in milk products. In *Advanced Dairy Chemistry* Fox PF (Ed.), London, UK: Chapman & Hall, Vol. 3: Lactose, Water, Salts and Vitamins, pp. 303–346, 1997.

Salameh A.K.A., Mauer L.J.L., Taylor L.S.L. Deliquescence lowering in food ingredient mixtures. *Journal of Food Science* 71(1) (2006) E10–E16. https://doi.org/10.1111/j.1365-2621.2006.tb12392.x.

Schuck P., Dolivet A., Méjean S., Zhu P., Blanchard E., Jeantet R. Drying by desorption: A tool to determine spray drying parameters. *Journal of Food Engineering* 94(2) (2009) 199–204.

Schuck P. Lactose and oligosaccharides | Lactose: Crystallization. In *Encyclopedia of Dairy Sciences* (2nd ed.) Fuquay JW, Fox PF and McSweeney PLH (Eds.), San Diego: Academic Press, vol. 3, pp. 182–195, 2011a.

Schuck P. Dehydrated dairy products | Milk powder: Types and manufacture. In *Encyclopedia of Dairy Sciences* (2nd ed.) Fuquay JW, Fox PF and McSweeney PLH (Eds.), San Diego: Academic Press, vol. 2, pp. 108–116, 2011b.

Schuck P. Dehydrated dairy products | Milk powder: Physical and functional properties of milk powders. In *Encyclopedia of Dairy Sciences* (2nd ed.) Fuquay JW, Fox PF and McSweeney PLH (Eds.), San Diego: Academic Press, vol. 2, pp. 117–124, 2011c.

Schuck P., Dolivet A., Jeantet R. *Analytical Methods for Food and Dairy Powders*, Oxford, UK: Wiley-Blackwell, 2012.

Schuck P. Implications of non-equilibrium states and glass transition. In *Non-Equilibrium States and Glass Transitions in Foods: Processing Effects and Product-Specific Implications* Bhandari BR and Roos YH (Eds.), Woodhead Publishing Series in Food Science, Technology and Nutrition, Duxford, UK: Woodhead Publishing, pp. 303–321, 2016.

Simeão M., Da Silva C.R., Stephani R., De Oliveira L.F.C., Schuck P., de Carvalho A.F., Perrone I.T. Lactose crystallization in concentrated whey: The influence of vat type. *International Journal of Dairy Technology* 70(1) (2017) 1–6.

2.4 *Homogenization: a key mechanical process in interaction with product to modulate the organization of fat in spray-dried powders*

Christelle Lopez and Pierre Schuck

2.4.1 *Introduction*

Many food products are oil-in-water (O/W) emulsions, for example milk, cream and infant milk formula. One way to increase their shelf-life is to remove the water and to reduce water activity (a_w) by using a drying process, since dehydration will reduce the chemical, physical and biological degradation of the food products. The drying of food products also ensures easy handling and shipping as well as a reduction of transportation and storage costs. Spray-drying is the most widely used and effective technological solution in the food industry to prepare powders due to its low cost and the ready availability of equipment. Dairy powders can be classified depending on the amount of fat they contain: skim (0% fat), whole (26% fat) and fat-filled (above 26% fat) milk powders, including infant milk formula (20 to 25% fat) powders obtained by spray-drying that are of increasing economic importance worldwide.

The fat-filled dairy powders are dried O/W emulsions where the fat, e.g. vegetable oils or milk fat, plays a significant role. Dietary fats are mainly composed by triacylglycerols (TAG; esters of fatty acids and glycerol) which are hydrophobic molecules. They can be formulated in their anhydrous form (e.g. blend of vegetable oils in the case of standard infant milk formula). In order to avoid the phase separation of fat from the other components, fat needs to be dispersed in individual droplets that will be distributed in the volume of the product. The dispersion of TAG in the mix that will be dried requires the use of energy that is generally provided by the mechanical process of homogenization. Fat can also be naturally dispersed in an O/W emulsion such as in the case of milk where fat globules have a size distribution ranging from 1 to 10 μm and a mean diameter around 4 μm. In this latter case, decreasing the size of fat globules using homogenization will improve the physical stability of the O/W emulsion during the manufacture of the powder and its long-time storage. Homogenization is therefore a key technological step aiming at dispersing fat in the form of small emulsion droplets.

The quality of dried O/W emulsions depends on the composition of the product in interaction with processing (e.g. concentration and spray-drying) and storage conditions. The sensorial (e.g. off-flavors) and nutritional (e.g. oxidative degradation) quality of the spray-dried O/W emulsion powders and subsequent re-hydrated products are affected by the fatty acid composition, the fat content and the organization of fat (e.g. size of fat droplets, free fat content). Fat is also involved in the functional properties of the spray-dried powders, such as the flowing and rehydration properties (wettability, sinkability, dispersibility and solubility). However, at the industrial scale, the control of the processing and of the powder properties in relation with the organization of fat is still rather empirical. The control of the functional properties and quality of fat-filled spray-dried powders requires an understanding of the relationships between composition, processing, particle microstructure and product properties.

This book chapter, dedicated to the mechanical process of homogenization involved in the manufacture of spray-dried powders, is structured in five main parts, with the following objectives:

1) To present the mechanical process of homogenization that is commonly used to disperse fat in individual droplets
2) To discuss the position of the homogenization process in the technological scheme of spray-dried powder production
3) To examine the possible organizations of fat within spray-dried powders
4) To evaluate the potentialities of homogenization in the preparation of nanostructured lipid carriers
5) To illustrate the key role played by homogenization in the preparation of dairy products such as infant milk formula and to highlight the potentialities of homogenization for innovation in the near future.

2.4.2 Homogenization: the mechanical process commonly used to disperse fat in droplets

The mechanical process of homogenization was designed by Auguste Gaulin and first presented at the Paris World's Fair in 1900. Gaulin invented his homogenizer for the processing of milk, i.e. to reduce the milk fat globules in size with the final objective of retarding phase separation and the formation of a cream layer on the top of the bottle. The main objective of homogenization is therefore to form and stabilize an O/W emulsion in order to avoid the phase separation of fat.

Homogenization is able (1) to sub-divide the relatively large poly-disperse fat droplets of a coarse O/W emulsion into a large number of fat droplets of smaller size, but also (2) to fractionate continuous anhydrous fat into individual droplets dispersed in another liquid containing emulsifiers, and then to create an O/W emulsion during emulsification (Figure 2.4.1). In the case of the mixture containing anhydrous fat, the shear stress and inertial forces applied by a pump located before the homogenization step can create a coarse O/W emulsion. Then, the size of the fat droplets will be reduced under pressure in the homogenizer.

A homogenizer is a machine consisting of a positive displacement pump and a homogenizing valve. The principle of the Gaulin-type homogenizer consists in forcing a liquid under pressure through a small adjustable gap between the valve and the seat, causing turbulence and intense mixing (Figure 2.4.1). Homogenization results in the reduction of fat droplet size due to shear stress, inertial forces and cavitation. Homogenizers can have a single stage, or two stages. The main pressure is applied in the first stage. The second stage generally applies a lower pressure than the first stage, in general about 10 to 20% of the pressure applied in the first stage. The objective of this second stage is to dissociate the aggregates of fat droplets that are formed after the first stage and to decrease the viscosity of the product. The pressure conditions of the homogenizers vary according to the apparatus and valve type.

Figure 2.4.2 shows the impact of the homogenization pressure on the size distribution of fat droplets. Homogenization of full-fat milk leads to a bimodal size distribution of fat globules, with the formation of small fat globules around 0.15 μm in diameter that are

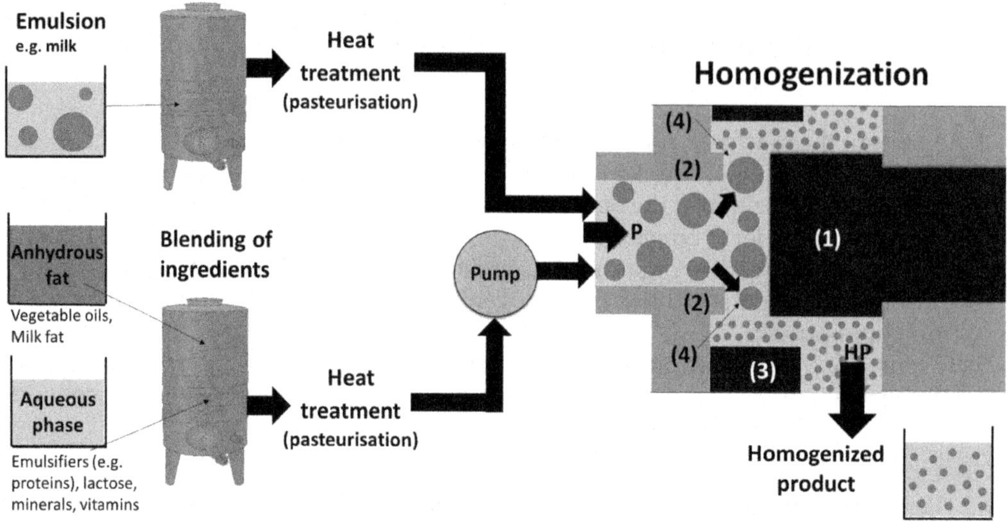

Figure 2.4.1 Processing involving a coarse oil in water emulsion (e.g. milk) or the blending of the ingredients (anhydrous fat, aqueous phase containing surface-active molecules and other ingredients), heat treatment and homogenization. During homogenization, the fluid product (P) containing coarse fat droplets enters the valve seat. As the product flows through the adjustable area between the homogenizing valve (1) and seat (2), the intense energy transition occurring in microseconds produces turbulent three-dimensional mixing layers that disrupt the droplets at the discharge from the adjustable gap (4). The homogenized product (HP) impinges on the impact ring (3) and exits from the homogenizer.

mainly covered by milk casein micelles (Lopez et al., 2015). The increase in the homogenization pressure leads to a fractionation of fat into an increased number of fat droplets with a decrease in the size distribution of the fat droplets and an increase in the surface area. According to Stoke's law, the decrease in fat droplet size decreases the cream separation rate due to the density difference between fat and the aqueous phase. This decrease in size is also responsible for a higher physical stability of the O/W emulsion and then to an increase of its shelf-life. In general, homogenization is performed in the food industry with pressures ranging from 0.2 to 50 MPa. In the preparation of fat-filled spray-dried dairy powders, the homogenization pressure is commonly in the range 15 to 25 MPa.

Homogenization leads to changes in the structural organization of components in the product. Indeed, after homogenization, the tension-active molecules or emulsifiers that are initially exclusively solubilized in the aqueous phase (e.g. milk proteins) partition between the aqueous phase and the fat/water interface. The fat and the proteins considered separately as well as the fat/protein ratio are involved in the interfacial mechanisms occurring during the formation of the O/W emulsion by homogenization of the product. Emulsifiers are able to stabilize the fat/water interface created under pressure during the homogenization.

As an example, we can consider the partitioning of proteins in an infant milk formula containing 25% fat and 12% milk proteins (i.e. g per 100 g powder). Authors generally consider that the amount of milk proteins adsorbed at the surface of fat droplets is around 1.5 to 3 mg proteins per m^2 surface, but it can vary as a function of the type of proteins. Since the surface of fat droplets in an infant milk formula ranges from 21 to 38 m^2/g fat ((Lopez et al., 2015); Figure 2.4.2B), we calculated that about 7 to 24% of the milk proteins could be

(A) Homogenization of milk: impact of the pressure on the fat globules size distribution

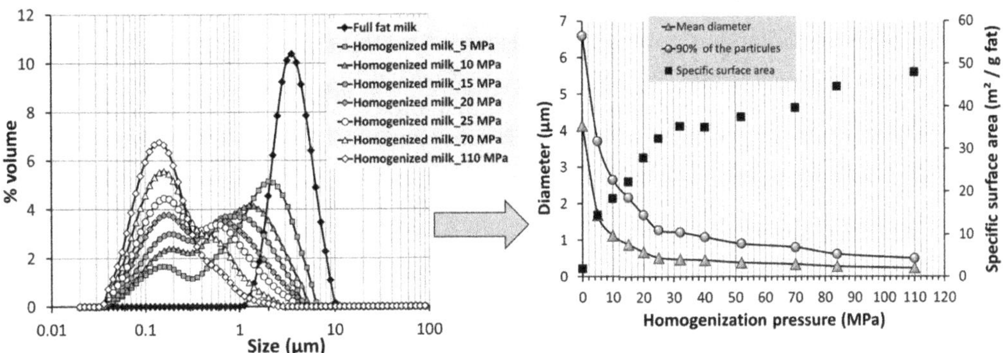

(B) Homogenization to prepare O/W emulsions such as infant milk formula

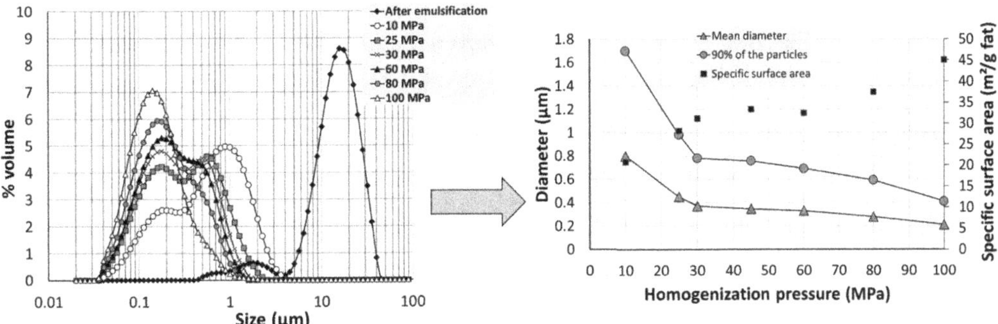

Figure 2.4.2 Impact of the homogenization pressure on the size distribution of fat droplets in O/W emulsions. (A) Homogenization of full-fat milk. (B) Homogenization in the case of the preparation of infant milk formula, i.e. blends of vegetable oils with an aqueous phase containing milk proteins, lactose and minerals.

adsorbed at the surface of the fat droplets in an infant milk formula, leading to the formation of protein-coated fat droplets. The other fraction of the proteins remains in the aqueous phase of the product.

Many consumers are interested in decreasing their consumption of animal products, such as bovine milk, because of health, environmental and ethical reasons. The food industry is therefore developing a range of plant-based milk alternatives. These milk substitutes should be affordable, convenient, desirable, nutritional and sustainable (McClements et al., 2019). Plant-based emulsifiers (proteins, polysaccharides, phospholipids, colloidal particles) differ in their effectiveness at forming and stabilizing O/W emulsions. Proteins from plants (e.g. from pea) are often organized as non-soluble globular structures of large size and have a low capacity to stabilize emulsion droplets. The functional performance can be improved by mixing milk proteins and plant-based proteins. The milk proteins will adsorb at the fat/water interface to stabilize the emulsion droplets while the non-soluble plant-based proteins will remain in the aqueous phase of the product. Changing the composition of the products requires an adaptation of the parameters used for the homogenization process, mainly the homogenization pressure as regards to the size distribution of the fat droplets.

The spreading of proteins adsorbed at the fat/water interface after homogenization could affect the mechanisms of their digestion in the gastrointestinal tract as compared to the proteins remaining in the aqueous phase. Further research studies are required in this field.

High-pressure homogenization (HPH) is based on the same design principle as conventional homogenization techniques but is operated at higher pressure, i.e. above 50 to 200 MPa. The high pressure applied to the product aims at strongly decreasing the size of fat droplets in order to produce a stable O/W emulsion, and/or to prepare products with appropriate rheological properties. The HPH process is used also to reduce the bacterial microflora, to reduce the activity and inactivate enzymes. Results obtained for single- or two-stage HPH-treated milk samples at similar pressures suggest that two-stage treatment has a greater destructive effect on milk microflora. The majority of the killing effect is probably achieved by the rapid pressure drop and its associated forces experienced by the milk at the primary homogenizing valve. The synergistic effect of the similar but significantly reduced forces experienced by microbes at the second stage during two-stage treatment may kill sub-lethally damaged cells, which may otherwise recover after single-stage treatment. Cell disruption may also be achieved by high-velocity collisions between bacterial cells and other milk components or solid components of the homogenizer. Sudden pressure drops, torsion and shear stresses, cavitation shock waves resulting from imploding gas bubbles, turbulence, impact with solid surfaces and viscous shear are all mechanisms that have been proposed for cell disruption during HPH.

Homogenization is generally performed at about 50 to 60°C to ensure the complete melting of fat. Indeed, some dietary fats such as milk fat or palm oil exhibit melting temperatures around 40°C. The high temperature also helps to reduce the viscosity of the product, which facilitates the encapsulation of fat by the tension-active molecules (e.g. proteins). Applying pressure on the products during homogenization leads to an increase in temperature. Floury et al. (2000) reported a temperature increase of 0.164°C per MPa on HPH-treatment of 10% O/W emulsion samples at 20 to 300 MPa. A mean temperature increase of 0.176°C per MPa, in the pressure range 50–200 MPa, was calculated for HPH-treatment of raw whole milk. Some of the observed temperature increase of the product is probably due to adiabatic heating, while the majority may be due to the high velocity at which the fluid exits the primary homogenizing valve. The fluid will be exposed to high turbulence, shear and cavitation forces during HPH, a large part of which may be transformed into thermal energy in the product. For a similar homogenization pressure, the temperature increases as a function of the increase in fat content in the product. For example, milk outlet temperature increased in a linear manner (0.5°C per 1% fat) as milk fat content increased from 0 to 10% in a two-stage HPH operating at 150 MPa (Hayes and Kelly, 2003). The observed increase in heating during HPH at higher milk fat contents may be a direct result of viscous dissipation or of the increased number of fat globules both initially and post-primary homogenization. This larger population of fat globules increases the probability of collisions between particles which may, in turn, exert greater shear and other forces upon each other. Milk samples of varying fat contents may also absorb different amounts of thermal energy due to adiabatic heating.

Differences in valve construction can lead to variation in the mechanisms of fat droplet disruption under pressure in the homogenizer and then can affect the size distribution of the fat droplets. For a similar homogenizer, the size distribution of the fat droplets depends on technological parameters such as the homogenization pressure applied to the product (Figure 2.4.2), and on the physico-chemical characteristics of the product such as its viscosity, and the amount and type of tension-active molecules able to stabilize to interface formed during homogenization. If the number of tension-active molecules is sufficient to cover the surface created during homogenization, the size of fat droplets decreases as the pressure increases (Figure 2.4.2).

2.4.3 *Position of the homogenization process in the technological scheme of spray-dried powder production*

The wet-mixing spray-drying process is currently the most widely used method of producing fat-filled dairy powders. One of the advantages of the wet-mix process is that the successive individual technological steps (i.e. preparation of the wet mix containing the ingredients, heat treatment, concentration by evaporation under vacuum, homogenization and spray-drying) can be controlled and adapted as a function of the chemical composition of the product. This results in improved quality powder in terms of microbiological, physical and chemical properties.

For the production of fat-filled spray-dried dairy powders, the technological step of homogenization, aiming at dispersing fat (e.g. vegetable fat, milk fat) into small individual droplets, can be located at different stages (Figure 2.4.3):

1. After concentration of the fat-free blend of ingredients and introduction of fat in line.
2. After the preparation of the product containing all the ingredients, including fat, at the required dry matter.
3. After blending all the ingredients, including fat, and pasteurization and before the concentration step. This third solution is preferred at the industrial scale when the product needs to be stored for several hours before the concentration step. It is also possible for a second homogenization step to be performed before spray-drying of the product to ensure a good dispersion of fat in the product and then to avoid phase separation and free fat formation.

In the two first cases, homogenization is performed in a product with high dry matter, i.e. from 48 to 52%, while in the third case, homogenization is performed with lower dry matter (i.e. 30%).

For similar parameters of homogenization (i.e. apparatus, pressure, temperature), the efficiency of the homogenization in terms of fat droplet size distribution can be affected by the concentration, i.e. the amount of dry matter, and then the viscosity of the product.

During the mechanical process of homogenization, the ingredients are submitted to pressure (from 15 to 25 MPa) that can alter their structure and functional properties. Hence, some ingredients such as the thickening agents (e.g. starch) are introduced by dry blending in order to avoid their alteration during the homogenization step. Other ingredients may not be added before the homogenization step; this is, for example, the case for oils rich in n-3 and n-6 long-chain polyunsaturated fatty acids (LCPUFA; e.g. docosahexaenoic acid (DHA), eicosapentaenoic acid (EPA), arachidonic acid (AA)) that are highly sensitive to oxidation. The LCPUFA-rich oils that are essential bioactive components in infant milk formula are mainly prepared from fish oil (e.g. tuna) or from unicellular organisms: ARA from *Mortierella alpina* oil and DHA from *Crypthecodinium cohnii*. In order to ensure their chemical protection from oxidation, the LCPUFA-rich oils can be encapsulated in a matrix composed of modified starch or caseins, together with antioxidant molecules (e.g. vitamin C, tocopherols), and used as spray-dried powder ingredients. These encapsulated LCPUFA-rich oils can be added (1) in the concentrated and homogenized product just before the spray-drying step in order to limit the duration of their

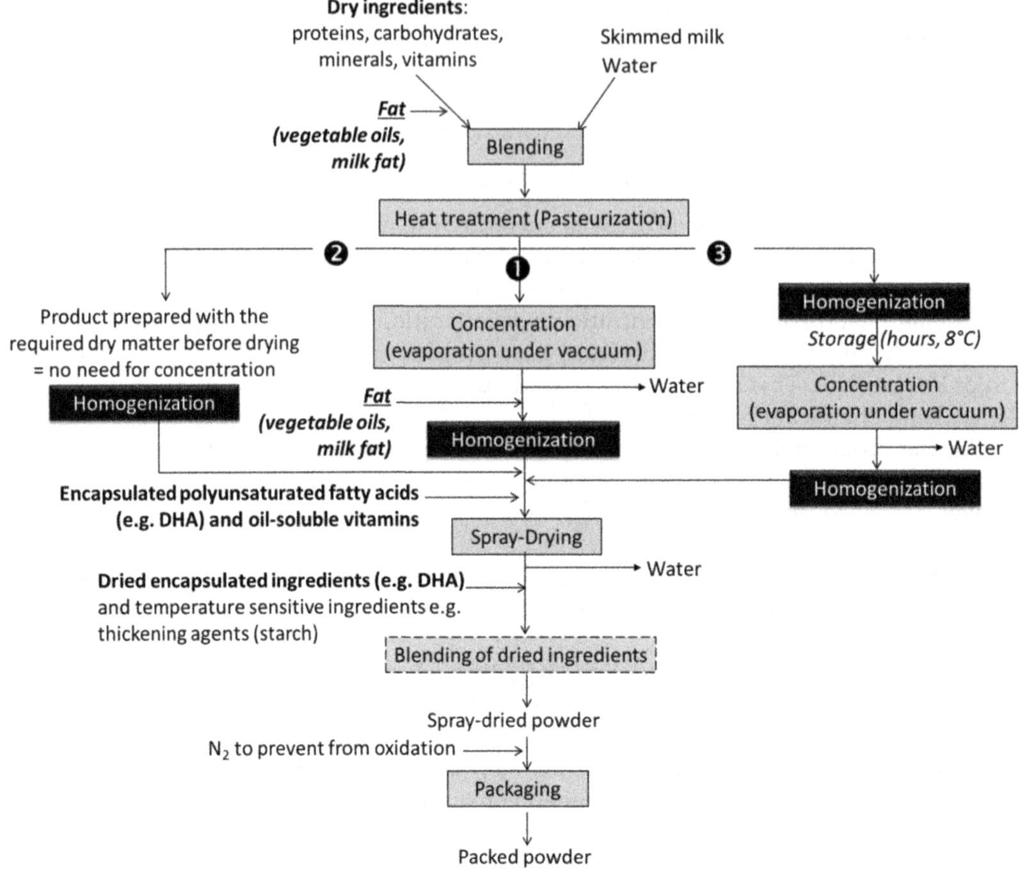

Figure 2.4.3 Flow chart of fat-filled spray-dried powder production. The step of homogenization is essential to disperse and stabilize fat.

contact with pro-oxidant molecules (e.g. iron, copper) dispersed in the aqueous phase, or (2) by dry blending after the spray-drying process. The final infant milk formula powder is always packaged in an N_2 atmosphere in order to prevent oxidation of the LCPUFA during the storage of the product.

2.4.4 *Organization of fat within spray-dried powders*

2.4.4.1 *Physical stability of O/W emulsions*

From a physical point of view, fat-filled dairy powders are dried O/W emulsions. The O/W emulsions are thermodynamically unstable systems in which the fat is dispersed as small droplets within the continuous phase, the two phases being immiscible. The fat droplets dispersed in dairy powders are often called encapsulated fat or oil (see Vignolles et al., 2007). The main physical instabilities of the dried O/W emulsion in powders correspond to aggregation and coalescence. Aggregation or flocculation of fat droplets is mainly the result of the interfacial properties of the droplets (the amount and type of tension-active molecules present at the surface of the fat droplets). Coalescence corresponds to the irreversible fusion of two or several fat droplets in the liquid state to form droplets of a larger size. Coalescence can occur when fat droplets are closely organized, as is the case during the concentration process (evaporation under vacuum) and during the spray-drying process. Coalescence can also occur during the storage of the spray-dried powder. For example, the presence of partially crystallized fat can enhance the disruption of the interfacial film especially as a mechanical stress is applied. The disruption of the dried O/W emulsion leads to the formation of non-emulsified fat, also called free fat. Many publications still define free fat as the solvent-extractable fat under controlled conditions. The following definition has been proposed: free fat is the fat (1) which is not entirely coated and stabilized by amphiphilic molecules, e.g. adsorbed proteins, or (2) which is not entirely protected by a matrix composed of amorphous carbohydrate (e.g. lactose) and proteins during drying (Vignolles et al., 2007). The analytical methods for the quantification of free fat are detailed elsewhere (Schuck et al., 2012; Vignolles et al., 2007).

2.4.4.2 *Microstructure of fat-filled dairy powders: focus on fat organization*

In-situ structural observations of fat-filled spray-dried dairy powders performed by confocal microscopy revealed that fat can be dispersed in small spherical droplets homogeneously distributed in the powder particle (Figure 2.4.4A). Fat can also be present in non-spherical large fat droplets and in the non-emulsified organization also called free fat (Figure 2.4.4B–D). Free fat was observed at the joining points of agglomerated powder particles as observed in Figure 2.4.4B. Free fat can also be located within the powder particle,

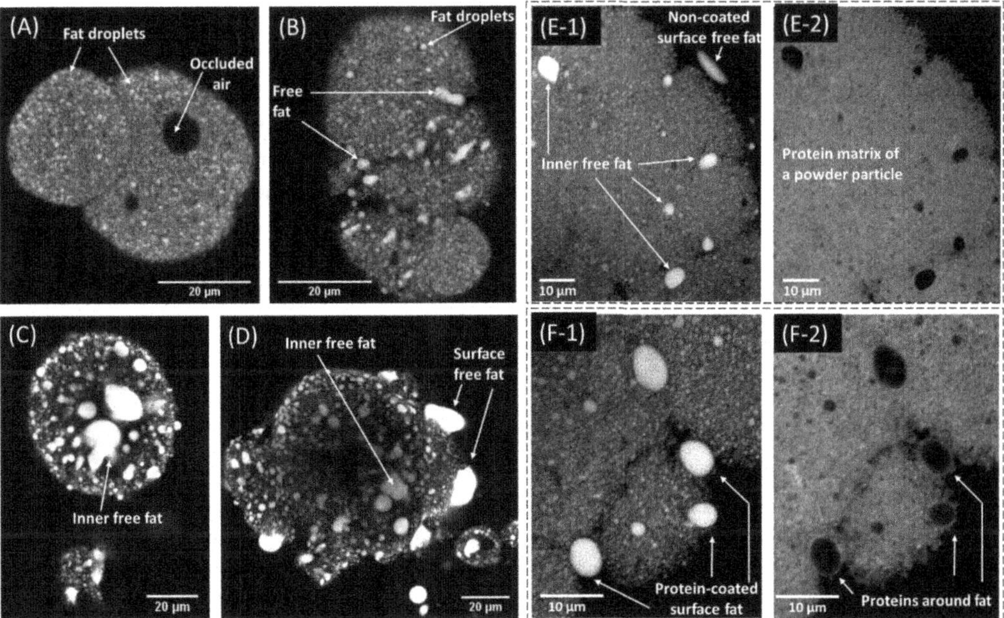

Figure 2.4.4 Microstructure of industrial infant milk formula spray-dried powders showing differences in the organization of fat. Confocal laser scanning microscopy images of individual powder particles; A–D, E-1 and F-1: labeling of fat performed with Nile red fluorescent dye, E-2 and F-2: labeling of proteins using fast green FCF. E and F show both the labeling of fat (E-1, F-1) and proteins (E-2, F-2) in the same powder particle.

in the pores and capillaries created during the drying process and at the surface of the powder particles. Surface free fat non-coated by proteins has been observed using confocal microscopy with the double labeling of fat and proteins (Figure 2.4.4E). Non-spherical large particles of fat, coated by proteins, can be found at the surface of powder particles (Figure 2.4.4F). The surface properties of fat located at the surface of powder particles are important in the functional properties of the powder, mainly its capacity for rehydration. The formation of free fat has been attributed to the physical instability of the emulsion that has evolved through the mechanisms of coalescence of fat droplets and then disruption of the emulsion droplets. Similar organizations of emulsified fat and free fat located both at the surface and within the powder particles have been previously reported (Vignolles et al., 2007). The composition of the surface of powder particles can be investigated by using X-ray photoelectron spectroscopy (XPS) (Vignolles et al., 2007).

Small spherical fat droplets homogeneously distributed in the volume of the powder particles have been related to low free-fat powders. In contrast, high free-fat powders showed a surface free fat layer and an inner degradation of the integrity of the fat droplets (Figure 2.4.4C and D; (Vignolles et al., 2007)). The correlation between large fat droplet diameters and high free fat content in the spray-dried powder has been established. From experimental results it is admitted that the fat droplet size distribution in the concentrate before spray-drying has to be smaller than 1 μm to achieve low free fat content in the powder (Vignolles et al., 2009a).

Authors reported that the free fat does not have the same composition as the total fat present in the powder. It is then possible that the composition of fat is not homogeneous in the powder. This highlights mechanisms involving fat fractionation within the spray-dried powder probably as a function of the physical state (liquid, crystallized) of the individual TAG. However, discrepancies exist between authors. Kim et al. (2005) reported that free fat was slightly more concentrated in saturated fatty acids from six to eight atoms of carbon chain length, and less concentrated in unsaturated fatty acids (C18:1 n-9, C18:2 n-6, C18:3 n-3). They supposed that free fat was enriched in high-melting-point TAG and that the latter tended to accumulate at the powder particle surface. On the other hand, Vignolles et al. (2009a) showed that the free fat removed from powder particles was enriched in low-melting-point fatty acids (i.e. fatty acids that are more unsaturated and/or fatty acids with short chains) as compared to the fat remaining in the powder particles. A high concentration of free fat enriched in low-melting-point unsaturated TAG may favor oxidation, while the dispersion of these TAG within lipid droplets may favor their chemical protection. Since the possible fractionation of fat between free fat and encapsulated fat may affect the functional properties of spray-dried powders, the underlying mechanisms need further investigation.

2.4.4.3 *Main parameters involved in the organization of fat within spray-dried powders*

Two main groups of parameters that are involved in the physical organization of fat in fat-filled spray-dried powders and leading to the formation of free fat have been identified: the composition of the product and processing. Furthermore, the interactions between the product and the process are highly important for the formation of the O/W emulsion during homogenization and its subsequent drying and storage.

2.4.4.3.1 Composition The chemical composition of the spray-dried powders is often complex. Fat-filled spray-dried powders contain fat but also other components such as the proteins, lactose, and minerals.

The amount of fat is an important parameter to consider since a high fat content in the product is often related to technological difficulty in physically stabilizing the fat and thus related to high free fat content. As a consequence, high-fat-content powders can exhibit poor rehydration properties and poor flowability. These powders are also more prone to sensory defaults such as off-flavors due to oxidation.

The fatty acid composition of the fat (e.g. amount of saturated and unsaturated fatty acids) that governs its melting point can impact the organization of fat in spray-dried powders. Fats that contain a high amount of saturated fatty acids (e.g. palmitic acid C16:0, stearic acid C18:0), such as milk fat and palm oil, have a high melting point and then are partially crystallized at the temperature of spray-dried powder storage. Fat crystals can contribute in the physical destabilization of the emulsion droplets. Note that the blending of the ingredients, including fat, and the homogenization step need to be performed above the final temperature of melting of the fat used. Also, the temperature of the powder particles during spray-drying ranges from 45 to 90°C, and thus the fat is in a liquid state throughout the process. The liquid state of fat favors its encapsulation during homogenization and during the spray-drying process.

The surface-active molecules contained in the encapsulating phase, and able to play a role at the fat/water interface during the technological step of homogenization and to stabilize the lipid droplets, are important to consider. These molecules must have a high solubility in the aqueous phase, have to form a coat around each fat droplet to ensure a good emulsion stability (good emulsifying properties) and must tolerate drying conditions for the protection of fat droplets. The amount of surface-active molecules (e.g. proteins, phospholipids from vegetable lecithin or from the milk fat globule membrane), the type of proteins (proteins from milk: casein micelles, sodium caseinate, whey proteins, isolates of milk proteins; proteins from a vegetable origin such as rice) and the structure of the proteins (e.g. heat-denatured proteins forming aggregates, hydrolyzed proteins) are involved in the stabilization of fat droplets during homogenization. For a similar amount of total proteins (e.g. 12 g per 100 g powder in infant milk formula), the milk proteins (i.e. caseins and whey proteins, 40/60 wt%) are able to efficiently stabilize the lipid droplets formed during homogenization, while hydrolyzed milk proteins exhibit a lower emulsifying and stabilizing capacity. As a consequence, fat-filled spray-dried powders formulated with hydrolyzed proteins for milk protein allergy reasons (e.g. hypoallergenic infant milk formula) often show a higher amount of free fat (Figure 2.4.4C and D) compared to infant formula prepared with regular proteins. In this case, increasing the pressure of homogenization and adding surface-active molecules to the product (e.g. soy lecithin) can contribute to the physical stabilization of fat droplets and thus to avoiding the formation of free fat.

The lactose, in its crystallized or amorphous states, can be involved in the physical stability of fat through several mechanisms: its role during homogenization and upon storage (physical constraints due to lactose crystals may damage the lipid droplets and lead to coalescence), its role in the viscosity of the product and its role as a protective agent favoring the emulsifying properties of proteins, especially with heat-denatured whey proteins as reported (Vignolles et al., 2009b). Lactose in its amorphous state has been reported to improve the stabilization of fat droplets and then contributes to a reduction of free fat content in the powder. Lactose, in its crystallized state, is known to improve the efficiency of spray-drying. Then, as well as the amount of lactose, the physical state of lactose is an important parameter to control in the production of fat-filled spray-dried powders.

2.4.4.3.2 Processing and storage The organization of fat in spray-dried powders can be affected by the homogenization parameters such as pressure and temperature. Moreover, the other steps of processing can cause a range of structural and physicochemical modifications of the emulsion, such as:

1) The heat treatment that can denature the proteins and affect their capacity to stabilize the emulsion.
2) The step of lactose crystallization which involves the puncturing of fat globule membranes and the generation of a capillary interstices network that stresses and causes fat droplet disruption and eventual migration towards the surface powder particles when the fat is in a melted form.
3) The parameters used for spray-drying (e.g. inlet and outlet temperatures) (Vignolles et al., 2010).

The conditions of storage of the fat-filled powders can affect the organization of fat, for example, changes in temperature may affect the crystallization properties of the fat. Moreover, changes in the fat droplet size and accumulation of free fat can occur during the storage of powders.

2.4.4.4 Consequences of the organization of fat on the quality of the powder

The amount of fat, its organization (dispersion in fat droplets, free fat) and its localization in the powder (inner fat or surface fat) may affect the functional properties and the quality of the powder (e.g. flow and wetting properties, poor rehydration).

The hydration ability of a spray-dried powder in water, including the steps of wetting, dispersion and solubilization, is an essential functional property. The surface composition of the powder, e.g. the amount and dispersion of fat or free fat that is highly hydrophobic, negatively affects the wetting capacity. The presence of fat on the surface of the powder particles provides hydrophobic layers that cause milk powder to become less flowable and less soluble in water. Surface fat may also form weak bridges between powder particles and promote agglomeration and caking, thus reducing the powder's functional properties. A solution to improve the wettability and thus the rehydration properties of high-fat-containing spray-dried powders is to perform coating with amphiphilic surface-active molecules such as lecithin (e.g. soy lecithin), mono- and diglycerides.

The presence of free fat and surface fat is often considered a quality defect.

2.4.5 Homogenization for the preparation of nanostructured lipid carriers able to encapsulate bioactive compounds

Homogenization can be used to prepare nanostructured lipid carriers (i.e. solid lipid nanoparticles, nano-emulsions, liposomes) for the encapsulation of bioactive compounds (e.g. n-3 fish oil or algal oil) and their protection from chemical degradation. The small size of the nanostructured lipid carriers has also been reported to increase the bio-accessibility of the bioactive compounds.

Homogenization has proven beneficial in comparison to solvent evaporation methods used in the pharmaceutical industry as this mechanical process does not require the use of organic solvents that are toxic for food applications.

The spray-drying of these encapsulation matrices is then a way to concentrate the bioactive ingredients and to extend their storage stability due to less molecular mobility. Various protective water-soluble biopolymers are used as wall materials to ensure the dispersion, the protection and the physical stability of the nanostructured lipid carriers within the powder, e.g. carbohydrates such as starches and maltodextrin, acacia gum and sodium caseinate proteins.

The use of nanostructured lipid carriers dispersed in spray-dried powders can also contribute to masking fishy smells and off-flavors due to oxidation (e.g. polyunsaturated fatty acids such as DHA).

2.4.6 *The key role played by homogenization in the preparation of infant milk formula*

2.4.6.1 *Processed fat droplets in infant milk formula vs. milk fat globules in human milk*

Exclusive breastfeeding is highly recommended during the first 6 months of a baby's life to provide the benefits of human milk components (WHO, 2011). Breast milk is a natural O/W emulsion that results from the mechanisms of milk fat globule secretion by the mammary epithelial cells. Breast milk fat globules are biological entities constituted by a core of triacylglycerols (TAG) enveloped by a biological membrane called the milk fat globule membrane (MFGM). Milk fat globules are efficient conveyers of energy (9 kcal/g fat; TAG represent 50% of energy in breast milk (Grote et al., 2016), lipid-soluble vitamins (A, D, E, K) and bioactive molecules (e.g. fatty acids, phospholipids, cholesterol, MFGM-specific proteins) to newborns. It is recognized that breast milk fat globules are of major significance in infant nutrition and that they are involved in the immunological protection, the growth, development and health of infants (Demmelmair and Koletzko, 2018; Hamosh et al., 1999; Koletzko et al., 2011).

When mothers cannot or do not want to breastfeed their infant or express their milk, infant milk formula (IMF) may be used. IMFs are manufactured foods designed for the bottle-feeding of babies from powder or liquid. The composition of IMF is designed to mimic the chemical composition of human mother's milk at approximately one to three months *post-partum* and is defined by the legislation. The most commonly used IMFs are enriched in cow's or goat's milk whey proteins and also contain casein as a protein source (or hydrolyzed proteins in case of allergy to milk proteins). Standard IMFs contain a blend of vegetable oils (e.g. palm oil, coconut oil, rapeseed oil, sunflower oil) or mixtures of vegetable oils with milk fat, mainly from bovine origin as a fat source. The different sources of fats are combined to mimic the fatty acid composition of human milk. IMFs are also composed of lactose as a carbohydrate source, vitamins, minerals and other ingredients depending on the manufacturer (e.g. prebiotics). Non-dairy emulsifiers and stabilizers can be used to prevent the separation of the oil phase from water in the reconstituted IMF. They include citric acid esters of monoglycerides and diglycerides, lecithins (from vegetable sources such as soya, from fish and krill), gums and maltodextrins. The production of IMF involves different technological steps such as mixing of the ingredients, heat treatment, homogenization, concentration and spray-drying in the case of IMF powders. Industrial processing procedures are used to produce stable and reproducible end-product IMF.

The microstructure of industrial IMF after their rehydration in a bottle has been investigated and compared to mature human milk (1 to 6 months *post-partum*) (Lopez et al., 2015). Figure 2.4.5 highlights the differences in fat droplet size and interface composition between breast milk fat globules and the processed fat droplets in IMF. The size distribution of mature breast milk fat globules spans from 0.4 to 13 μm with a mean diameter of 5 μm, corresponding to a surface of 2 m^2/g fat (Figure 2.4.5A) (Lopez and Ménard, 2011). The processed fat droplets found in the IMF have a significantly smaller size than breast milk fat globules (mean diameters: 0.3–0.8 μm; Figure 2.4.5A) and exhibit a higher surface (from about 20 to 40 m^2/g fat; Figure 2.4.5A). The small size of the fat droplets found in IMF results from the homogenization step involved in the manufacture of these processed products. As already detailed at the beginning of this chapter, the objective of homogenization is to create an O/W emulsion by mixing the blend of fats and the other components (proteins, minerals, lactose). The pressure applied upon homogenization is adjusted to form small-sized droplets (i.e. mainly <1 μm) in order to ensure the physical stability of the processed emulsion during long-term storage of the powder and after hydration of the powder in the bottle. The differences in the size distributions of processed lipid droplets observed in different industrial IMF (Figure 2.4.5A) result from variations in the technological parameters used by different manufacturers or for different markets (i.e. differences in the homogenization process, concentration, evaporation and drying parameters or in the chemical composition).

The composition and the architecture of the TAG/water interface are also different between breast milk fat globules and the processed lipids droplets in IMF (Figure 2.4.5D and E). The microscopy images and electrophoresis analysis clearly show the differences in interface composition: processed fat droplets IMF are mainly covered by bovine milk proteins (i.e. whey proteins and caseins Figure 2.4.5D), while breast milk fat globules are enveloped by membrane-specific proteins from the MFGM (Figure 2.4.5E). The surface of breast milk fat globules is covered by the MFGM, which is rich in phospholipids and cholesterol (Figure 2.4.5E).

Moreover, the structural analysis of various commercialized IMF revealed the presence of aggregates of proteins or of complexes formed between lipid droplets and proteins (Lopez et al., 2015). These protein aggregates and lipoprotein complexes may result from the thermo-induced denaturation of proteins occurring during the heat treatments performed for the microbial safety of IMF. Such lipoprotein complexes induced by the industrial process raise questions about the accessibility of TAG to the digestive enzymes in the gastrointestinal tract of newborns, and then about the nutritional and health impacts.

The structure of fat in IMF is different from that in human milk (size of the lipid droplets, interfacial composition and architecture of the lipid droplet surface). The most striking difference is that processed lipid droplets in IMF are not surrounded by a biological membrane and thus lack MFGM components (phospholipids, sphingomyelin, gangliosides, cholesterol, membrane-specific proteins). Giuffrida et al. (2013) estimated that the mean intake of total phospholipids per day in infants 4 weeks old is about 140 mg when the infant is fed exclusively with human milk, with milk sphingomyelin (e.g. the main sphingolipid found in the MFGM) accounting for about 40% of the polar lipids. However, these milk phospholipids are absent from standard IMF which usually contain soy lecithin (i.e. mainly PC). The functional effects of these differences are largely unknown. The potential consequences of the size of the lipid droplets (<1 μm in IMF vs. 5 μm in breast milk) and interfacial composition (proteins vs. MFGM) on the mechanisms of digestion and absorption of lipids as well as on the metabolic programming have been revealed recently (Oosting et al., 2012). Recent studies highlighted the benefits provided by the presence of MFGM in IMF.

2.4.6.2 *Future directions in which to improve the structure of fat in IMF: the key role played by homogenization*

The structural differences observed between breast milk fat globules and processed fat droplets in IMF, combined with the nutritional and health benefits of MFGM

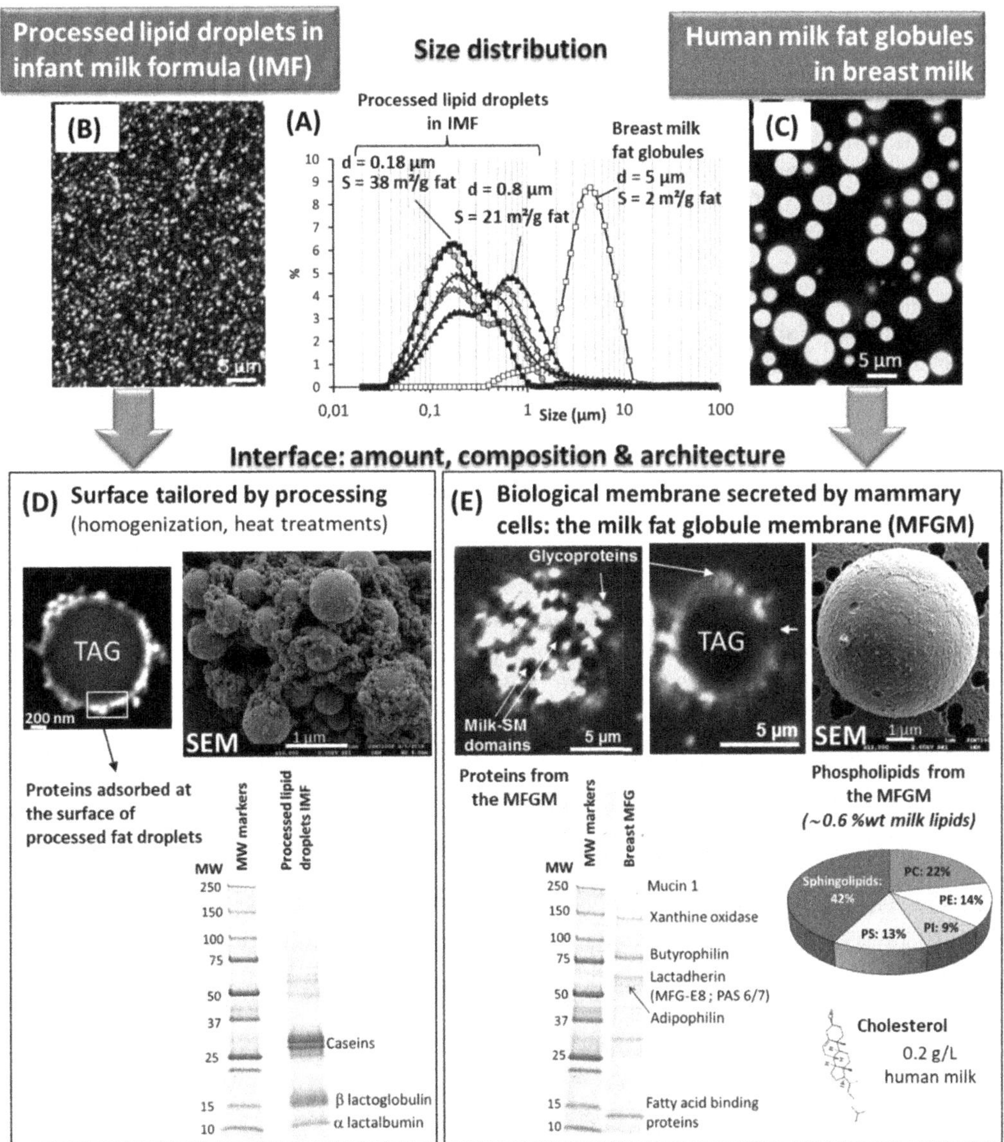

Figure 2.4.5 Role of processing in the organization of fat and fat/water interface architecture in standard industrial infant milk formula (IMF) vs. human milk. (A) Size distributions of processed lipid droplets in IMF and breast milk fat globules in human milk. (B) and (C) confocal laser scanning microscopy (CLSM) images showing the processed lipid droplets in IMF and breast fat globules in human milk, respectively (labeling of fat performed using Nile red fluorescent dye). Differences in the composition and structure of the interface: (D) and (E) show CLSM (left part) and scanning electron microscopy (SEM) images (right image) with electrophoresis to identify the proteins present at the surface, determined for the processed lipid droplets in IMF and breast fat globules in human milk, respectively. Breast milk fat globules are surrounded by the milk fat globule membrane (MFGM) that contains cholesterol and phospholipids. Abbreviations: PC = phosphatidylcholine, PE = phosphatidylethanolamine, PI = phosphatidylinositol, PS = phosphatidylserine, sphingolipids corresponds mainly to sphingomyelin but also to glycosphingolipids (e.g. gangliosides). Adapted from (Lopez et al., 2015).

components, provide opportunities to improve the qualitative lipid supply to infants through tailoring the emulsion in IMF. The manufacture of processed lipid droplets bioinspired by breast milk fat globules, with the size of the lipid droplets and the composition and architecture of their surface close to human milk fat globules, could improve the nutritional properties and health impact of IMF (Oosting et al., 2012). The improvement of the structure of the emulsion in IMF is of growing interest worldwide (Bourlieu et al., 2017). Technological solutions able to produce IMF containing processed fat droplets bio-inspired by breast milk fat globules have been proposed (Lopez et al., 2019).

A supplementation of IMF in MFGM can be performed with the addition of bovine milk fat globules concentrated in cream. In this case, MFGM components are added to IMF together with milk fat that represents 98% of milk lipids. Also, technologies have become available to obtain MFGM from cow's milk. Various dairy streams permit the recovery of MFGM and are available in sufficient amounts to produce ingredients at the industrial scale. Buttermilk, butter serum, beta serum and cheese whey are suitable sources because of their low cost and their relatively high content in MFGM components.

The development of improved IMF containing processed fat droplets bio-inspired by breast milk fat globules will therefore constitute a challenge in the near future. The science of the emulsions and the interactions product – process, more specifically the homogenization process in this case, constitute the heart of the future development in spray-dried IMF powders. However, several technical challenges will need to be overcome. The order of incorporation of the ingredients is of high importance to cover the surface of the fat droplets with MFGM components, since milk proteins have a high affinity for the fat/water interface. Then, the fat (blend of vegetable oils, mixture with milk fat) has to be first homogenized in the presence of a hydrated MFGM-rich ingredient to prepare the emulsion and tailor the surface composition of the fat droplets. In a second time, the other ingredients (i.e. proteins, lactose, minerals) will be incorporated in the emulsion. The size of the fat droplets can be modulated by changing the homogenization pressure, i.e. a decrease in the homogenization pressure leads to an increase in the mean diameter of the fat droplets. After the preparation of the emulsion, the challenge is the physical stability of the emulsion, since large fat droplets (i.e. around 5 µm) may favor the formation of free fat during the storage of spray-dried IMF powders and after their rehydration in the bottle.

2.4.7 Conclusion

In the technological scheme of fat-filled spray-dried powder production, homogenization plays a key role. The mechanical process of homogenization is able to disperse fat in individual droplets and to modulate their size for functional or nutritional reasons. The efficiency of homogenization depends on the parameters applied, for example, the pressure and the temperature, in interaction with the composition of the product. The organization of fat in spray-dried powders, which impacts the quality of the final product, depends not only on the homogenization but also on the other technological steps (concentration, spray-drying) and storage conditions. Used most of the time with empirical knowledge, homogenization will undoubtedly be a source of innovation in the near future, e.g. for the encapsulation of bioactive compounds and in the improvement of the structure of the emulsion in infant milk formulae.

References

Bourlieu, C., Deglaire, A., de Oliveira, S.C., Menard, O., Le Gouar, Y., Carriere, F., and Dupont, D. (2017). Towards infant formula biomimetic of human milk structure and digestive behaviour. *OCL - Oilseeds Fats Crops Lipids 24*(2), D206.

Demmelmair, H., and Koletzko, B. (2018). Lipids in human milk. *Best Pract. Res. Clin. Endocrinol. Metab. 32*(1), 57–68.

Floury, J., Desrumaux, A., and Lardières, J. (2000). Effect of high-pressure homogenization on droplet size distributions and rheological properties of model oil-in-water emulsions. *Innov. Food Sci. Emerg. Technol. 1*(2), 127–134.

Giuffrida, F., Cruz-Hernandez, C., Flück, B., Tavazzi, I., Thakkar, S.K., Destaillats, F., and Braun, M. (2013). Quantification of phospholipids classes in human milk. *Lipids 48*(10), 1051–1058.

Grote, V., Verduci, E., Scaglioni, S., Vecchi, F., Contarini, G., Giovannini, M., Koletzko, B., and Agostoni, C. (2016). Breast milk composition and infant nutrient intakes during the first 12 months of life. *Eur. J. Clin. Nutr. 70*(2), 250–256.

Hamosh, M., Peterson, J.A., Henderson, T.R., Scallan, C.D., Kiwan, R., Ceriani, R.L., Armand, M., Mehta, N.R., and Hamosh, P. (1999). Protective function of human milk: The milk fat globule. *Semin. Perinatol. 23*(3), 242–249.

Hayes, M.G., and Kelly, A.L. (2003). High pressure homogenisation of raw whole bovine milk (a) effects on fat globule size and other properties. *J. Dairy Res. 70*(3), 297–305.

Kim, E.H.-J., Chen, X.D., and Pearce, D. (2005). Melting characteristics of fat present on the surface of industrial spray-dried dairy powders. *Colloids Surf. B Biointerfaces 42*(1), 1–8.

Koletzko, B., Agostoni, C., Bergmann, R., Ritzenthaler, K., and Shamir, R. (2011). Physiological aspects of human milk lipids and implications for infant feeding: A workshop report. *Acta Paediatr. Oslo Nor. 1992 100*(11), 1405–1415.

Lopez, C., and Ménard, O. (2011). Human milk fat globules: Polar lipid composition and in situ structural investigations revealing the heterogeneous distribution of proteins and the lateral segregation of sphingomyelin in the biological membrane. *Colloids Surf. B Biointerfaces 83*(1), 29–41.

Lopez, C., Cauty, C., and Guyomarc'h, F. (2015). Organization of lipids in milks, infant milk formulas and various dairy products: Role of technological processes and potential impacts. *Dairy Sci. Technol. 95*(6), 863–893.

Lopez, C., Cauty, C., and Guyomarc'h, F. (2019). Unraveling the complexity of milk fat globules to tailor bioinspired emulsions providing health benefits: The key role played by the biological membrane. *Eur. J. Lipid Sci. Technol. 121*, 1800201.

McClements, D.J., Newman, E., and McClements, I.F. (2019), Plant-based milks: A review of the science underpinning their design, fabrication, and performance. *Compr. Rev. Food Sci. Food Saf. 18*, 2047–2067.

Oosting, A., Kegler, D., Wopereis, H.J., Teller, I.C., van de Heijning, B.J.M., Verkade, H.J., and van der Beek, E.M. (2012). Size and phospholipid coating of lipid droplets in the diet of young mice modify body fat accumulation in adulthood. *Pediatr. Res. 72*(4), 362–369.

Schuck, P., Jeantet, R., and Dolivet, A. (2012). *Analytical Methods for Food and Dairy Powders*. Coord. P. Schuck, R. Jeantet, A. Dolivet. Wiley-Blackwell, Oxford, UK, 268 pages.

Vignolles, M.-L., Jeantet, R., Lopez, C., and Schuck, P. (2007). Free fat, surface fat and dairy powders: Interactions between process and product. A review. *Le Lait 87*(3), 187–236.

Vignolles, M.L., Lopez, C., Madec, M.N., Ehrhardt, J.J., Méjean, S., Schuck, P., and Jeantet, R. (2009a). Fat properties during homogenization, spray-drying, and storage affect the physical properties of dairy powders. *J. Dairy Sci. 92*(1), 58–70.

Vignolles, M.-L., Lopez, C., Madec, M.-N., Ehrhardt, J.-J., Mejean, S., Schuck, P., and Jeantet, R. (2009b). Protein-lactose matrix effects on fat encapsulation during the overall spray-drying process of dairy powders. *Aust. J. Dairy Technol. 64*, 75.

Vignolles, M.-L., Lopez, C., Le Floch-Fouéré, C., Ehrhardt, J.-J., Méjean, S., Jeantet, R., and Schuck, P. (2010). Fat supramolecular structure in fat-filled dairy powders: A tool to adjust spray-drying temperatures. *Dairy Sci. Technol. 90*(2–3), 287–300.

WHO (2011). Exclusive breastfeeding for six months best for babies everywhere. WHO: World Health Organization. Available from: http://www.who.int/mediacentre/news/statements/2011/breastfeeding_20110115/en/.

2.5 Technology, modeling and control of the processing steps in spray drying

Meng Wai Woo, Cordelia Selomulya and Xiao Dong Chen

2.5.1 Principles of spray drying

Spray drying is a common process used to convert liquid feed (e.g. milk) into solid particles. A typical spray-drying process comprises the atomization of liquid droplets into hot air in the drying chamber and the collection of the resulting powder. The evaporation rate of the moisture is determined by the difference between the vapor pressure of water on the droplet surface and the humidity of surrounding air. Initially the droplet undergoes a constant-rate drying period limited by mass transfer, whereby moisture is evaporating and a saturated vapor film is forming on the droplet surface. The droplet surface temperature approaches the wet bulb temperature of the drying air (Masters, 1980) and eventually exceeds the wet bulb temperature with the build-up of solid on the surface. When the droplet reaches a critical moisture content, the second stage of falling-rate drying period occurs, as moisture diffusion within the particle becomes the rate-limiting step. As the solid concentration continues to increase, the evaporation rate decreases as more energy is required to remove the same amount of moisture and the droplet's temperature rises towards the dry bulb temperature. Increasing solid concentration increases the mass transfer resistance, and the structure of the shell (e.g. permeability) also influences the drying behavior. When the particle reaches the drying air temperature, the evaporation ceases at the equilibrium moisture content, which is dependent on the drying air temperature and relative humidity of the ambient air (Kentish et al., 2005). The dried particles are then collected via a cyclone separator that recycles the fines back to the drying chamber.

2.5.2 Spray-drying layout and operation

2.5.2.1 Typical multi-stage layout

Dairy powder spray drying typically involves three stages of drying. Figure 2.5.1 shows this typical configuration, in which the first stage of drying is imparted by the atomization and convective drying of the milk concentrate droplets. This stage is where the bulk of the moisture is removed and the droplet solidifies into non-sticky powder. Hot air (which may be dehumidified for better control) is used for the first stage of drying which is relatively short and instant for the spray-drying process. As the name of the process suggests, this can be deemed the "actual" spray-drying process.

It is very important for the powder to become relatively non-sticky as it approaches the second stage of drying which is imparted by the internal static fluidized bed typically positioned at and as part of the bottom of the drying chamber. The purpose of the static fluidized bed is to provide relatively milder evaporation but longer residence time to remove the bulk of the remainder of the moisture in the powder. At the beginning of this stage, the powder would have already been well into the falling-rate period of drying and would tend to leave the internal bed at about 2–3%wt moisture content.

Hot air from the drying process typically leaves the dryer at outlets at the top of the dryer or via the side of the drying chamber. The air leaving the chamber is a combination

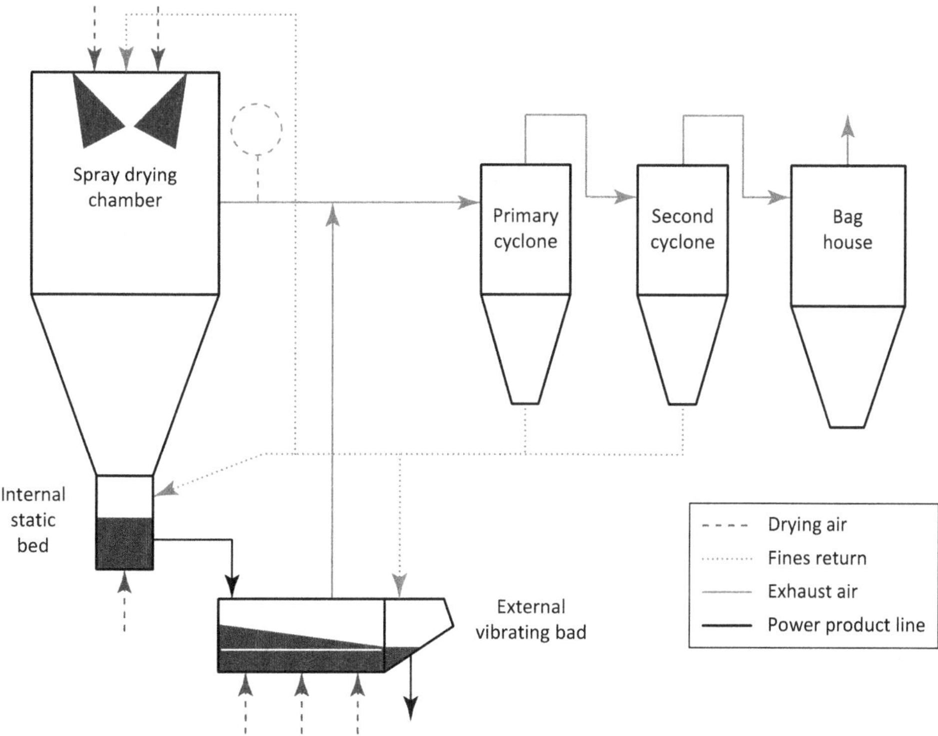

Figure 2.5.1 Typical three-stage spray-drying process for dairy powders.

of the air from the primary drying stage, the static fluidized bed and the cooling air from the cooling lances. Fine powders will then leave the chamber via the top or side outlets of the chamber. On the other hand, relatively larger powders will then mix with the powder at the internal static fluidized bed.

After the powder leaves the second-stage internal static fluidized bed, it will then undergo further fluidization, giving relatively longer residence time in the external vibrating fluidized bed. This third stage may sometimes function as a drying stage to further remove and control the moisture content of the powder or to function to provide mild cooling to the powder, depending on the temperature of the fluidization air used at the third stage. Beyond the third stage of the overall dairy powder production process is typically the sieving or the packing stage. Keeping this typical layout in mind, subsequent discussion in this part of the book chapter will focus on a few important control strategies for the production of dairy powders.

2.5.2.2 Outlet temperature-based control

It is common practice to take the outlet temperature as the basis for the control of the first two stages of the multi-stage drying process. The premise for this strategy is that the atomized droplets in the chamber actually experience rapid evaporation and do not fully undergo the high temperature condition input at the first-stage dryer. In contrast, as the drying process proceeds, the air within the primary drying chamber cools down (relative to the inlet thermal condition), and it is at this cooler temperature that the solidified powder will experience the bulk of its residence time in the first-stage dryer.

Therefore, if we consider the humidity and the temperature of the outlet stream from the chamber as a basis, we are indirectly using the outlet temperature as an indication of the lowest possible moisture achievable from the first drying stage. This basis for control inevitably assumes "quasi-equilibrium" between the powder and the conditions of the outlet drying air. On top of using this as an indirect control of the particle moisture content from the first drying stage, it is important to also ensure the particles enter the second stage of drying under non-sticky conditions. The sticky behavior of dairy powder (and in most spray-drying of food powder) is controlled by a combination of powder moisture and temperature. More detailed description is given in the references cited here (Ozmen and Langrish, 2002, 2003a). It is also important to maintain the outlet temperature at levels suitable to prevent product degradation, which are product-specific. A typical outlet temperature from the primary chamber for dairy powder may be around 60–70°C.

It is noteworthy that the outlet air temperature and humidity are a result of the drying process of both the first-stage and second-stage dryer, and thermal and humidity conditions at both first- and second-stage hot air inlets. In addition, cooling air at the atomizer lance and plenum chamber will also affect the overall thermal conditions at the outlet of the chamber. Depending on the specific system, the flow rate of cooling air required may be significant relative to the flow rate of the hot air supplied to the drying chamber. From mass and energy balance principles, higher hot air flow rates and higher inlet temperatures will increase the outlet temperature, while a higher spray rate will reduce the outlet temperature, and vice versa. These relations are certainly not linear. Based on these principles, some trial-and-error strategies for the operation of spray dryers are available in the literature (Woo et al., 2007). In using the outlet temperature as the baseline of spray-drying operation, it is important not to compensate high product (spray) throughput with an excessively higher primary-stage hot air inlet temperature. This may lead to excessively high initial temperatures, leading to undesired product degradation.

2.5.2.3 Fines control and agglomeration

The return of fines is an important aspect of the multistage spray-drying operation to control the amount of product loss (reaching the bag house) and to control the forced agglomeration of the product. Both factors will affect the bulk density of the product which is a critical product attribute controlled in the industry. Different fines return strategies can be applied as shown in the dashed green lines in Figure 2.5.1. If the fines are returned to the end of the external vibrating fluidized bed, its function is to mainly combine and mix the fines with the relatively larger powder in the final product. This is then mainly a product-loss minimization strategy. Compared to not returning the fines to the end of the external vibrating fluidized bed, in general if the powder is not excessively cohesive, this strategy may increase the bulk density of the final product as the fines will make the product more compact. Fines may also be returned to the internal static bed to give more contact with the relatively warmer and wetter powder (relative to the powder at the end of the external bed).

If the fines are returned to the top of the drying chamber in the vicinity of the nozzles to contact with semi-dried droplets, forced agglomeration will then be induced. This will in effect reduce the amount of fines in the spray-drying system. Different configurations are available depending on the scale of the spray dryer. Some fine-return configurations resemble a central "waterfall" of fines from which the nozzles spray inwards towards the "waterfall" of fines. In some other configurations, the fines are returned towards the nozzle sprays. Regardless of the configuration, the key is to force the fine particles to contact semi-dried (sticky) particles, forming stable and raspberry-type agglomerates. This will improve the dissolution behavior and reduce the cohesiveness of the product. If the particles contact wet droplets too close to the nozzles, large undesired clumps may form, leading to products with undesired dissolution behavior. In order to directly prevent the formation of fines, spray nozzles may be directly toward each other to induce forced agglomeration of semi-dried (sticky) particles. At the moment, the amount of overlap in the nozzle spray to induce such forced agglomeration is still based on trial and error in commercial operations.

Another strategy in the control of the bulk density of the product is to actually reduce the suction at the exhaust of the external vibrating fluidized bed. In some operations, the exhaust from the external bed can be independently controlled from the main exhaust line leaving from the primary chamber going into the cyclone. With the reduction in the exhaust suction, more fines will remain in the final product which will increase the bulk density of the product.

2.5.2.4 Adjusting the feed viscosity

Spray drying is an energy-intensive process, and thus the feed into the spray dryer is often pre-concentrated via heat-driven evaporation under vacuum which consumes less energy (Singh and Singh, 2015). The increase in the total solid content of dairy liquid increases its viscosity, which influences the drying process and the properties of the final powder. Some of the unwanted effects include reduced solubility (Baldwin et al., 1980) and the development of off-flavor in whey protein concentrate (Park et al., 2014), milk protein concentrate and skim milk powders (Park et al., 2016).

Different methods can be used to adjust the feed viscosity at higher solid contents. High-pressure homogenization (HPH) has been used to improve the stability of emulsions by modifying the structure and functional properties of proteins to prevent phase separation (Ye and Harte, 2014), and reduce viscosity (Floury et al., 2000). Hydrodynamic cavitation generates cavitation bubbles when the liquid is forced through a constriction so that the pressure decreases below the vapor pressure of the medium at that temperature. When the liquid expands and the pressure increases, the cavitation bubbles collapse and the cavitation shockwave causes microscopic mixing that can be scaled up without excessive heating from the friction between the rotor and liquid (Tao et al., 2016). Li et al. (2018) demonstrated that the hydrodynamic cavitation reduces the viscosity of milk protein concentrate (MPC80) up to 56% compared to the solutions typically used for spray drying, and has a strong impact on the final powder's properties. It was found that although the particle size and bulk/tapped density of the resulting powder slightly increased compared to that of the un-treated feed, the solubility and wettability of the powder were relatively unchanged. The reduction in viscosity was attributed to the temporary breakdown of the protein structure reducing the elastic modulus of the protein due to the small eddies generated during the hydrodynamic cavitation (Patist and Bates, 2008). If enough energy is applied, permanent viscosity reduction (Doona, 2010) and protein denaturation (Villamiel and de Jong, 2000) might occur which affect the rheological properties of the solution. Another method to reduce viscosity is by direct injection of carbon dioxide, which has been used in milk processing to prolong the shelf life of raw milk by reducing microbial activity (Hotchkiss et al., 2006). Marella et al. (2015) showed that the injection of CO_2 before and during the ultrafiltration of skim milk to produce milk protein concentrate resulted in up to two-fold reduction in the viscosity of the retentate. The dissolved CO_2 adsorbs onto the casein micelles and generates carbonic acid to bind with the calcium ions, solubilizing some of the calcium phosphate in casein micelles. The reduction in viscosity was attributed to the lower amount of calcium in the retentate, although a reduction in membrane flux was also observed due to the presence of solubilized calcium phosphate which is an important salt foulant in the membrane system. The spray-dried MPC retained its solubility such that the carbon dioxide treatment can be used to produce MPC80 powders with reduced calcium content and improved functionality.

2.5.3 *Modeling of spray dryers*

The simplest form of spray-dryer modeling is to treat the drying chamber as a black box, which in essence follows the conventional flowchart-based heat and mass balance calculations. These approaches are well-described in the literature, and interested readers are directed to the references cited here (Straatsma et al., 1991, Ozmen and Langrish, 2003, Schuck et al., 2009). While most of the black-box model only accounts for conventional energy balances surrounding the dryer, the SD2P© platform proposed by Schuck et al. (2009) accounts for the extra energy requirement for milk solute drying in the mass and energy balance of the spray-drying system. The frameworks cited above are mainly steady-state models assuming "equilibrium" conditions in the outlets of the spray dryer. There are also black-box modeling approaches reported in the literature to account for the dynamic control of spray dryers (Petersen et al., 2017).

The focus of this section of the chapter, however, will be on non-black-box approaches in the modeling of the spray dryer. One key advantage of such an approach is that the model will be able to describe the local conditions within the dryer, which may be more useful for the scoping of the dryer (sizing) or in detailed design or troubleshooting of the dryer. Such more detailed modeling of the spray dryer will be particularly useful in evaluating what-if situations.

2.5.3.1 *One-dimensional approach*

A one-dimensional simulation of the spray dryer assumes a simplified plug flow within the spray dryer. In all the works using this approach, the spray-drying chamber is treated as a long-scale spray dryer where the droplets move downwards in a plug flow manner towards the exit, despite possible recirculation of the particles (Patel et al., 2010, Langrish, 2009, Truong et al., 2005). A key assumption of this approach is that the droplets are assumed to be uniformly distributed radially across each section of the "plug flow" within the spray-drying chamber. The setting up of a one-dimensional framework typically involves only the simultaneous solution of six differential equations presented here below to describe fundamentally the spray-drying process within the chamber.

Drying air energy balance:

$$\frac{dT_b}{dL} = \frac{\left(\frac{dm_w}{dt}\frac{\theta}{v_p}\right)\left[\Delta H_L - C_{p,v}\left(T_b - T_p\right)\right] - \frac{\theta}{v_p} h A_p \left(T_b - T_p\right) - U(\pi D)\left(T_b - T_{amb}\right) - \dot{V}\rho_b \Delta H_L \frac{dY}{dL}}{\dot{V}\rho_b C_{p,b}} \tag{2.5.1}$$

Drying air mass (absolute humidity) balance:

$$\frac{dY}{dL} = \frac{\frac{dm_w}{dt}\frac{\theta}{v_p}}{\dot{m}_{b,dry}} \tag{2.5.2}$$

Droplet drying or mass change:

$$\frac{dm_w}{dt} = h_m A_p \left(\psi \rho_{v,\text{surface}} - \rho_{v,\text{bulk}}\right) \tag{2.5.3}$$

Droplet energy change:

$$\frac{dT_p}{dt} = h A_p \left(T_b - T_p\right) - \frac{dm_w}{dt} \Delta H_L \tag{2.5.4}$$

Droplet momentum change:

$$\frac{dv_p}{dt} = C_D \frac{18\mu_b Re}{24\rho_p d_p^2}\left(v_b - v_p\right) + \frac{g}{\rho_p}\left(\rho_p - \rho_b\right) \tag{2.5.5}$$

These equations can be solved using a forward method with respect to the length of the tower corresponding to Equations 2.5.1 and 2.5.2. The time-based differential equation for the droplet phase can be solved with respect to the length scale of the tower by considering the velocity of the droplet at each step of the solution. These solutions can be implemented directly on a spreadsheet (Microsoft Excel) or using commercially available mathematical tools (MATLAB). For the benefit of the reader, Figure 2.5.2 describes graphically the physical meaning of each differential equation and how they are linked. It can be seen that in this theoretical framework, two-way heat and mass coupling between the air and the droplet phase is assumed. Only a one-way momentum coupling is assumed for simplicity (the air influencing the movement of the droplet). This is because the airflow behavior was modeled based on the conservation of mass in the air phase without an explicit solution to any momentum equation of the air.

As the humidity and temperature of the air and the moisture content of the particles are tracked throughout the simulation, this predictive approach does not necessitate the assumption of equilibrium outlet powder moisture content, as in the black-box approach to spray-dryer modeling. One main advantage of this predictive approach is that it can now be used as an approximate scaling tool to obtain a rough functional size of the spray tower. The knowledge of such a rough estimate of the chamber can then be used as a basis for more detailed design of the spray chamber. Proper setup of the one-dimensional framework will also provide quick computation, making it suitable as a training tool for operators.

A few important notes are given here, highlighting the current limitations or further areas for improvement for such a one-dimensional predictive framework. It can be underlined that most of them pertain to numerical challenges and not to the lack of a physical mathematical model for the spray dryer:

1. Current reported simulation works only involve the modeling of the drying process using a single representative droplet size. During the preparation of this chapter, the authors were not aware of any work involving a whole spectrum of droplet sizes. The

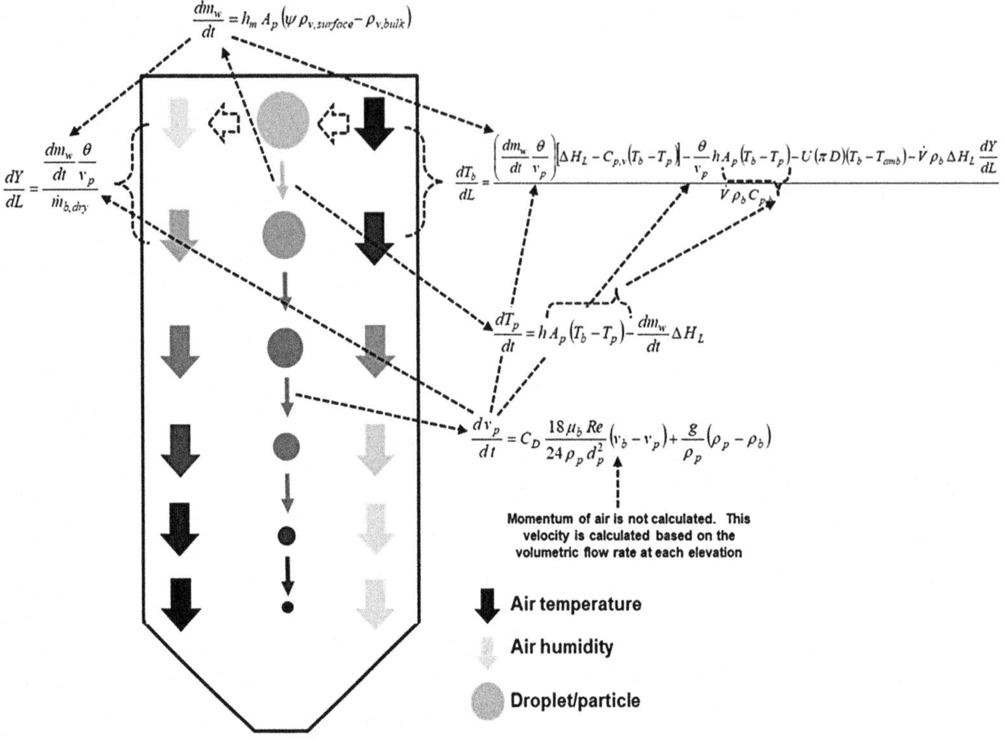

Figure 2.5.2 Graphical representation of the one-dimensional theoretical framework.

implementation of droplet size distribution in the framework may require additional solutions of the droplet momentum, droplet drying and droplet temperature change equation, for each droplet size class.

2. Current reported simulation studies do not account for the recirculation of airflow within the chamber. The term "recirculation" here refers to coherent upward bending of airflow within the chamber such as those illustrated in Figure 2.5.1, rather than small-scale turbulent recirculation within the chamber. The outlet conditions for such upward recirculated airflow within the chamber may be different when compared to that of a single plug flow.
3. The application of such a plug flow framework for a counter-current spray tower (should this be explored for dairy powder processing) may involve significant numerical complexity. This is because the equations regarding the heat, mass and momentum transfer of the droplet will need to be solved in the opposite direction as the energy and mass balance of the air phase. The propagation of the effect of the two-way coupling between the droplet and the air will then be in opposite directions. For this reason, iterations are required to fully develop the solution until it reaches a converged powder drying history. Such iteration schemes can be implemented by Microsoft Excel using the Visual Basic macros (author's experience – unpublished results). Ali et al. (2014) have shown that this is possible to iterate with MATLAB too.
4. For the counter-current spray drying, there is always a possible movement reversal for particles. Capturing this phenomenon and incorporating the heat and mass coupling for the recirculated particles in a one-dimensional modeling framework is

numerically not trivial. Nevertheless, this is an important phenomenon as particle re-entrainment may denote the maximum possible counter-current inlet airflow rate (should re-entrained be deemed a loss in product).

2.5.3.2 Computational fluid dynamics approach

The CFD technique is a widely established technique in the modeling of engineering systems involving fluid flow (air and/or liquid flow). For the interested readers, a few useful general references are cited here (Versteeg and Malalsekera, 2007, Patankar, 1980). In essence, the technique involves resolving and predicting the flow pattern and its associated energy transfer within a particular engineering system. Figure 2.5.3 illustrates an example of the flow field prediction within a short-form spray dryer using the CFD technique. The path lines of the predicted flow field may be colored to exhibit the temperature, turbulence behavior or even the velocity of the flow field depending on the needs of the analysis. Droplets and particles are usually then modeled as discrete parcels of particles transported within the simulation domain, interacting with the flow field accounting for turbulence dispersion. The drying process is then captured by the implementation of suitable droplet drying models (with material-specific kinetics) for each parcel of particles in the simulation. Details and guidelines on the setting up and interpretation of a CFD model for spray dryers are beyond the scope of this chapter and can be found in a handbook cited here (Woo, 2016). This section of the book chapter will aim to highlight the advantages/disadvantages of the CFD technique and to focus on a few common questions in evaluating the use of this technique for spray-drying modeling.

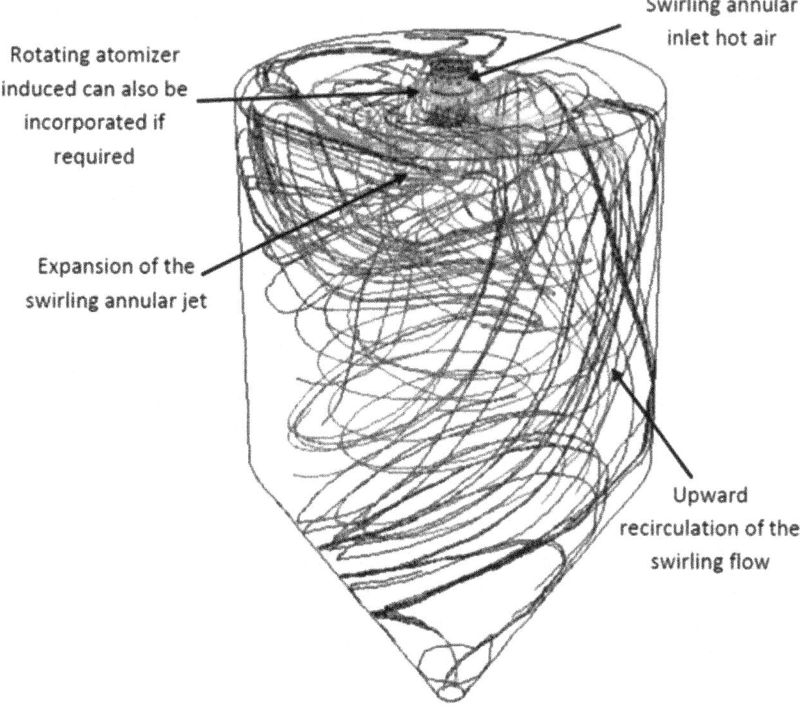

Figure 2.5.3 Detailed flow behavior and droplet/particle movement feature in a CFD simulation of spray dryers (modified from Woo et al., 2016).

The main advantage of using the CFD technique is that it allows very detailed analysis within the dryer without any overarching assumptions such as the uniform radial droplet distribution, as in the one-dimensional models. Therefore, this can be used for detailed design and operation control, for example: to examine suitable positioning of the spray nozzles, to understand and control the deposition of powder, to enhance the turbulent mixing within the chamber, for the design and the optimization of swirls in the chamber to enhance mixing, etc. There are numerous reports utilizing this technique for industrial-scale spray dryers (Gabites et al., 2010, Jin and Chen, 2009, Ullum et al., 2010, Ali et al., 2017). On the other hand, if the need of the analysis is mainly to know suitable inlet or outlet operating conditions for a spray-drying operation, the CFD technique may be overkill due to the complexity in setting up the model and the long computing time required (days) even for a simple analysis. In this respect, the one-dimensional modeling technique may be preferable. The complexity in setting up a CFD model of a spray dryer does not just lie in the numerical aspect of the model but is also contributed by the difficulty in obtaining reliable boundary conditions (required inputs) for the model. These properties may be different from measured operating conditions in the spray-drying facility, and more information on this area is given in (Woo, 2016). In particular, challenges in obtaining the atomization characteristics for simulations are discussed in the next section of this book chapter. It is noteworthy that the bulk of the resources required for a CFD modeling project of a spray dryer is in obtaining these input properties, rather than in the actual numerical implementation and computation time.

In view of the complexity of such a model and the possible uncertainties in obtaining the required inputs for the CFD framework, analysis based on a CFD framework should be used for evaluating comparative differences between different spray-dryer designs and operations, rather than with the expectation of obtaining an "absolute" dryer prediction. For example, to put this into perspective, the CFD technique should not be used as a process control tool to predict spray-dryer operations for the control of tight outlet powder moisture content. In contrast, it can be used for what-if situations, for example for evaluating if larger vane angles (swirls) in the chamber will or will not improve the overall drying process.

Figure 2.5.4 illustrates the structure of a CFD model of a spray dryer. At the core of the CFD framework is the general fluid flow simulation technique, which is well-established in the literature. Key elements which are unique in a spray-dryer CFD model are the sub-models incorporated into the overarching fluid flow prediction framework. Over the past decades, there has been significant development in the droplet drying, particle-tracking and deposition sub-models. On that note, the particle-tracking sub-models reported so far in the literature concern the tracking of entrained powders transported within the drying chamber. To the best knowledge of the authors, there is yet to be a reported simulation accounting for both entrained (first-stage drying) and fluidized powders (second-stage internal static bed) within a single simulation. In current reported simulations, only the airflow from the internal static bed was considered, without capturing the particle collision behavior in the fluidization regime. Considering the increasing importance of powder agglomeration in the spray-drying industry, there is now significant interest in the development of the agglomeration sub-model. Most of the sub-models available and implemented for pilot- or commercial-scale spray dryer, reported in the literature, focus on the prediction of size enlargement of the powder or on the regions suitable for agglomeration within the drying chamber (Verdumen et al., 2004, Jaskulski et al., 2015, Malafronte et al., 2015); on the other hand, they are not able to predict the structure of the final agglomerate.

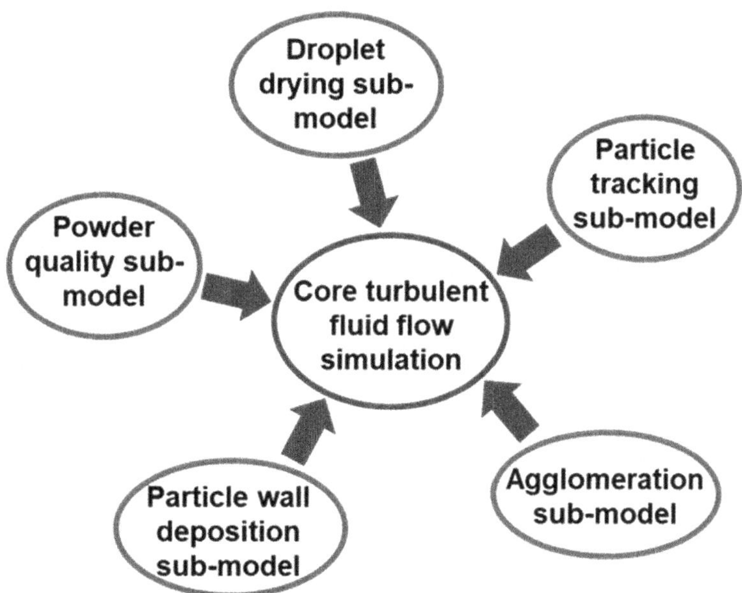

Figure 2.5.4 Structure of a CFD predictive framework of a spray dryer.

2.5.3.3 Modeling of droplet drying

Regarding the spray dryer prediction, for the implementation using the one-dimensional or, alternatively, the lumped droplet drying model is normally applied. As the class of the model suggests, this approach provides a prediction of the average moisture content and temperature of the particle without distinguishing the possible moisture gradient within each particle. One mathematical form to capture the heat and mass transfer in droplets/particles is to use Equations 2.5.3 and 2.5.4 for each droplet in the simulation. These mathematical forms were introduced mainly due to the familiarity and experience of the authors, and other forms are also available in the literature.

The crux in droplet drying modeling lies in mathematically capturing the falling-rate period as solidification progresses. This should be reflected in the progressive reduction of the particle surface vapor concentration, leading to the progressive reduction in the driving force for evaporation. We will review and compare two approaches regarding modeling the falling rate period of drying: the reaction engineering approach (REA) and the characteristic drying curve (CDRC) method. Starting with the REA approach, the falling-rate period is captured by modeling the drying process as a reaction with a fractionality term in Equation 2.5.3 as a function of the activation energy for evaporation E_v (Chen, 2008),

$$\psi = \exp\left(-\frac{E_v}{RT_p}\right) \tag{2.5.6}$$

The activation energy is then a function of moisture content within the particle, and it can be described by a material-specific master activation energy curve. A compilation of experimentally determined material-specific master activation energy curves, relevant for dairy products, is given in (Woo, 2016). In addition to being material-specific, the master activation energy curves are also initial solid concentration-specific.

For the CDRC approach, Equation 2.5.3 is typically reformulated into the following form (Langrish and Kockel, 2001),

$$-\frac{dm_w}{dt}=\left[\frac{X-X_{eq}}{X_{cr}-X_{eq}}\right]^N\frac{hA_p\left(T_b-T_{p,wb}\right)}{\Delta H_L} \tag{2.5.7}$$

The square bracket on the right side empirically describes the falling-rate period (relative to the wet bulb drying rate), in which the characteristics of the falling rate are governed by the parameter N in Equation 2.5.7. In this formulation, the wet bulb period of evaporation is either ignored or mathematically captured by empirically denoting the critical moisture content beyond which the falling-rate drying begins. Prior to the critical moisture content, wet bulb evaporation is applied. A compilation of experimentally determined material-specific N parameters and the associated critical moisture content is given in (Woo, 2016). There is minimal dairy product-relevant characterization done for the CDRC framework. For skim milk with an initial solid %wt of 20, N was found to be 1 (linear falling rate) while the initial moisture content can be taken as the critical moisture content (Langrish and Kockel, 2001).

2.5.3.4 *Modeling of atomization*

One main uncertainty faced in the operation of spray dryers to obtain concentrated dairy products is the selection of nozzles. For the efficient operation of spray dryers, it is important that the nozzle (which includes its internal accessories) provides the required droplet size distribution leading to the desired dehydrated powder size range. Another aspect is that the nozzle configuration should provide the optimal spray angle, particularly if efficient agglomeration control is required. These two atomization parameters are highly affected by the viscosity of the concentrated milk (typically when >50%wt in solid). In contrast, most nozzles available in the industry are characterized only by water. How do we extrapolate the performance of commercially available nozzles to actual concentrated milk atomization?

There are conventional correlations available in the literature which can be used to account for the spray of different viscosities. These correlations are available in standard handbooks about spray drying, and interested readers are directed to the suitable references cited here (Masters, 1979). The highly empirical nature of these nozzles may impose limitations to the correlations to only nozzle configurations (in particular in the internal designs of the nozzle) similar to those used in the derivation of the correlations. Such highly empirical forms can also be delineated from the correlations themselves as they are normally linked to the atomization parameters such as the feed flow rate, surface tension, viscosity, solid concentration and atomization pressure, etc. Even if the orifice size is incorporated into the correlation, all these parameters do not reflect the internal design of the nozzle configuration, which is the key to distinguish nozzle characteristics. The authors encountered discrepancies when using some of the correlations available in the literature.

While spray-drying operators may have an intuitive guess about the actual atomized droplets based on the available water-based characterization of the nozzle, this unknown atomization parameter (the initial droplet size distribution) seemed to be a significant challenge in the modeling of spray dryers, as it is a critical input to the model. A precise estimation of the atomization parameters typically requires laser or image-based measurements, that are quite difficult to obtain in the industrial environment. On the other end of the spectrum, making such measurements in the laboratory is also difficult due to the absence of a high-pressure pumping system and of an evaporator system required to achieve high solid concentration in the milk.

There have been attempts by the authors to overcome this bottleneck in nozzle characterization by using volume-of-fluid (VOF)-based CFD simulations to model the flow of concentrated milk within the nozzle and the subsequent dispersion of the milk upon leaving the nozzle. This kind of simulation is potentially able to shed light on the effect of different nozzle/feed conditions as well as of the internal design of the nozzle on the atomization behavior of the fluid. However, the limitation of this approach is that only the development of the high-velocity thin concentrated milk films leaving the atomizer, which precedes the breakup leading to the formation of the droplets, can be predicted. Due to this limitation, only the initial film velocity and spray angle can be estimated from this type of simulation. The reason behind this limitation is that the VOF-based simulations require the tracking of the boundary between the immiscible fluids (milk and air) in the simulation. To clearly distinguish a typical spray-dried droplet, mesh sizes in the length scale order of 10^0 microns are required. On the contrary, the geometry of a nozzle and its internal components is of the order of millimeters. Taking into account both length scales in the system will mean impractically high meshing and computational requirements.

Therefore, the authors conclude that the direct use of the CFD technique to model the atomization process of a conventional industrial-scale nozzle is not feasible. Nevertheless, the VOF-based CFD technique may be advantageous in the prediction of film characteristics produced from a nozzle. Regardless of the myriad combinations of nozzle design and operation, it is the characteristics of the film leaving the nozzle which determine the final droplet size distribution from the nozzle. Therefore, one possible framework which can be adopted in the future is to "move" the boundary of the correlations to that between the film characteristics and the final droplet size distribution. The VOF-CFD predictive framework can then be used to predict the film characteristics from nozzle-specific designs and operations. This will then represent a more fundamental approach, albeit still empirical in nature, to predict the atomizer performance for highly concentrated milk product. It is foreseen that such correlations can then be used across different forms of atomizers available in the industry.

2.5.4 *Unique spray-drying technologies and explorations*

2.5.4.1 *Single-droplet drying experiments to obtain drying kinetics and morphological changes*

It is usually not possible to investigate the process of particle formation (morphology and appearance) during spray-dryer operations. It is also difficult to obtain drying kinetics for droplets *in situ* in a spray dryer. The latter is the most important point when designing and optimizing spray-drying processes, sometimes in the dryer-wide simulations including computational fluid dynamics. The drying kinetics includes the drying rate of droplets which is influenced by drying conditions and the compositions of the solutes and/or solid materials. The knowledge of this kinetics, once coupled with the momentum transport and energy transport equations, is very powerful for scaling up or down the processes. To establish the drying kinetics, one must have a way to suspend a droplet of solution in a controlled drying gas flow (hot air of known humidity for instance) and hence to calculate heat and mass transfer correlations (see Figure 2.5.5); moreover, it would be necessary to record the weight loss and size change over time. One possible approach is to use a glass filament technique. As drying proceeds, due to the flexibility of the glass filament, it would move up relative to its original positioning (Figure 2.5.5). This displacement can be nicely correlated to weight loss. More details of this technique can be found in PhD theses by Lin (2004) and Fu (2011), and many published papers.

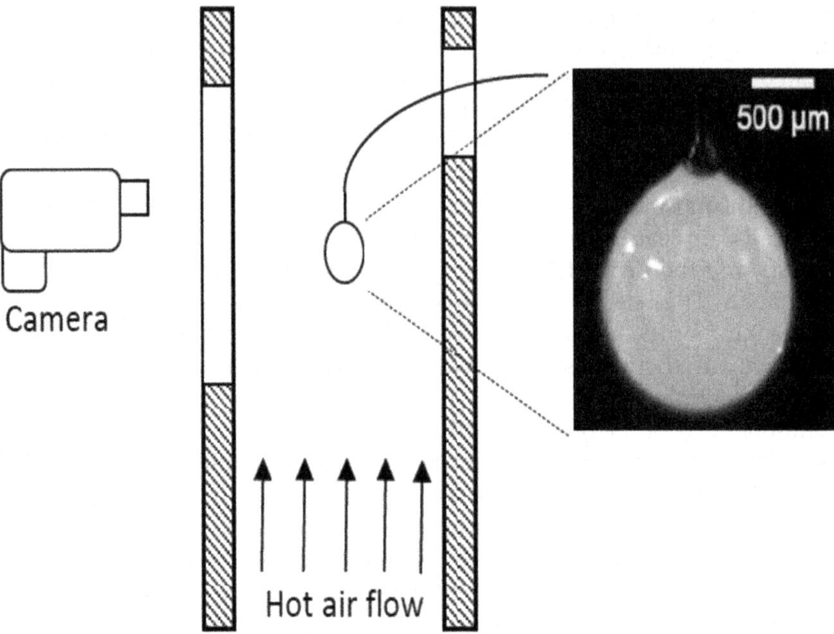

Figure 2.5.5 "Hanging" of a droplet in a controlled drying gas flow to obtain drying kinetics (modified from Fu, 2011).

To establish drying kinetics, one must also measure the droplet temperature over time (usually due to the small *Biot-Chen* number (Chen and Peng, 2005) for evaporation, a single average temperature measurement inside the droplet would be enough). To hang a droplet, Sano and Keey (1982) described a comprehensive study on skim milk droplet drying. Basically, a glass filament with a knob at the end of it is used. The part of the filament above the knob is coated with hydrophobic material to prevent the droplet climbing beyond the knob. There is another method which "freely" suspends a droplet using acoustic waves. However, the more common method is still the glass filament method (Lin, 2004; Fu, 2011). As drying proceeds, the droplet shrinks over time, and the size change is recorded using a video camera and interpreted through an imaging software such as Image-J. The device can be made simple and operated manually as shown by Sano and Keey (1982), and a version of that was implemented for factory operation (Stevenson, 1999). A commercially available single-droplet dryer is illustrated in Figure 2.5.6. This system has been incorporated with morphological observations and wetting/dissolution tests on-line (Fu, 2011).

2.5.4.2 Mono-disperse droplet spray-drying experiments to validate CFD and to obtain particles under well-controlled operating conditions

Based on his extensive experience, as an engineer working in Fonterra to produce milk powders and later as an academic working on industry-funded projects on spray drying in the period of 1991–2003, Chen realized that it was difficult to draw definitive conclusions from the research results that were obtained using the conventional laboratory- or pilot-scale spray dryers at university or in companies, especially when the intention for these

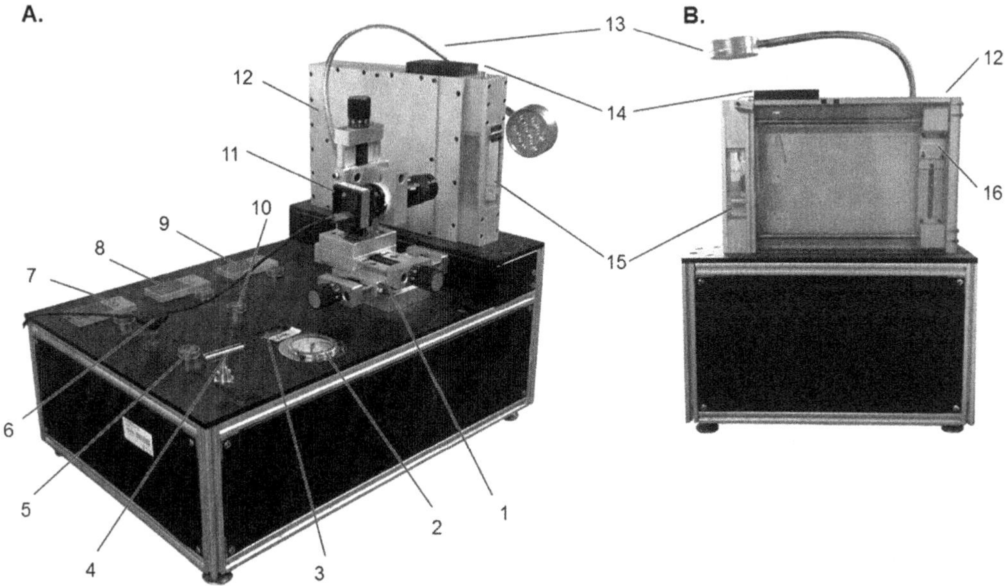

Figure 2.5.6 Single droplet drying apparatus. (A) Axonometric view; (B) right side view. 1: Adjustable stand to control the position of high-definition camera, 2: pressure gauge, 3: mass flowmeter, 4: needle valve, 5: switch of the gooseneck lamp, 6: pressure regulator, 7: temperature controller for the constant room chamber containing mass-measuring glass filament, 8: temperature controller for drying air flow, 9: over-heat temperature controller, 10: main power switch, 11: high-definition camera and charge-coupled device, 12: constant room chamber containing mass-measuring glass filament, 13: gooseneck lamp, 14: removable holder of diameter-measuring glass filament, 15: drying chamber, 16: adjustable holder for mass-measuring glass filament, which could move the glass filament to the drying chamber for mass-measuring experiment. (Courtesy of Nantong Dong-Concept New Material Technology Ltd, Jiangsu Province, China.)

studies was to scale up and to optimize industrial operations. The small-scale dryers commercially available for the laboratory usually can only produce very small particles, and they show polydisperse sizes and different morphology even within the same experimental run. It is usually not practical to carry out research on industrial-scale dryers, which are of the capacity of several to tens of tons of powder product produced per hour. In fact, the complications encountered in the small laboratory dryers and the large industrial ones are similar in many ways. Therefore, the purpose was to simplify the real situation so that it could be reproduced in a small laboratory. It is usually very difficult to discern the effects of operating parameters on particle morphology development. To understand fundamentally what is going on in spray drying, a better-controlled scenario was needed. In the 2002–2003 period, at the University of Auckland, Chen proposed the making of a mono-disperse droplet spray dryer. This was realized through several PhD studies (Patel, 2004, 2008; Wu, 2010; Rogers, 2011). Although many modifications have been made in recent years for the better performance of the atomization system, the main components of the design have remained. One key aspect of the mono-disperse atomizer is the use of the piezoelectric element to generate the droplets. Wu (2010) performed the most significant study on the atomization behavior of mono-disperse droplet generators, proving that there

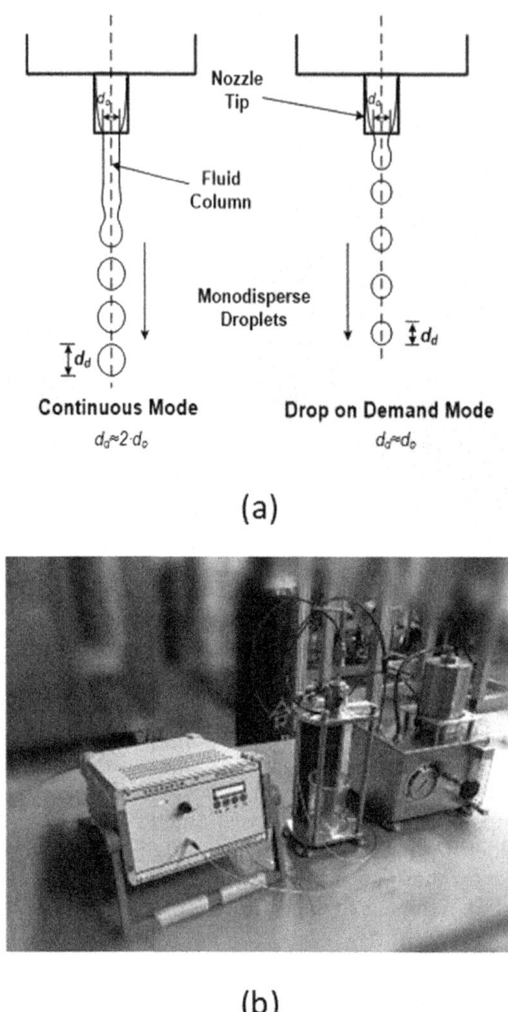

Figure 2.5.7 (a) The principles of mono-disperse droplet generation and (b) the mono-disperse droplet generation and demonstration unit available commercially. (Courtesy of Nantong Dong Concept New Material Ltd, China.)

is a wide range of frequencies for the piezoelectric material to operate at that can generate uniform droplets. There are two modes of droplet generation through the piezoelectric mechanism shown in Figure 2.5.7a. A commercially available unit for continuous droplet generation mode is shown in Figure 2.5.7b. Figure 2.5.8 displays a commercially available mono-disperse droplet spray dryer.

The air flow pattern is simple and mostly straight downwards to make the particle trajectories simple to interpret. The particles are easily collected under the dryer as the "rain fall" of the particles can be easily captured due to the simple flow path. The uniformity of the particles made is excellent (refer to Figure 2.5.9), and the impact of operation parameters and feed characteristics can be more effectively studied.

Figure 2.5.8 Mono-disperse droplet spray dryer available commercially. (Courtesy of Nantong Dong Concept New Material Ltd, China.)

2.5.5 *Superheated steam spray drying*

The control of surface fat on spray-dried dairy powder has been receiving on-going attention in spray drying. A series of experimental investigation using the single-droplet drying technique and laboratory-scale superheated steam spray drying was initiated to evaluate if the use of superheated steam as the drying medium may help in addressing this challenge (Lum, 2019, Lum et al., 2017). The initial hypothesis in that series of work was that superheated steam, which in essence is composed of water molecules, may exhibit a "hydrophilic"-like drying environment to "push" the fat into the droplet during spray drying. While the reports revealed that the superheated steam spray-dried skim milk and full cream milk powder indeed exhibited higher wettability, unfortunately, there was no significant change in the composition of the surface fat, disproving the initial hypothesis of that series of work. In contrast, it was found that superheated spray drying produced dairy powder with significantly more pores (which may have led to the higher wettability). Further analysis also showed that there was minimal increase in insolubility in the powder, an aspect very critical in the evaluation of the use of superheated steam in the processing of dairy products. This series of work is still at the preliminary phase, and further exploration is required to evaluate systematically the application of superheated steam in producing dairy powder, in terms of dryer design and operations. An important strategy in designing the equilibrium moisture content of powders from the droplet convective drying process is given in the following article (Lum et al., 2018). More work is also required to evaluate the long-term stability of superheated spray-dried dairy powder. It is well-known that superheated steam drying is advantageous in terms of energy saving and in providing an oxygen-free environment

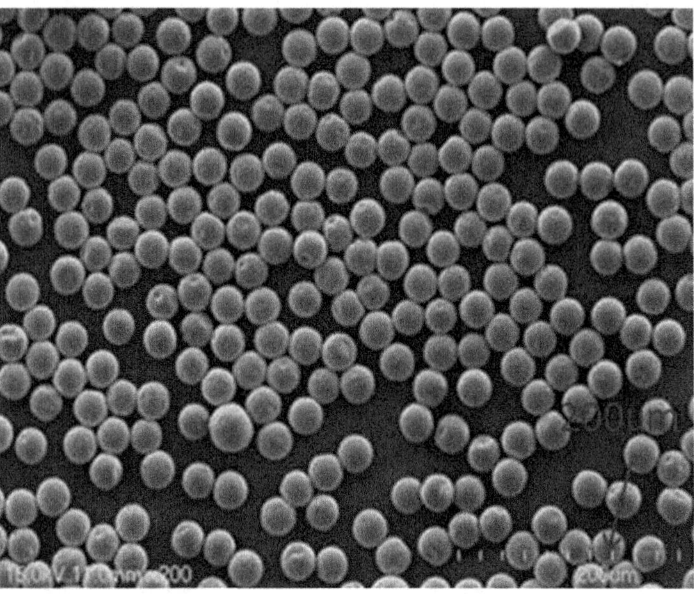

Figure 2.5.9 Chitosan micro-particles produced using a mono-disperse droplet spray dryer.

to minimize oxidation degradation in the dried product. On a separate note, it was found that superheated steam spray drying is also a very useful avenue for controlling fast-to-crystallize materials in spray drying. This may include fast-to-crystallize materials such as edible table salt or mannitol. This unique capability of superheated steam drying may find applications in spray drying individual components of milk or its micronutrients or minerals.

2.5.6 *Nomenclature*

Symbols	
v	Velocity (m s^{-1})
t	Time (s)
ρ	Density (kg m^{-3})
g	Gravitational acceleration (m s^{-2})
T	Temperature (°C)
C_p	Specific heat capacity (J °C^{-1} kg^{-1})
$\dot{m}$	Mass flow rate of bulk air (kg s^{-1})
$\dot{m}_{b,dry}$	Mass flow rate of dry air (kg s^{-1})
ΔH_L	Enthalpy of vaporization (J kg^{-1} vapor/water)
Y	Bulk air absolute humidity (kg vapor kg^{-1} dry air)

$\frac{dm_w}{dt}$	Evaporation rate from a single droplet (kg water droplet^{-1})
θ	Rate of number of droplets (s^{-1})
A	Area (m^2)
h	Heat transfer coefficient (J s^{-1} m^{-2} °C^{-1})
U	Heat loss coefficient (J s^{-1} m^{-2} °C^{-1})
D	Diameter of the chamber (m)
dL	Length of the control volume (m)
$\dot{V}$	Bulk air volumetric flow rate (m^3 s^{-1})
h_m	Mass transfer coefficient (kg s^{-1} m^{-2} °C^{-1})
R	Universal gas constant (8314 J kg^{-1} K^{-1})
E_v	REA activation energy (J kg^{-1})
X	Particle moisture content (kg water/kg dry solid)
C_D	Drag constant
N	CDRC falling-rate constant
Ψ	REA fractionality term
Subscripts	
p	Particle
b	Bulk air
v	Vapor
amb	Ambient
w	Water
eq	Equilibrium

References

Ali, M., Mahmud, T., Heggs, P.J., Ghadiri, M., Bayly, A., Ahmadian, H., Martin de Juan, L. 2017. CFD modeling of a pilot-scale countercurrent spray drying tower for the manufacture of detergent powder. *Drying Technology*, 35(3), 281–299.

Ali, M., Mahmud, T., Heggs, P.J., Ghadiri, M., Djurdjevic, D., Ahmadian, H., Martin de Juan, L., Amador, C., Bayly, A. 2014. A one-dimensional plug-flow model of a counter-current spray drying tower. *Chemical Engineering Research and Design*, 92(5), 826–841.

Baldwin, A.J., Baucke, A.G., Sanderson, W.B. 1980. The effect of concentrate viscosity on the properties of spray dried skim milk powder. *Dairy Science and Technology*, 15, 289–297.

Chen, X.D. 2008. The basics of a reaction engineering approach to modelling air-drying of small droplets or thin layer materials. *Drying Technology*, 26(6), 627–639.

Chen, X.D., Peng, X.F. 2005. Modified *Biot* number in the context of air drying of small moist porous objects. *Drying Technology*, 23(1–2), 83–103.

Doona, C.J. 2010. *Case Studies in Novel Food Processing Technologies: Innovations in Processing, Packaging, and Predictive Modelling*, Elsevier, Cambridge, UK.

Floury, J., Desrumaux, A., Lardienes, J. 2000. Effect of high pressure homogenization on droplet size distribution and rheological properties of model oil - in - water emulsions. *Innovative Food Science & Emerging Technologies*, 1, 127–134.

Fu, N. 2011. Single droplet drying of food and bacterium containing liquids and particle engineering. PhD thesis, Chemical Engineering, Monash University, Melbourne, Australia.

Gabites, J.R., Abrahamson, J., Winchester, J.A. 2010. Air flow patterns in an industrial milk powder spray dryer. *Chemical Engineering Research and Design*, 88(7), 899–910.

Hotchkiss, J.H., Werner, B.G., Lee, E.Y.C. 2006. Addition of carbon dioxide to dairy products to improve quality: A comprehensive review. *Comprehensive Reviews in Food Science and Food Safety*, 5(4), 158–168.

Jaskulski, M., Wawrzyniak, P., Zbiciński, I. 2015. CFD model of particle agglomeration in spray drying. *Drying Technology*, 33(15–16), 1971–1980.

Jin, Y., Chen, X.D. 2009. A three-dimensional numerical study of the gas/particle interactions in an industrial-scale spray dryer for milk powder production. *Drying Technology*, 27(10), 1018–1027.

Kentish, S., Davidson, M., Hassan, H., Bloore, C. 2005. Milk skin formation during drying. *Chemical Engineering Science*, 60(3), 635–646.

Langrish, T.A.G. 2009. Multi-scale mathematical modeling of spray dryers. *Journal of Food Engineering*, 93(2), 218–228.

Langrish, T.A.G., Kockel, T.K. 2001. The assessment of a characteristic drying curve for milk powder for use in computational fluid dynamics modelling. *Chemical Engineering Journal*, 84(1), 69–74.

Li, K., Woo, M.W., Patel, H., Metzger, L., Selomulya, C. 2018. Improvement of rheological and functional properties of milk protein concentrate by hydrodynamic cavitation. *Journal of Food Engineering*, 221, 106–113.

Lin, S.X.Q. 2004. Drying of a single milk droplet. PhD thesis, Department of Chemical and Materials Engineering, The University of Auckland, New Zealand.

Lum, A., Cardamone, N., Beliavski, R., Mansouri, S., Hapgood, K., Woo, M.W. 2018. Unusual drying behaviour of droplets containing organic and inorganic solutes in superheated steam. *Journal of Food Engineering*, 36, 1802–1813.

Lum, A., Mansouri, S., Hapgood, K., Woo, M.W. 2017. Single droplet drying of milk in air and superheated steam: Particle formation and wettability. *Drying Technology* (doi: 10.1080/07373937.2017.1416626, Accepted 10 December 2017).

Malafronte, L., Ahrné, L., Innings, F., Jongsma, A., Rasmuson, A. 2015. Prediction of regions of coalescence and agglomeration along a spray dryer—Application to skim milk powder. *Chemical Engineering Research and Design*, 104, 703–712.

Marella, C., Salunke, P., Biswas, A.C., Kommineni, A., Metzger, L.E. 2015. Manufacture of modified milk protein concentrate utilizing injection of carbon dioxide. *Journal of Dairy Science*, 98(6), 3577–3589.

Masters, K. 1979. *Spray Drying Handbook*, George Godwin Limited, London.

Masters, K. 1980. *Spray Drying Handbook*, John Wiley & Sons, New York.

Ozmen, L., Langrish, T.A.G. 2002. Comparison of glass transition temperature and sticky point temperature for skim milk powder. *Drying Technology*, 20(6), 1177–1192.

Ozmen, L., Langrish, T.A.G. 2003a. A study of the limitations to spray dryer outlet performance. *Drying Technology*, 21(5), 895–917.

Ozmen, L., Langrish, T.A.G. 2003b. An experimental investigation of the wall deposition of milk powder in a pilot-scale spray dryer. *Drying Technology*, 21(7), 1253–1272.

Park, C.W., Stout, M.A., Drake, M.A. 2014. The effect of feed solid concentration and inlet temperature on the flavour of spray-dried whey protein concentrate. *Journal of Food Science*, 79(1), C19–C24.

Park, C.W., Stout, M.A., Drake, M.A. 2016. The effect of spray drying parameters on the flavour of nonfat dry milk and milk protein concentrate 70%. *Journal of Dairy Science*, 99(12), 9598–9610.

Patankar, S. 1980. *Numerical Heat Transfer and Fluid Flow*, McGraw-Hill, New York.

Patel, K.C. 2004.A novel concept of spray drying: Modelling and design. ME thesis, Chemical Engineering, University of Auckland, New Zealand.

Patel, K.C. 2008. Production of uniform particles via single stream drying and new applications of the reaction engineering approach. PhD thesis, Chemical Engineering, Monash University, Melbourne, Australia.

Patel, K., Chen, X.D., Jeantet, R., Schuck, P. 2010. One-dimensional simulation of co-current, dairy spray drying systems – pros and cons. *Dairy Science and Technology*, 90(2–3), 181–210.

Patist, A., Bates, D. 2008. Ultrasonic innovations in the food industry: From the laboratory to commercial production. *Innovative Food Science and Emerging Technologies*, 9(2), 147–154.

Petersen, L.N., Poulsen, N.K., Niemann, H.H., Utzen, C., Jorgensen, J.B. 2017. An experimentally validated simulation model for a four-stage spray dryer. *Journal of Process Control*, 57, 50–65.

Rogers, S. 2011. Developing and utilizing a mini food powder production facility to produce industrially relevant particles for functionality testing. PhD thesis, Chemical Engineering, Monash University, Melbourne, Australia.

Sano, Y., Keey, R.B. 1982. The drying of a spherical particle containing colloidal material into a hollow sphere. *Chemical Engineering Science*, 37(6), 881–889.

Schuck, P., Dolivet, A., Mejean, S., Zhu, P., Blanchard, E., Jeantet, R. 2009. Drying by desorption: A tool to determine spray drying parameters. *Journal of Food Engineering*, 94(2), 199–204.

Singh, P.K., Singh, H. 2015. *Dry Milk Products. Dairy Processing and Quality Assurance*, John Wiley & Sons, Ltd, Iowa.

Stevenson, M. Computational modelling of drying milk droplets. Master Thesis. University of Auckland, New Zealand.

Straatsma, J., van Houwelingen, G., Meulman, A.P., Steenbergen 1991. Dryspec2: A computer model of a two stage dryer. *Journal of the Society of Dairy Technology*, 44, 107–111.

Tao, Y., Cai, J., Liu, B., Huai, X., Guo, Z. 2016. Application of hydrodynamic cavitation to wastewater treatment. *Chemical Engineering and Technology*, 39(8), 1363–1376.

Truong, V., Bhandari, B., Howes, T. 2005. Optimization of co-current spray drying process of sugar-rich foods. Part 1 – Moisture and glass transition temperature profile during drying. *Journal of Food Engineering*, 71(1), 55–65.

Ullum, T., Sloth, J., Brask, A., Wahlberg, M. 2010. Predicting spray dryer deposits by CFD and an empirical drying model. *Drying Technology*, 28(5), 723–729.

Verdumen, R.E.M., Menn, P., Ritzert, J., Blei, S., Nhumaio, G.C.S., Sorenson, T.S., Gunsing, M., Straatsma, J., Verschueren, M., Sibeijn, M., Schulte, G., Fritsching, U., Bauckhage, K., Tropea, C., Sommerfeld, M., Watkins, A.P., Yule, A.J., Schonfeldt, H. 2004. Simulation of agglomeration in spray drying installations: The EDECAD project. *Drying Technology*, 22(6), 1403–1146.

Versteeg, H.K., Malalsekara, W. 2007. *An Introduction to Computational Fluid Dynamics: The Finite Volume: Method*, Pearson Education Limited, Harlow.

Villamiel, M., De Jong, P. 2000. Influence of high-intensity ultrasound and heat treatment in continuous flow on fat, proteins, and native enzymes of milk. *Journal of Agricultural and Food Chemistry*, 48(2), 472–478.

Wu, W.D. 2010. A novel micro-fluidic-jet-spray-dryer equipped with a micro-fluidic-aerosol-nozzle: Equipment development and applications in making functional particles. PhD thesis, Chemical Engineering, Monash University, Melbourne, Australia.

Woo, M.W. 2016. *Computational Fluid Dynamic Simulation of Spray Dyers – An Engineer's Guide*, CRC Press, Boca Raton, Florida.

Woo, M.W., Daud, W.R.W., Tasirin, S.M., Talib, M.Z.M. 2007. Optimization of the spray drying parameters – A quick trial and error method. *Drying Technology*, 25(10), 1741–1747.

Ye, R., Harte, F. 2014. High pressure homogenization to improve the stability of casein - hydroxypropyl cellulose aqueous systems. *Food Hydrocolloids*, 35, 670–677.

2.6 Agglomeration processes of dairy powders

Mathieu Person, Romain Jeantet, Pierre Schuck,
Cécile Le Floch-Fouéré and Bernard Cuq

2.6.1 Introduction

The specificities of powders are particularly well-adapted to meet the functional needs of dairy products: storage, handling, preservation, transport, dosage and uses. For dairy products, manufacturers and consumers desire a quick dissolution in water, without the formation of lumps. Agglomeration processes, applied to combine fine primary particles into larger ones, have been widely used by the dairy industry for many years to produce instant powders with improved rehydration properties (Gaiani et al. 2007; Williams 2007; Barkouti et al. 2013; Ji et al. 2016; Chever et al. 2017). Good wettability of agglomerated dairy powders is favored by the large and porous structures. Agglomerated dairy powders exhibit large voids between the primary particles which allow water to easily penetrate into the porous structures and solubilize the solid bridges between the particles, releasing them into solution. The released particles reach similar sizes and thus similar solubilization rates, that depend on the properties of the dairy components.

The agglomeration processes of food powders have been widely studied for 20 years (Iveson et al. 2001; Palzer 2011; Cuq et al. 2013). The wet controlled growth agglomeration of dairy powders is based on the addition of water to generate stickiness between primary particles. Mechanical energy input is necessary to disperse the liquid over the particles and promote growth by generating collisions and contact between the sticky particles. Agglomeration mechanisms depend on the opposite contributions of cohesion forces between particles and of rupture forces with shearing effects. The agglomeration of dairy powders depends on the physicochemical reactivity of their components, and more particularly on the amorphous lactose, which could have sticky behavior. The process conditions, including time, temperature and water content, control the agglomeration mechanisms. A final drying operation is applied to stabilize the agglomerates. Wet agglomeration of the dairy powder is mainly conducted using three different types of pneumatic mixing agglomerators: spray drying, fluidized-bed and steam-jet. Pneumatic mixing agglomerators use an air stream to agitate the particles under low shear conditions. The present chapter gives an overview of the processes used to agglomerate dairy powders by describing the principles, equipment characteristics, agglomeration mechanisms, the impact of process parameters and properties of agglomerated powders.

2.6.2 Agglomeration by spray drying

Spray drying is the standard technique to produce dairy powders, by introducing small droplets of concentrated milk into a hot air stream. The particles are formed by the rapid water elimination from the droplets due to high exchange surface. Recently, the spray-drying technologies were adapted to generate additional agglomeration of dairy powders by recycling the small primary dried particles (Williams 2007; Palzer 2011; Chever et al. 2017). During spray drying, the droplets of milk develop a sticky surface due to the large amount of amorphous lactose. Collisions between sticky droplets and dried particles lead to adhesion, promote agglomeration mechanisms and lead to the formation of large structures directly inside the spray dryer. The specific conditions of spray drying concern the fast reaction rates and low mechanical forces acting on the particles. The spray-drying agglomeration of dairy products leads to small (diameter between 130–300 μm), highly porous and brittle agglomerates, with low tapped and bulk densities. Dairy powders (and also instant coffee or cacao beverages, infant formulas, maltodextrins, dextrose syrup or powdered flavors) are agglomerated by spray drying. Spray-drying agglomeration of skim milk powders produces agglomerates with good dispersibility in water, but has no effect on the solubility of agglomerates obtained from whole milk powders. The addition of small amounts of surfactant molecules (e.g. lecithin) is then required to overcome the hydrophobic nature of the fatty surface of the powder.

Agglomeration of dairy powders by spray drying is conducted using specific spray driers (Figure 2.6.1). To promote the agglomeration mechanisms, the small dried particles (called the "fines") are collected after the exit from the tower and injected into the drying chamber, close to the zone of initial spraying of the milk. The liquid droplets can collide with the fines to form agglomerates. The spray drier equipment can be differentiated by the position of the injector of the fines (Verdurmen et al. 2005; Williams 2007). When the fines particles are introduced into the atomization zone of the liquid, agglomerates are formed by droplet coalescence and contact between liquid droplets and the small particles. These agglomerates present good mechanical stability, but poor instant properties. When the fines are introduced into the intermediary drying zone, the small particles can collide

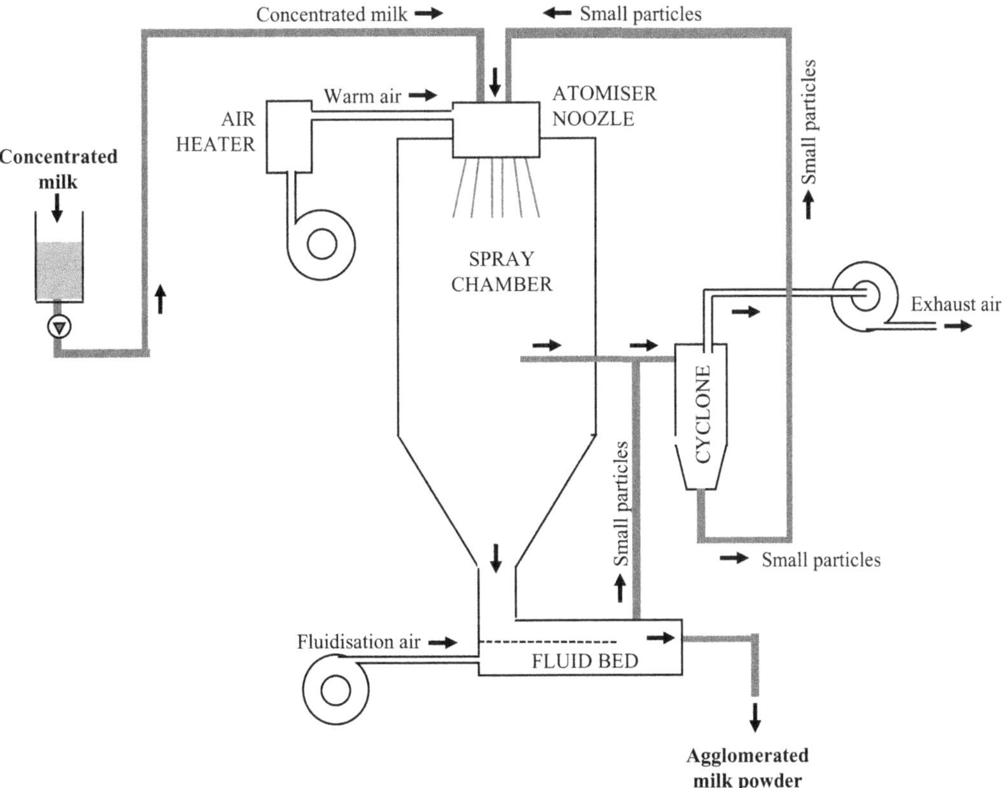

Figure 2.6.1 Schematic representation of the specific spray-dried technology used to agglomerate powders.

with the sticky particles to form agglomerates. This zone is the most effective in promoting highly porous agglomerates with good mechanical stability and instant behavior. The recycling flow of fines particles means that a part of the powder is processed several times at high temperatures, which can impact the heat-sensitive components.

A combination of spray- and belt-drying technology has been proposed to promote agglomeration for sticky or fatty dairy powders (Palzer 2011). The liquid is atomized into a short spray tower and the particles fall on a moving belt where they agglomerate to form a powder cake. The cake is dried with hot air, cooled, milled down and sieved. This process is used for dairy blends, coffee whiteners, whey and whey protein, cheese powder, maltodextrins and glucose syrup, infant formulas, fruit and vegetable powders, soy sauce powder and instant coffee.

The agglomeration by spray-drying of dairy powders involves a succession of several mechanisms (Williams 2007; Palzer 2011). During the atomization of milk, a large number of droplets are produced in a very small volume near the top of the tower. During flight, water evaporation reduces the diameter of the droplets over time. The rate of drying and the diameter of the droplets are important factors to control. During the drying, the solute and dispersed phases of milk accumulate at the surface of droplets, reducing the water content and increasing the viscosity. High evaporation flux can cause the differential migration of fat and protein, compared to the liquid phase containing dissolved lactose. This results in a surface composition unlike the bulk of the particles. Drying causes the surface to become sticky.

The spray parameters are important factors affecting the agglomeration mechanisms during spray-drying (Chen 1992; Panda et al. 2001; Shakeri and Chandra 2002; Williams 2007; Williams et al. 2009). The formation of liquid droplets is influenced by the nozzle type, spray pressure, number of sprays and their direction relative to each other. Increasing the atomization pressure or the spray flow rate increases the presence of small droplets in the agglomeration zone and favors the agglomeration. The key mechanisms are the frequency of collisions and the likelihood of adhesion. Two types of collisions can be considered. Primary agglomeration is caused by collisions between primary droplets due to spray streams overlapping. These agglomerates are characterized by high mechanical strength, but poor dispersibility in water. Secondary (or forced) agglomeration is caused by collisions of primary droplets with the recycled fines particles. After the droplets have traveled some distance, they enter a stream of fines particles and may collide, depending on their physicochemical surface characteristics and air stream conditions. High agglomeration is obtained with multiple passes of the fine powder through cyclones until the particles grow large enough to fall to the bottom of the dryer.

Adhesion between particles depends on the process conditions, and on the existence of attractive forces and collision energy (Williams 2007; Malafronte et al. 2016). Coalescence dominated by adhesion forces occurs at high water contents when, after contact, a droplet/particle merges with another droplet/particle. Stickiness occurs between moist particles when, after contact, necking appears through the formation of a liquid bridge due to the low viscosity of the surface liquid at the particle–particle interface. For successful adhesion, the kinetic energy of the particles has to be dissipated in the liquid bridge developed during the viscous flow mechanism. If the kinetic energy is too large or viscous forces in the liquid bridges are too weak, particle rebound will occur. The location of the nozzles inside the drier significantly impacts the residence time and adhesion mechanisms (Williams 2007; Chever et al. 2017). Shorter residence times and powders with uniform particle size were observed when fines were returned to the top of the dryer because of greater efficiency of agglomeration. The flow rate of the recycled fines particles can range between 5 and 50% of the mass ratio of sprayed liquid flow rate (Williams et al. 2009). Forced agglomeration was optimal at high fines flow rates for generating high densities of fines in the collision zone. The frequency of collisions between droplets and particles is also enhanced with a high velocity of liquid droplets and fines particles entering the collision zone.

For dairy powders, amorphous lactose contributes to the formation of viscous bridges. The temperature inside the spray dryer is critical to control the sticky behavior of amorphous lactose (Palzer 2011; Malafronte et al. 2016). Particles are sticky when the difference between process temperature and glass transition temperatures is greater than about 30°C. Using highly concentrated skim milk favors direct natural agglomeration at the top of the spray dryer (Williams et al. 2009). When considering forced agglomeration with recycled small particles, low-concentrated milk is optimal as it leads to the formation of small droplets, increases the density of droplets in the collision zone and increases the probability of collision, but reduces the diameter of the agglomerates. The characteristics of the recycled fines particles impact the mechanisms of forced agglomeration (Williams et al. 2009). Fines with low diameters and water contents above 5% promote optimal forced agglomeration mechanisms.

Drying mechanisms occur constantly during all the successive steps of the agglomeration process by spray drying. The liquid bridges formed between the droplets and fines particles after collisions are quickly dried. The increasing viscosity leads to the formation of viscous and finally solid bridges that give cohesion to the agglomerates.

2.6.3 *Fluidized-bed agglomeration*

Fluidized-bed agglomeration consists of fluidizing the native powder by an upwardly directed air stream and spraying a binder solution onto the moving particles (Figure 2.6.2). It is well-adapted to manage the successive agglomeration mechanisms (wetting, colliding, growing and drying). The continuous heat and mass transfers between the fluidizing air and particles allow short processing times. Fluidized-bed agglomeration is used to improve the instant and flow properties of different food powders: dairy powders, culinary powders for vending machines and vitamin mixes (Palzer 2011; Barkouti et al. 2013; Turchiuli et al. 2013; Ji et al. 2015, 2017). Water is used as a binder for the fluidized-bed agglomeration of particles with amorphous components. Viscous liquids are used for crystalline components. The agglomerated dairy powders are characterized by a large range of diameters (200–2500 μm), irregular shape (with low values of circularity and convexity and high values of elongation), low bulk density and a porous structure due to the lack of kinetic energy during the collision between particles in the agglomeration zone. The agglomerated skim or whole milk powders had a very good wettability and were nearly instant.

The fluidized-bed granulator (Figure 2.6.2) contains several functional zones (Burggraeve et al. 2013). An air-handling system with filtering, heating, cooling and drying functions is used for managing the temperature and relative humidity of the airflow introduced inside the equipment. The airflow is forced through the bed of powder using the air distributor plate, in order to fluidize the particles in the granulation zone. The fluidized-bed process provides intense mixing conditions. The airflow leaves the equipment through an air filter system, removing the residual small particles. In the air expansion chamber, the largest particles are withdrawn and filter bags can be periodically shaken to reintroduce the collected fines into the fluidized-bed area. In the granulation zone, the binder liquid is sprayed onto the fluidized particles via a nozzle system. Different locations

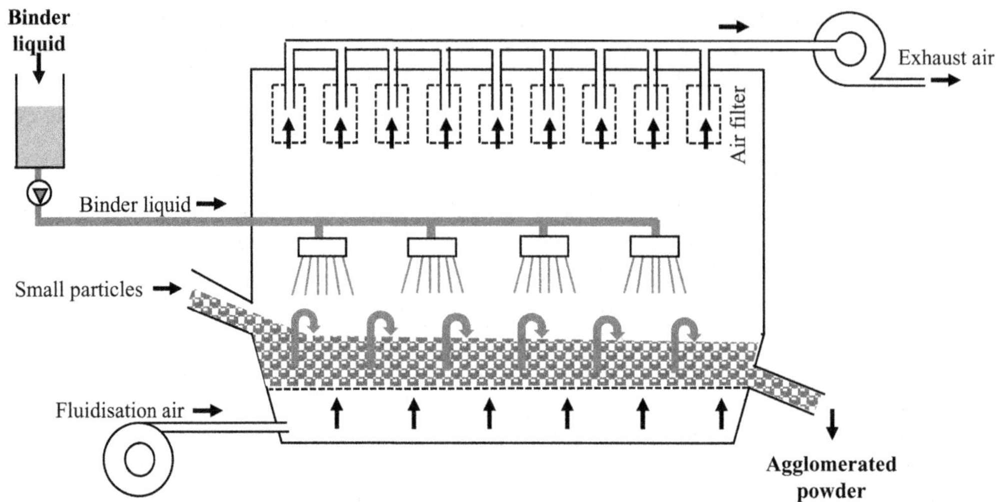

Figure 2.6.2 Schematic of a fluidized-bed granulator.

of the spray nozzle are possible (Jones 1985; Srivastava and Mishra 2010). In the top-spray granulator, the nozzle is located at the top of the chamber. The binder liquid is sprayed from the top down onto the fluidized bed, counter-currently to the fluidizing air. This technology favors porosity and wettability of the granules. In the bottom-spray granulator, the nozzle is positioned at the base of the chamber, in the middle of the distributor plate. The liquid is sprayed in the same direction as the fluidizing air. This technology is used in the pharmaceutical industry for producing active layering and for coating to control drug release. In the tangential-spray granulator, the nozzle is located at the side of the chamber, embedded in the powder bed during granulation. The tangential spray method is used for granulating and pelletizing with subsequent coating.

Three active zones can be described in a fluidized-bed granulator (Heinrich et al. 2005; Jiménez Munguia 2007). The wetting-active zone is located below the spraying nozzle at the topmost part of the bed. It is a low-temperature and high-humidity zone, with large gradients of relative humidity and temperature, due to the wetting of the fluidized particles by the sprayed liquid, and to the evaporation of the liquid. The isothermal zone is located near the walls and around the wetting-active zone. There is equilibrium between heat and mass transfer. The air temperature is homogeneous. The heat transfer zone is located above the bottom air distributor plate. The temperature of the air stream decreases due to the energy absorbed by the cold particles coming from the upper zones.

The vibro-fluidized-bed has been specifically developed to process sticky powders (Banjac et al. 2009). The residence time is then controlled by managing vibration parameters without affecting the gas flow rates.

A succession of simultaneous and competitive mechanisms are considered to describe the fluidized-bed agglomeration of dairy powders (Palzer 2011; Barkouti et al. 2013; Turchiuli et al. 2013). Whatever their position in the chamber, the powder is fluidized by an upward directed air stream. The temperature of the fluidizing air ranges typically between 70 and 100°C. The large agglomerates can be fluidized since their velocity is superior to the maximum air velocity in the chamber. The particles with intermediate diameters are fluidized up to heights depending on their diameter and do not enter the wetting active zone.

The compositions of the dairy powders impact the agglomeration mechanisms (Forny et al. 2011; Barkouti et al. 2013; Ji et al. 2015, 2016). The content of amorphous lactose in primary particles is a key factor controlling the stickiness of particles during processing. The protein composition of primary particles can impact the agglomeration behavior and rehydration properties of milk powders. When the dispersibility is the rate-limiting step, the agglomeration process does not impact the rehydration behavior. When the wettability is the rate-limiting step, the agglomeration process improves the rehydration of powders. For milk protein isolates, both the wetting and dispersing processes are rate-limiting steps. The presence of fat (whole or skim milk powders) does not have a significant impact on the growth mechanisms, with similar diameters, porous microstructures and irregular shapes. The flowability of skim milk powders was not impacted by the agglomeration. On the other hand, the flowability of whole milk powders was significantly improved by the agglomeration.

To agglomerate water-soluble amorphous substances, pure water or water-solution at low viscosity is sprayed as a binder (Turchiuli et al. 2005; Palzer 2011; Szulc and Lenart 2013). High binder liquid concentration resulted in high compactness of agglomerates due to thicker bridges between particles. Different binder solutions (water, lactose solutions or sucrose solutions) were tested for the fluidized-bed agglomeration of milk protein isolates powders (Gaiani et al. 2007; Ji et al. 2015). When using pure water, the agglomerates are characterized by a poor spherical shape, large surface area, apparent density close to that of native particles and low solubilization. Adding lactose or sucrose in the binder solutions significantly gives agglomerates with similar values of porosity, lower values of apparent densities and higher wettability. The presence of hydrophilic lactose and sucrose on the surface of the particles allows the development of more hydrophilic bridge surfaces and favors wetting. However, using lactose or sucrose solutions had no impact on the solubilization rate, because it is controlled by the primary particles.

Several process parameters impact the fluidized-bed agglomeration of milk powders (Srivastava and Mishra 2010; Barkouti et al. 2013; Szulc and Lenart 2013; Ji et al. 2015, 2016, 2017). The conditions of liquid spraying (liquid flow rate, particle load and droplets size) impact the growth mechanisms. Agglomerate growth depends on the amount of water sprayed, whatever the water flow rate and the particle load. The strength of inter-particle forces depends on the size of the primary particles. A small amount of liquid at the particle surface is enough for the successful coalescence of small particles, due to less shear stress acting at the liquid bridges. In the case of large particles, only particles wetted by the large droplets (or a larger number of small droplets at the same time) can make permanent bonds due to the need for stronger bonds. The morphology of agglomerates is difficult to predict, because it depends on the droplet deposition, droplet spreading, bridge breakage and the amount of and the way that binders are introduced.

Water droplets colliding with porous particles containing water-soluble substances can penetrate immediately into the particles. Due to the absorbed water, the glass transition temperature of the amorphous substance strongly decreases, and the amorphous matrix becomes sticky when reaching the rubbery state. Due to the air stream, the dry and wetted particles can impact together. Depending on the colliding conditions, they can sinter together with the plasticized components present at the surface of the wetted particles. Plastic deformation of the plasticized regions of the wetted particles can occur. Van der Waals forces between colliding particles contribute, increasing adhesion. In the initial phase of the wetting stage, the concentration of polar amorphous solids present on the particle surface is low, the liquid viscosity is low, and true liquid bridges stabilized by capillary forces are formed between colliding particles. With increasing solid content by the dissolving of polar amorphous substances, the viscosity increases and adhesion becomes a time-dependent process. At high viscosities, adhesion is based on viscous bridges between the particles.

The granule growth during the wetting period can be described in two stages (Barkouti et al. 2013). A rapid increase of the diameter is first observed when starting the liquid spraying and then a slower increase in diameter till reaching a plateau, that corresponds to equilibrium between the growth and breakage mechanisms. The agglomeration of dairy powder was described by considering class by class the growth mechanisms of three classes of particles (primary particles, nuclei and agglomerates): the association of initial particles into compact intermediate nuclei structures, and the association of the intermediary structures into large porous agglomerates.

The wet agglomerates are dried to reach low water content (1–5%) and cooled below their glass transition temperature to avoid caking during storage. The amorphous components involved in the viscous bridges might then crystallize to promote solid bridges. A decrease of the drying temperature causes a higher mean diameter of the agglomerates, because the drying mechanisms are slowed down during processing (Banjac et al. 2009).

2.6.4 *Steam-jet agglomeration*

Steam-jet agglomeration is a special kind of agglomeration process particularly well-adapted for water-soluble materials (Schuchmann et al. 1993; Schuchmann 1995; Hogekamp et al. 1994; Hogekamp et al. 1996; Hogekamp 1999a, 1999b, 2004; Palzer 2011). It takes advantage of the ability of steam to quickly penetrate primary particles, driven by the difference between the vapor pressure in the gas phase and the equilibrium vapor pressure above solid surfaces of lower temperatures. In the steam-jet agglomeration process, free-falling particles are wetted instantly and quite homogeneously by steam flow in the wetting chamber. The low steam consumption and low drying requirements are major advantages of the jet agglomeration process. The low thermal load applied to the product prevents thermal degradation. As size enlargement takes place in a low concentration swarm under the influence of small forces in the steam-jet process, the resulting agglomerates have high porosity and good dispersibility in water. Mechanical stability is lower compared to agglomerates produced in a fluidized bed. In the food industry, the applications of steam-jet agglomeration started in the 1950s with the production of instant beverage powders, and different food materials are processed: milk powders, powdered milk constituents, sugars, water-soluble extracts and cocoa powders. The steam-jet agglomeration of dairy powders has been investigated recently (Person 2018).

The steam-jet agglomeration equipment is a tower (Figure 2.6.3). Primary particles are fed at the top of the tower and enter the wetting zone in free fall. Condensing steam is ejected from steam nozzles towards the particle swarm in the closed wetting zone. Steam provides the necessary binder liquid to increase the adhesion forces between free-falling and colliding particles in the agglomeration zone. The residence time of the particles in the agglomeration zone is short. A subsequent drying step directly in the agglomeration tower or in an external fluidized drying bed is applied to stabilize the agglomerates. Fine volatile particles in the process air can be exhausted through a separation device. Martins and Kieckbusch (2008) described a piece of steam-jet equipment in which particles fall in front of a steam nozzle to promote agglomeration, reach an inclined vibratory screen, are dried on a rotary drier and sieved.

The steam-jet agglomeration process can be described as a succession of several mechanisms that are impacted by ingredient characteristics and process conditions (Schuchmann 1995; Hogekamp et al. 1996; Hogekamp 1999a; Martins and Kieckbusch 2008; Takeiti et al.

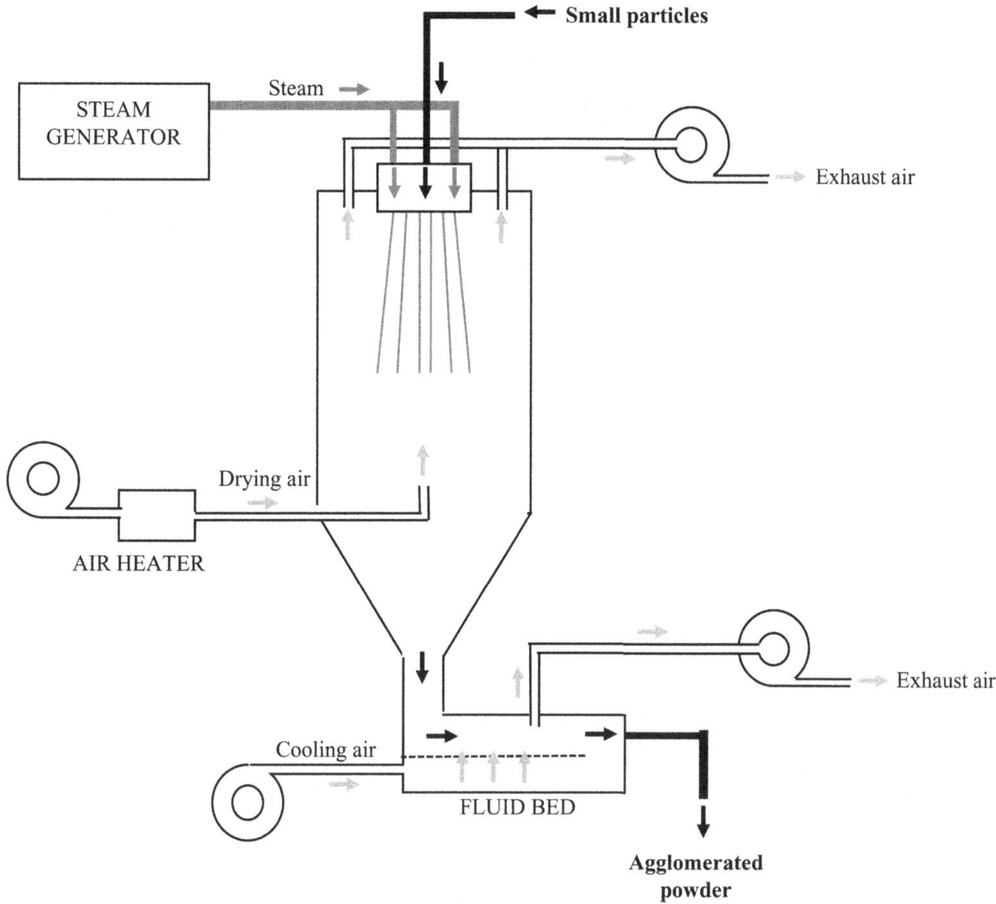

Figure 2.6.3 Schematic of steam-jet agglomeration equipment.

2008; Vissotto et al. 2010; Palzer 2011; Person 2018). The composition of the primary particles (lipid content, sucrose or glucose content) impacts the agglomeration mechanisms, through their water absorption capacity and glass transition temperature. The content of amorphous lactose of primary particles is a key factor of the agglomeration mechanisms. The particle size distribution of the primary particles also impacts the agglomeration mechanisms. For the agglomeration of sugar-based powders, the reduction of the content of fines in native powders was found to be favorable to obtain instant products.

Condensing steam is ejected towards the particle swarm in the wetting zone. Approximately 25–50% of the water contained in the steam condenses in the agglomeration zone. The entry temperature of particles is critical to the condensation process. Wetting by steam condensation can only occur as long as the particle surface temperature lies below the equilibrium temperature corresponding to the prevailing relative humidity. The lower the entry temperature, the faster the condensation on the particle surface. In addition to condensation, droplets are formed by the condensation of vapor in the gas phase. After water condenses on the particle surface, steam migrates into the particles and the particle surface temperature increases up to 80–90°C. The increase in temperature and in water content plasticizes the particle surface and solubilizes solutes, creating a viscoelastic front. The viscosity at the particle surface decreases, whereas the particle inside remains dry.

Steam provides the necessary binder liquid to increase the adhesion forces between free-falling and colliding particles in the agglomeration zone. The collision frequency is low since the difference in velocity between particles and the particle concentration in the agglomeration zone are low. The particles have different sinking velocities depending on their size. Turbulence caused by steam jets promotes collisions by intensifying the relative motion of the particles. The effective length of the agglomeration zone can be adjusted by varying the steam input. An increase in steam pressure favors the growth mechanisms, leading to the production of large granules, due to the formation of a greater number of inter-particle liquid bridges. An increase in the solid feed rate at constant steam flow had a negative impact, with the formation of granules of smaller diameter.

After collision, the surface of colliding particles is deformed and sintering increases the adhesion forces between particles. Liquid bridges develop at the contact points between primary particles and promote the formation of the agglomerates. The turbulence caused by the steam jets can provide the destruction of fragile agglomerates by shear forces.

Drying of the agglomerates causes crystallization of soluble substances in the liquid bridges joining the primary particles and turns the liquid bridges into solid ones. The use of low drying temperatures could be advantageous by preventing browning reactions and nutrient loss. When applying high temperatures during the first drying stage when the agglomerates are still moist, supplementary agglomeration mechanisms may occur.

References

Banjac, M., Stamenić, M., Lečić, M. and Stakić, M. 2009. Size distribution of agglomerates of milk powder in wet granulation process in a vibro-fluidized bed. *Brazilian Journal of Chemical Engineering* 26(3):515–525.

Barkouti, A., Turchiuli, C., Carcel, J.A. and Dumoulin, E. 2013. Milk powder agglomerate growth and properties in fluidized bed agglomeration. *Dairy Science and Technology* 93(4–5):523–535.

Burggraeve, A., Monteyne, T., Vervaet, C., Remon, J.P. and De Beer, T. 2013. Process analytical tools for monitoring, understanding, and control of pharmaceutical fluidized bed granulation: A review. *European Journal of Pharmaceutics and Biopharmaceutics* 83(1):2–15.

Chen, X.D. 1992. Whole milk powder agglomeration – Principles and practice. In *Milk Powders for the Future,* ed. X.D. Chen. Palmerston North, New Zealand: Dun More Press.

Chever, S., Mejean, S., Dolivet, A., Mei, F., Den Boer, C.M., Le Barzic, G., Jeantet, R. and Schuck, P. 2017. Agglomeration during spray drying: Physical and rehydration properties of whole milk/sugar mixture powders. *LWT - Food Science and Technology* 83:33–41.

Cuq, B., Mandato, S., Jeantet, R., Saleh, K. and Ruiz, T. 2013. Food powder agglomeration. In *Handbook of Food Powders,* ed. B. Bhandari, N. Bansal, M. Zhang, and P. Schuck, 150–177, Woodhead Publishing Series in Food Science, Technology and Nutrition No. 255.

Forny, L., Marabi, A. and Palzer, S. 2011. Wetting, disintegration and dissolution of agglomerated water-soluble powders. *Powder Technology* 206(1–2):72–78.

Gaiani, C., Schuck, P., Scher, J., Desobry, S. and Banon, S.. 2007. Dairy powder rehydration: Influence of protein state, incorporation mode, and agglomeration. *Journal of Dairy Science* 90(2):570–581.

Heinrich, S., Henneberg, M., Peglow, M., Drechsler, J. and Mörl, L. 2005. Fluidized bed spray granulation: Analysis of heat and mass transfers and dynamic particle populations. *Brazilian Journal of Chemical Engineering* 22(2):181–194.

Hogekamp, S. 1999a. Steam jet agglomeration – Part 1: Production of redispersible agglomerates by steam jet agglomeration. *Chemical Engineering and Technology* 22(5):421–424.

Hogekamp, S. 1999b. Steam-jet agglomeration - Part 2: Modelling agglomerate growth in a modified steam-jet agglomerator. *Chemical Engineering and Technology* 22(6):485–490.

Hogekamp, S. 2004. Agglomeration by use of steam. PhD thesis, Karlsruhe, Germany.

Hogekamp, S., Schubert, H. and Wolf, S. 1996. Steam jet agglomeration of water soluble material. *Powder Technology* 86(1):49–57.
Hogekamp, S., Stang, M. and Schubert, H. 1994. Jet agglomeration and dynamic adhesion forces. *Chemical Engineering and Processing* 33(5):313–318.
Iveson, S.M., Litster, J.M., Hapgood, K. and Ennis, B.J. 2001. Nucleation, growth and breakage phenomena in agitated wet granulation processes: A review. *Powder Technology* 117(1–2):3–39.
Ji, J., Cronin, K., Fitzpatrick, J., Fenelon, M. and Miao, S. 2015. Effects of fluid bed agglomeration on the structure modification and reconstitution behaviour of milk protein isolate powders. *Journal of Food Engineering* 167:175–182.
Ji, J., Fitzpatrick, J., Cronin, K., Fenelon, M. and Miao, S. 2017. The effects of fluidised bed and high shear mixer granulation processes on water adsorption and flow properties of milk protein isolate powder. *Journal of Food Engineering* 19:219–227.
Ji, J., Fitzpatrick, J., Cronin, K., Maguire, P., Zhang, H. and Miao, S. 2016. Rehydration behaviours of high protein dairy powders: The influence of agglomeration on wettability, dispersibility and solubility. *Food Hydrocolloids* 58:194–203.
Jimenez Munguia, M.T. 2007. Agglomération de particules par voie humide en lit fluidisé. PhD thesis, ENSAIA Massy, France.
Jones, D.M. 1985. Factors to consider in fluid bed processing. *Pharmaceutical Technology* 9:50–62.
Malafronte, L., Ahrné, L., Robertiello, V., Innings, F. and Rasmuson, A. 2016. Coalescence and agglomeration of individual particles of skim milk during convective drying. *Journal of Food Engineering* 175:15–23.
Martins, P.C. and Kieckbusch, T.G. 2008. Influence of a lipid phase on steam jet agglomeration of maltodextrin powders. *Powder Technology* 185(3):258–266.
Palzer, S. 2011. Agglomeration of pharmaceutical, detergent, chemical and food powders—Similarities and differences of materials and processes. *Powder Technology* 206(1–2):2–17.
Panda, R.C., Zank, J. and Martin, H. 2001. Modelling the droplet deposition behaviour on a single particle in fluidized bed spray granulation process. *Powder Technology* 115(1):51–57.
Person, M. 2018. Étude multi-échelle des relations procédés-processus-produits lors de l'agglomération de poudres de lait. PhD thesis, Agrocampus Ouest, France.
Schuchmann, H. 1995. Production of instant foods by jet agglomeration. *Food Control* 6(2):95–100.
Schuchmann, H., Hogekamp, S. and Schubert, H. 1993. Jet agglomeration processes for instant foods. *Trends in Food Science and Technology* 4(6):179–183.
Shakeri, S. and Chandra, S. 2002. Splashing of molten tin droplets on a rough steel surface. *International Journal of Heat and Mass Transfer* 45:4561–4575.
Srivastava, S. and Mishra, G. 2010. Fluid bed technology: Overview and parameters for process selection. *International Journal of Pharmaceutical Sciences and Drug Research* 2:236–246.
Szulc, K. and Lenart, A. 2013. Surface modification of dairy powders: Effects of fluid-bed agglomeration and coating. *International Dairy Journal* 33(1):55–61.
Takeiti, C.Y., Kieckbusch, T.G. and Collares-Queiroz, F.P. 2008. Optimization of the jet steam instantizing process of commercial maltodextrins powders. *Journal of Food Engineering* 86(3):444–452.
Turchiuli, C., Eloualia, Z., El Mansouri, N. and Dumoulin, E. 2005. Fluidised bed agglomeration: Agglomerates shape and end-use properties. *Powder Technology* 157(1–3):168–175.
Turchiuli, C., Smail, R. and Dumoulin, E. 2013. Fluidized bed agglomeration of skim milk powder: Analysis of sampling for the follow-up of agglomerate growth. *Powder Technology* 238:161–168.
Verdurmen, R.E.M., Verschueren, M., Gunsing, M., Straatsma, J., Bleib, S. and Sommerfeld, M. 2005. Simulation of agglomeration in spray drying installations: The EDECAD project. *Lait* 85:343–351.
Vissotto, F.Z., Jorge, L.C., Makita, G.T., Rodrigues, M.I. and Menegalli, F.C. 2010. Influence of the process parameters and sugar granulometry on cocoa beverage powder steam agglomeration. *Journal of Food Engineering* 97(3):283–291.
Williams, A.M. 2007. Instant milk powder production: Determining the extent of agglomeration. PhD thesis, Massey University, New Zealand.
Williams, A.M., Jones, J.R., Paterson, A.H.J. and Pearce, D.L. 2009. Effect of fines on agglomeration in spray dryers: An experimental study. *International Journal of Food Engineering* 5(2):1–36.

2.7 Product modification in a fluidized bed dryer

Nima Yazdanpanah

2.7.1 Introduction

Spray drying is a very significant part of dairy powder production where most of the powder's specifications and properties are determined to a large extent, while post-processing in fluidized beds could modify them [1]. Although the main morphological structure forms in the spray-dryer chamber, a fluidized bed dryer has the potential to improve powder properties.

Traditionally, fluidized bed dryers are used for drying, agglomeration and conditioning the powders after the spray dryer. In an integrated process, the spray dryer is placed upstream of the fluidized bed dryer and powdered material continuously flows to the fluidized bed, followed by a sieve (sifter) and a material collection bin. The dry and hot gas enters from the bottom of the fluidized bed dryer to pass through the powders and fluidize them for better heat and mass transfer. Large-scale, continuous, industrial fluidized bed dryers normally incorporate few baffles to maintain the residence time of the particle; however, the dryer is inclined toward the material outlet. In addition, circular vibration or reciprocal shaking of the entire equipment facilitates the powder flow toward the outlet and also assists the fluidization by breaking up the lumps and mixing the powder bed.

Inlet air temperature and humidity is selected to dry powders as much as possible. In advanced equipment designs and high-end models, some sensors are placed in the outlet air stream to measure the air outlet temperature and humidity. Calculating for heat loss from the body and the inlet moisture content, operators can estimate the drying performance and final moisture content of the final product. Near-infrared (NIR) sensors can also be utilized at the outlet to measure the final moisture content of the materials.

Depending on the spray dryer setup and the environmental and processing conditions, the produced powder might contain a "high moisture content." The acceptable moisture content for dried milk powder is 3–4% $w.w^{-1}$ dry basis [2,3]. Variation in the feed concentration, the drying conditions such as air temperature or flow rate, and the inlet air condition could increase the outlet moisture content. Large spray dryers mainly use "raw" air from the environment for drying; therefore, daily weather variations affect the inlet air moisture content, hence, the product could end up with a higher than desired moisture content.

Due to being exposed to slightly varied drying conditions, the particles might not have a uniform moisture content. The particles also might have a heterogeneous moisture profile from surface to core, where the bulk moisture content is higher than the dried surface [4]. Since the mass transfer is across the interface or surface of the particle, the surface of the particle could be closer to the equilibrium state with the drying gas; however, the core of the particle could maintain the moisture for a longer time.

The extra residence time in a fluidized bed dryer provides the "conditioning" opportunity for the powders. The powders have extra time to release internal moisture and attain bulk equilibrium. The surface of the particles (a thin layer of 5–10 micron) at the outlet of the spray dryer is in equilibrium with the water activity of the spray-dryer air. The heterogeneity of moisture distribution in the particle (radial distribution of moisture inside the particle) diminishes during the fluidized bed drying stage and the particles'core reaches to a homogeneous spatial moisture distribution state. This conditioning performance of the fluidized bed dryer can be controlled by the residence time of the powder via manipulating the equipment design (slop, length, baffles number and distance and air distributor design) and by the operating parameters (vibration, air velocity and pulsing).

Agglomeration is also performed in the fluidized bed dryer by spraying additional water on the surface of the powder bed [1]. This agglomeration, also called instantization, creates larger particles that are "heavier" which helps sinking the powder during rehydration for final customers. The agglomeration process doesn't change the weight of the particles and doesn't make them heavier. However, the surface to bulk ratio change in larger particles helps to overcome the surface tension of water during sinking, allowing the particles to sink and submerge faster.

These combinational effects mostly create amorphous particles with a uniform microstructure that are susceptible to deteriorative changes. The processing, storage time and handling of these spray-dried powders are very sensitive to the storage conditions, due to the presence of amorphous lactose in spray-dried dairy powders that are produced by conventional drying facilities [5–8]. The physical and thermodynamical states of dairy powders, such as the crystallinity of the powders, strongly depend on the process conditions, which can cause unstable powders. The unstable state of powders (amorphicity) causes some changes during storage, such as stickiness, caking, degradation and non-enzymatic browning [9–12].

The technology described in this chapter aims to address these challenges, and use the fluidized bed step opportunity to modify the microstructure of particles [12,13]. The improved performance of the powder, as demonstrated here, makes this advancement on the conventional fluidized bed dryer a viable investment.

2.7.2 Underlying physics

Among the different ingredients in dairy powders, amorphous lactose is the most hygroscopic and unstable material. By sorbing moisture, the amorphous lactose becomes sticky and forms bridges with other particles that lead to caking, forming a product that is non-free-flowing and difficult to handle. The particle formation in conventional spray drying is a very fast process. There is not enough time for molecules to organize in an ordered structure and create a crystalline form. The conditions in conventional fluidized bed dryers are also not adequate for phase change. Changes in ambient conditions, such as daily variations in the relative humidity, can affect the performance of the unit operations and change the final moisture content of the spray-dried powders.

Many researchers have recommended that the amorphous lactose fraction could be treated in a crystallization facility after spray drying to crystallize lactose-containing powders and thus limit the caking tendency of the powder [12,14–16]. The post-crystallization process in a fluidized bed dryer depends on the processing conditions, such as air temperature, relative humidity, time and initial moisture content of the spray-dried powders [17].

Different layers of the dairy powder particle may contain different components, such as crystalline lactose and amorphous lactose, and hence different water activities. The phase change, the crystallization of amorphous lactose, then takes place in different layers sequentially, when the elevated moisture content of the layer drags the glass transition of the materials below the temperature of the layer (which was increased by heat transferring from the air to the surface, then conducting the heat to the core). Therefore, the crystallinity of the different layers increases as heat and moisture penetrate toward the center. The presence of plasticizers, moisture in this case, reduces the glass-transition temperature, and the crystallization of lactose occurs at temperatures that are above the glass-transition temperature [18,19]. The glass-transition temperature is affected by various factors, of which the composition of the material, the molecular weight and the presence of plasticizers are most important. Since amorphous sugars are very hygroscopic, if they are exposed to high humidity in the rubbery state, they sorb moisture, leading to the glass-transition temperature falling below the ambient temperature, causing changes such as crystallization to take place. The glass-transition concept and state diagrams are useful for describing the physical and structural stability of food systems at specific conditions. Interested readers are referred to textbooks and extensive publications on the state diagrams and glass transition [20–27]. The glass-transition temperature (T_g) can be estimated by the Gordon–Taylor equation (Eq. 2.7.1) [28] and the Couchman–Karasz equation (Eq. 2.7.2) [29].

$$T_g = \frac{W_1 T_{g1} + kW_2 T_{g2}}{W_1 + kW_2} \tag{2.7.1}$$

where W_1 and W_2 are the weight fractions of the two components; T_{g1} is the glass-transition temperature of one component; T_{g2} is the glass-transition temperature of the other component (in this case, the sorbed moisture in the particle); and k is a curvature constant.

$$T_g = \frac{W_1 \Delta C_{P_1} T_{g1} + W_2 \Delta C_{P_2} T_{g2} + W_3 \Delta C_{P_3} T_{g3}}{W_1 \Delta C_{P_1} + W_2 \Delta C_{P_2} + W_3 \Delta C_{P_3}} \tag{2.7.2}$$

where T_{gi} is the glass-transition temperature of a component i (K); ΔC_{Pi} is the change in the heat capacity of this component at T_{gi} (J kg^{-1} K^{-1}); and W_i is its weight fraction.

With the presence of sufficient moisture in powders, the phase change and transition can take place during spray drying and/or fluidized bed drying when the process temperature is much higher than the glass-transition temperature. Therefore, the powders can be dried and crystallized simultaneously. The Williams–Landel–Ferry (WLF) equation (Eq. 2.7.3) can be applied to conditions that both dry and crystallize, where the rate of crystallization is related to the temperature difference (T–T_g).

Williams et al. (1955) described the temperature dependence of all mechanical relaxation processes using an empirical function [30]. According to the WLF equation, the ratio (r) of the crystallization time (q_{cr}) at any temperature (T) to the time for crystallization (q_g) at the glass-transition temperature (T_g), can be correlated by

$$\log_{10} r = \log_{10}\left(\frac{\theta_{cr}}{\theta_g}\right) = \frac{-17.44\left(T - T_g\right)}{51.6 + \left(T - T_g\right)} = \log_{10}\left(\frac{k_g}{k_{cr}}\right) \tag{2.7.3}$$

where q_{cr} is the crystallization time at any point in time (s); q_g is the time for crystallization at the glass-transition temperature (s); k_{cr} is the rate of crystallization (s^{-1}) at the particular local conditions (T–T_g); and k_g is the rate of crystallization at the glass-transition temperature (T_g). The WLF equation shows that the rate of crystallization can be increased by increasing the temperature difference (T–T_g). This temperature difference can be raised by increasing the inlet gas temperature to the fluidized bed dryer or increasing the moisture content of the powders by supplying more moisture with high-humidity fluidization air.

The three-parameter Guggenheim–Anderson–de Boer (GAB) equation has been used in estimating the moisture sorption of the three main components [31,32]. The GAB equation can be written as

$$\frac{X_j}{X_{0,j}} = \frac{C_j K_j a_w}{\left(1 - k_j a_w\right) - \left(1 - K_j a_w + C_j K_j a_w\right)} \tag{2.7.4}$$

The values of X_0, C and K for amorphous lactose, crystalline lactose and protein have been reported in the literature. The water activity (a_w) is the vapor pressure in the headspace of the materials in relation to the vapor pressure of pure water and is a measure of how effectively the water present can take part in a chemical (physical) reaction. The average moisture content of the layer is the sum of the dry mass fraction of the ingredient multiplied by the associated moisture content of the ingredient in the layer, which can be calculated from Equation 2.7.5:

$$\overline{X_i} = a_{as,i} X_{as,i} M_{as,i} + b_{cs,i} X_{cs,i} M_{cs,i} + \left(1 - a_{as,i} - b_{cs,i}\right) X_{pr,i} M_{pr,i} \tag{2.7.5}$$

In this case, during the crystallization and moisture redistribution, the water activities of the components need to be equal within each layer inside the particle.

Moisture transfer in the particle is in unsteady state, meaning the concentrations of water in each layer are changing with time (as a source or sink term), through diffusion, moisture sorption and crystallization. Although it is a difference in the chemical potential, or water activity, that defines the driving force for moisture migration, mass transfer is commonly characterized based on the moisture content and Fick's law. The transport phenomena are described by Equation 2.7.6:

$$\frac{\partial \overline{X}_i}{\partial t} = \frac{\partial}{\partial z}\left(D_{av,i}\left(\overline{X}_i, T_i\right) \frac{\partial \overline{X}_i}{\partial z} \right) + \dot{S} \tag{2.7.6}$$

The initial and boundary conditions required for solving Equation 2.7.6 are

IC: $t=0,\ 0 \le z \le R_{part}$, $\overline{X_i} = \overline{X_0}$ and $T_i = T_0$

BC1: $t>0,\ z=0$ $\quad D_{av,i} \dfrac{\partial \overline{X}_i}{\partial z} = 0$ (at center)

BC2: $t>0,\ z=R_{part}$, $\quad D_{av,1} \dfrac{\partial \overline{X}_i}{\partial z} = \dfrac{k_{ap}\left(pv_1 - pv_{air}\right)}{\rho_m}$ (at surface)

The first boundary condition represents the symmetry at the center of the particle. The second boundary condition states that the amount of water sorbing/desorbing on the particle surface is equal to the diffusive flux at the surface. Water is sorbed from air to the surface of the particle through the difference in the water activity (vapor pressure) of the components and the vapor pressure of water in air. Fluxes are defined as positive away from the particle and away from the center of the particle.

A similar analogy could be applied to heat transfer. Details of the model and numerical simulation can be found in the literature [33].

2.7.3 *Process integration*

Figure 2.7.1 shows a proposed system integration containing a spray dryer and a fluidized bed dryer. The additional heaters and humidifiers in the fluidized bed air inlet line adjust the air temperature and humidity. The sizes of the fluidized beds depend on the outlet powder flow rate from the spray dryer and the residence time needed for crystallization. The stages and the combination of humidity and temperature depend on the material properties and the powder bed behavior.

Different fluidization ability may occur for different amounts of crystallinity in the powders, so in the fluidization/crystallization process, the humidity and temperature can be increased as a function of time while the crystallinity of powders is developing. Figure 2.7.2 shows the step-wise process path in respect of the degree of amorphicity of materials. The two dashed lines between the upper and lower limits are representative limits for partially crystallized powder with some percentage improved crystallinity during the process, at different sections of the fluidized bed dryer. The safe process conditions of temperature and humidity to keep the bed well fluidized and avoid cake formation should be below these upper limits [12,17]. The process path shows the initial conditions for processing and the changes in different stages.

2.7.4 *Mathematical model of the process*

A multi-component diffusion model has been developed to describe moisture sorption onto a spray-dried amorphous skim milk particle, with the accompanying processes of moisture-induced crystallization and phase change, moisture redistribution within new components and moisture transport between different internal layers [33]. The separate components of amorphous lactose, crystalline lactose, protein and moisture have been included in each layer, where the ratio of crystalline to amorphous lactose may change in each layer during the crystallization time. The glass-transition temperature has been affected by the diffusion of moisture between the layers. The crystallization rate has been increased by raising the difference between the material temperature and the glass-transition temperature in each layer [22], by heat and mass transfer processes. The model has simulated moisture sorption for a series of different temperature and humidity conditions in a particle. Moisture profiles and crystallization rates show trends in the changes in moisture diffusion and the permeability of the layers within the particles accompanying the change in the amount of crystallinity in each layer. More detail on the model and results was reported in the literature [33]. Figure 2.7.3A,B shows the mass fraction of the amorphous and crystalline materials in each layer. The amount of amorphous lactose decreased during the process, while the amount of crystalline material increased. The

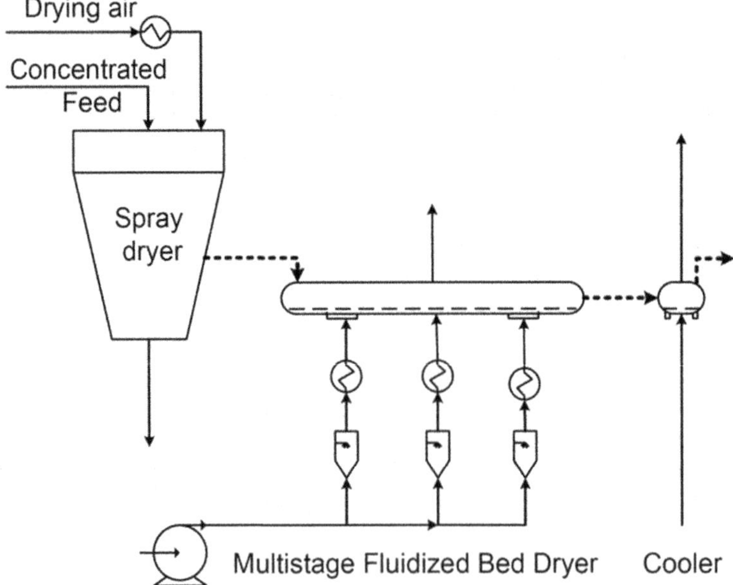

Figure 2.7.1 Process integration schematic.

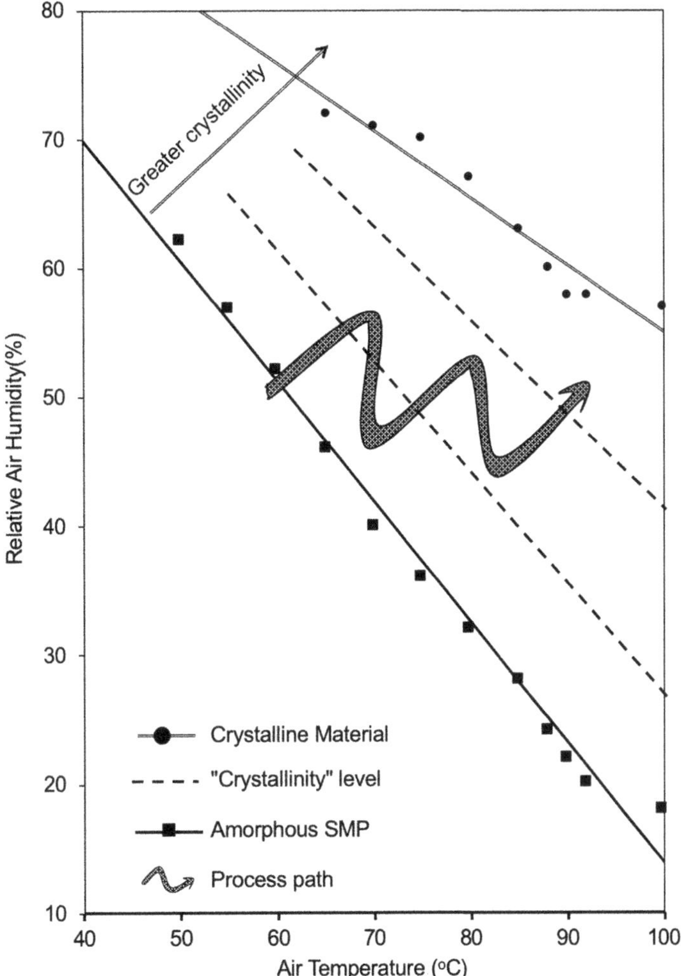

Figure 2.7.2 Process path at different stages of the fluidized bed.

fraction of amorphous material started at 0.5 and decreased to 0.05 at the end of the processing simulation (the fraction of protein had a value of 0.4, which remained constant; the value for crystalline lactose was assumed to be 0.1 as the initial condition). The results for the time evolution of moisture sorption and amorphicity show similar trends with the same timing. The values for the amorphous content of all the layers start from 0.5 and decrease gradually with a time delay between internal layers. The first layer starts to become crystalline immediately, and the other layers follow with a short delay that is equal to the corresponding moisture content results (the trends are similar to the moisture sorption results). The crystallinity results (Figure 2.7.3B) clearly show this difference and include a close-up view of the initial stage of the process when the layers start to change.

2.7.5 *Particle structure*

Most of the physical properties of spray-dried milk powder, such as flowability, stickiness, agglomeration and caking, are surface-dominated properties that involve overcoming the surface attraction between the particles [4,13]. Crystalline particles with crystalline surfaces have very weak tendencies to form bridges with other particles due to the high thermodynamic stability of the crystalline state. Even a thin layer of modified molecular structures on the surfaces of the particles can greatly improve their physical properties by preserving the desirable characteristics of the amorphous core. The lower amount of moisture on the surface causes fewer moisture-induced changes inside the particle.

The novel particle structure that is created by this method has a modified surface with crystalline lactose but maintains instead the amorphicity of the core in the fluidized bed apparatus used earlier (Figures 2.7.4–2.7.7). The new powder structure has been called an egg-shell structure because it has a crystalline surface and an amorphous core. The role of the thin crystalline layer on the surface of the particles was instead investigated and compared with raw commercial powders to assess changes in the stability, flowability and physical properties of freshly processed powders with egg-shell structures while maintaining the same functionality [13,34]. Also, the water-induced changes in raw commercial skim milk powder during storage was instead compared with processed powder, which has an egg-shell structure, and the effects of the crystalline surface layer on the deteriorative changes in milk powder will be investigated.

Particles with egg-shell structures have the combined characteristics of stable crystalline materials and bioavailable amorphous materials. The performance of this modified

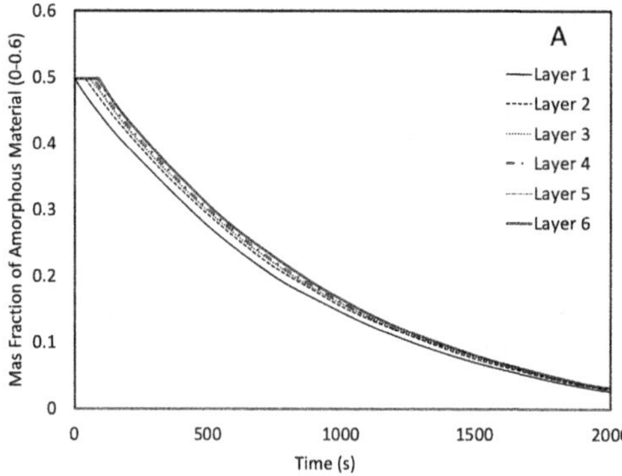

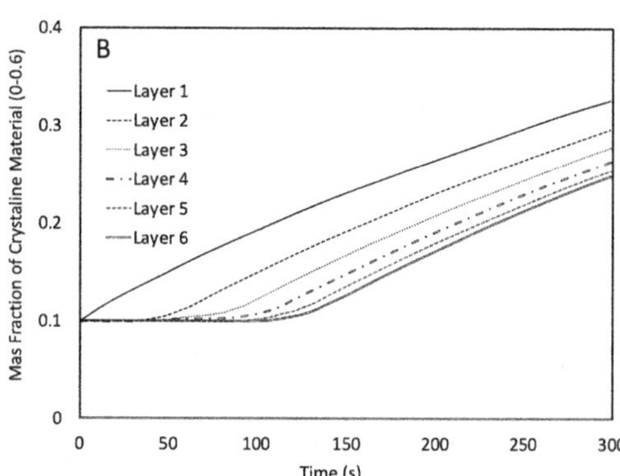

Figure 2.7.3 (A) Change in the mass fraction of the amorphous lactose in the layers during the simulation. (B) Improvement in the crystallinity of the layer during the simulation (early stage) [33].

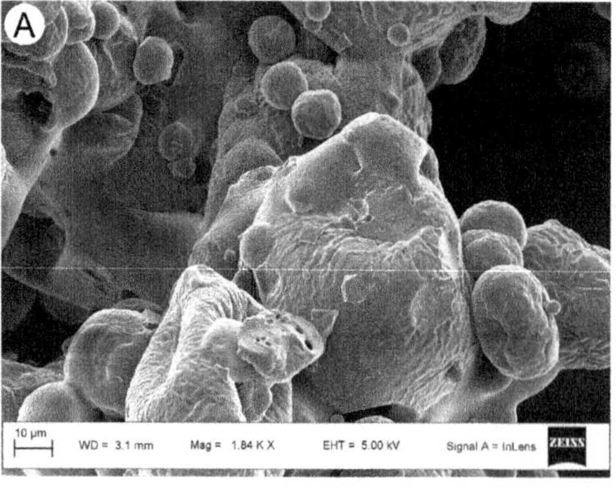

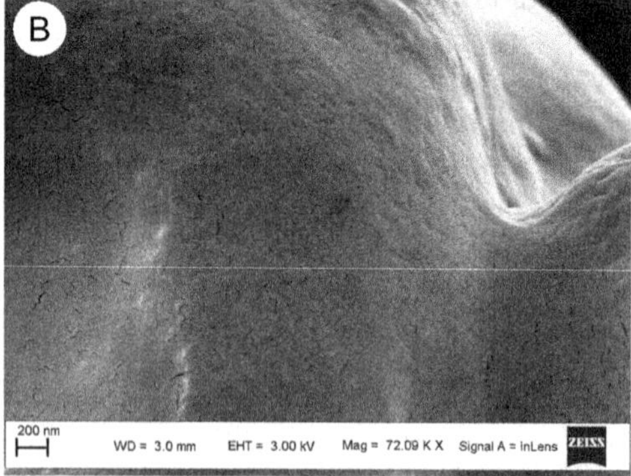

Figure 2.7.4 (A) Surface of raw milk powder at low magnification. (B) Close-up of the surface of raw particles.

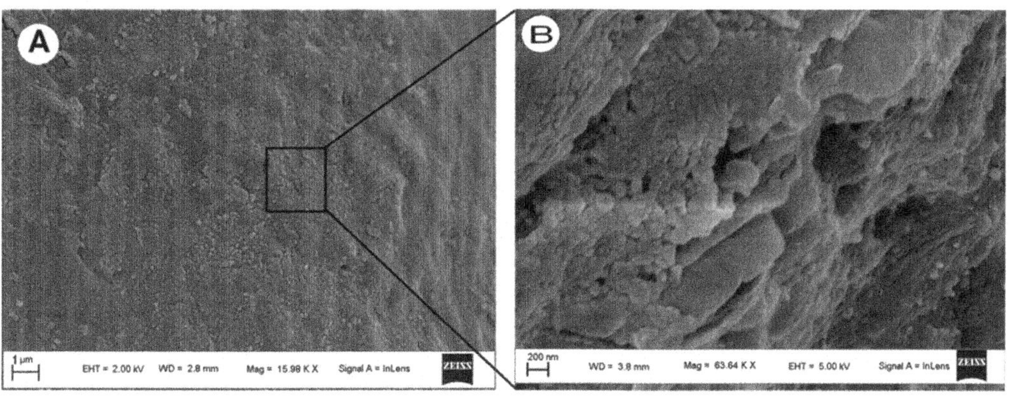

Figure 2.7.5 (A) Surface of processed milk powder at low magnification. (B) Islands of lactose crystals formed on the surface of processed powders; the crystal sizes are nanoscale.

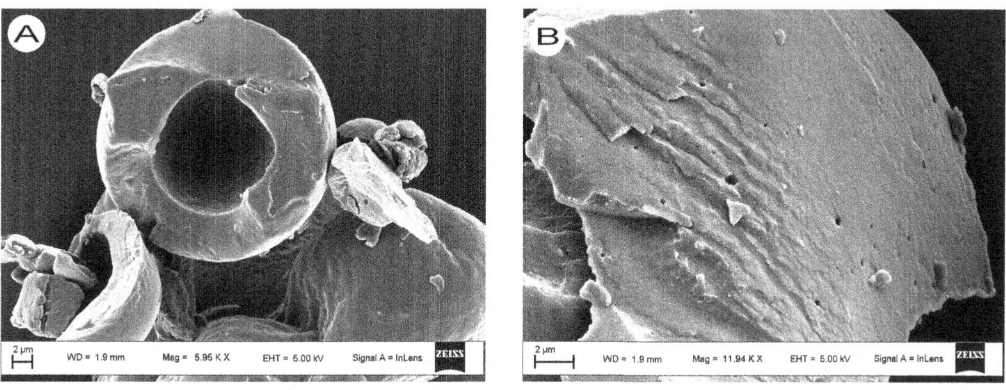

Figure 2.7.6 (A) Cross section of processed milk powder (crushed by mortar and pestle). (B) No lactose crystals appear in the sub-surface region of the processed powder.

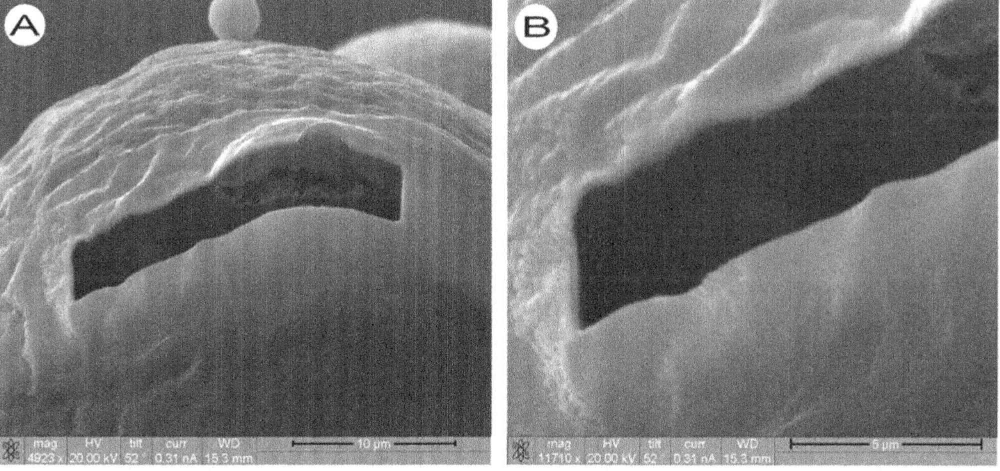

Figure 2.7.7 (A) Cross section of processed milk powder (ion-beam milled). (B) No lactose crystals appear in the sub-surface region of the processed powder.

structure on improving the shelf life [34] and powder behavior, such as flowability [4,13,35], has been demonstrated.

Figure 2.7.8 shows a schematic diagram of the different heterogeneous particle structures. A particle with a crystalline surface and an amorphous core is called an egg-shell structure. A structure where the crystalline domain is located in the core of the particle, and the surface layer is amorphous, is called an egg-yolk structure. The reason for the formation of these heterogeneous structures is the difference between the water activity of the processing air and the particle. This difference governs the moisture transfer direction through a sorption and desorption process into/from the particle. These phenomena were described in detail in the modeling section. When the dry particle is processed at high water activity, the moisture sorbs onto the surface of the particle and causes crystallization of the surface layer. For the case where the moisture content of the particle is higher than the water activity of the processing air, moisture diffuses toward the surface, and eventually to the air, and the particle dries out to reach an equilibrium state with the air. Therefore, the core of the particle, which has a low glass-transition temperature in the moist state, transforms to the crystalline state, but the surface layer that is in contact with dry air stays amorphous [4].

More analytical results and investigation of the powder using x-ray powder diffraction (XRD), Raman, Fourier-transform infrared spectroscopy (FTIR) and differential scanning calorimetry (DSC) demonstrate the modification of the particle microstructure [4,13,35]. The thin crystal layer around the particle has less permeability and moisture sorption and can protect the internal active ingredients of the dairy powder against deteriorative changes influenced by increasing the water activity during storage.

A protein content analysis and monitoring the protein unfolding during storage showed lower changes in the surface composition, protein modification and crystallinity during long-term storage at 25–30°C and 35% RH [34]. Agglomeration, large lactose crystal formation on the surface, surface composition changes and protein modifications were studied. The changes between the raw powder and the processed powder were compared after aging. The effect of the non-hygroscopic crystalline surface layer showed significant benefit in maintaining the physicochemical qualities of the powders over long storage times. The aged raw powder showed a 24% change in crystallinity, a 3% change in the

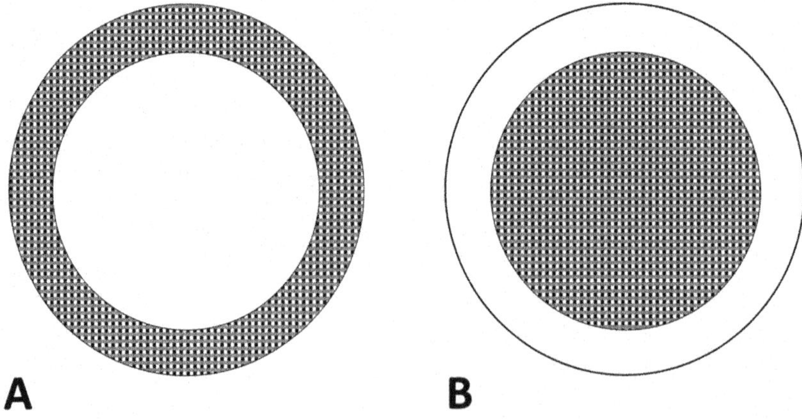

Figure 2.7.8 Schematic diagram of heterogeneous particle structures. (A) Egg-shell structure with crystalline surface. (B) Egg-yolk structure with crystalline core (the grey-patterned sections represent the crystalline domains).

lactose/protein ratio on the surface composition and 6% protein denaturation in comparison with the aged processed powder with a 4% change in crystallinity, a 1.5% change in the lactose/protein ratio on the surface composition and 2% protein denaturation, respectively, after 30 weeks storage.

2.7.6 Conclusion

The fluidized bed dyer, a significant step in dairy powder processing, is mostly used for secondary drying of powder after spray drying. The residence time and ample heat and mass transfer capacity in the fluidized bed make it a potential unit operation to modify particle microstructure and powder properties. One of the new advancements in the fluidized bed dryer, as introduced in this chapter, is modifying the particle surface and microstructure and inducing partial crystallization.

Processed dairy powders show less moisture sorption and more lactose crystals, leading to an improvement in the degree of amorphicity of spray-dried milk powder by post-processing in a fluidized bed dryer. Crystallized lactose appeared to form, and the processed milk powders were heavily textured (crystallized on the surface). The improvement in the surface texture and the lower moisture sorption from the environment mean that the powders have more stability and better flowability, which could potentially be a feasible solution for flow problems associated with handling, bulk transfer and hopper design in relation to these powders. Surface analysis reports, crystallinity, composition and protein modification all indicated that there were much smaller changes in the aging of the processed powder in contrast to significant changes in the aged raw powder.

References

1. Písecký, J., *Handbook of Milk Powder Manufacture*. 1997, Niro A/S, Copenhagen.
2. Birchal, V.S., et al., Effect of spray-dryer operating variables on the whole milk powder quality. *Drying Technology*, 2005. **23**(3): p. 611–636.
3. Kim, E.H.J., X.D. Chen, and D. Pearce, Surface composition of industrial spray-dried milk powders. 1. Development of surface composition during manufacture. *Journal of Food Engineering*, 2009. **94**(2): p. 163–168.
4. Yazdanpanah, N. and T.A.G. Langrish, Heterogeneous particle structure formation during post-crystallization of spray-dried powder. *Particuology*, 2016. **27**: p. 72–79.
5. Anema, S.G., et al., Effects of storage temperature on the solubility of milk protein concentrate (MPC85). *Food Hydrocolloids*, 2006. **20**(2–3): p. 386–393.
6. Fitzpatrick, J.J., et al., Effect of composition and storage conditions on the flowability of dairy powders. *International Dairy Journal*, 2007. **17**(4): p. 383–392.
7. Listiohadi, Y., et al., Moisture sorption, compressibility and caking of lactose polymorphs. *International Journal of Pharmaceutics*, 2008. **359**(1–2): p. 123–134.
8. Fyfe, K.N., et al., Storage induced changes to high protein powders: Influence on surface properties and solubility. *Journal of the Science of Food and Agriculture*, 2011. **91**(14): p. 2566–2575.
9. Mauer, L.J., D.E. Smith, and T. Labuza, Effect of water content, temperature and storage on the glass transition, moisture sorption characteristics and stickiness of β-casein. *International Journal of Food Properties*, 2000. **3**(2): p. 233–248.
10. Vega, C. and Y. Roos, The state of aggregation of casein affects the storage stability of amorphous sucrose, lactose, and their mixtures. *Food Biophysics*, 2007. **2**(1): p. 10–19.
11. Shrestha, A.K., et al., Spray drying of skim milk mixed with milk permeate: Effect on drying behavior, physicochemical properties, and storage stability of powder. *Drying Technology*, 2008. **26**(2): p. 239–247.
12. Yazdanpanah, N. and T.A.G. Langrish, Fast crystallization of lactose and milk powder in fluidized bed dryer/crystallizer. *Dairy Science and Technology*, 2011. **91**(3): p. 323–340.

13. Yazdanpanah, N. and T.A.G. Langrish, Egg-shell like structure in dried milk powders. *Food Research International*, 2011. **44**(1): p. 39–45.
14. Nijdam, J., A. Ibach, and M. Kind, Fluidisation of whey powders above the glass-transition temperature. *Powder Technology*, 2008. **187**(1): p. 53–61.
15. Saito, Z., Particle structure in spray-dried whole milk and in instant skim milk powder as related to lactose crystallization. *Food Microstructure*, 1985. **4**: p. 333–340.
16. Hynd, J., Drying of whey. *International Journal of Dairy Technology*, 1980. **33**(2): p. 52–54.
17. Yazdanpanah, N. and T.A.G. Langrish, Crystallization and drying of milk powder in a multiple-stage fluidized bed dryer. *Drying Technology*, 2011. **29**(9): p. 1046–1057.
18. Roos, Y.H. and N. Silalai, Glass transitions: Opportunities and challenges. In *Food Engineering Interfaces*, J.M. Aguilera, et al., Editors. 2011, Springer, New York. p. 473–490.
19. Omar, A.M.E. and Y.H. Roos, Water sorption and time-dependent crystallization behaviour of freeze-dried lactose-salt mixtures. *LWT - Food Science and Technology*, 2007. **40**(3): p. 520–528.
20. Buera, M.D.P., et al., State diagrams for improving processing and storage of foods, biological materials, and pharmaceuticals (IUPAC Technical Report). *Pure and Applied Chemistry*, 2011. **83**(8): p. 1567–1617.
21. Omar, A.M.E. and Y.H. Roos, Glass transition and crystallization behaviour of freeze-dried lactose-salt mixtures. *Lebensmittel-Wissenschaft und -Technologie*, 2007. **40**(3): p. 536–543.
22. Elmonsef Omar, A.M. and Y.H. Roos, Glass transition and crystallization behaviour of freeze-dried lactose-salt mixtures. *LWT - Food Science and Technology*, 2007. **40**(3): p. 536–543.
23. Barham, A.S., et al., Crystallization of spray-dried lactose/protein mixtures in humid air. *Journal of Crystal Growth*, 2006. **295**(2): p. 231–240.
24. Laaksonen, T.J. and Y.H. Roos, Thermal, dynamic-mechanical, and dielectric analysis of phase and state transitions of frozen wheat doughs. *Journal of Cereal Science*, 2000. **32**(3): p. 281–292.
25. Jouppila, K. and Y.H. Roos, Glass transition and crystallization in milk powder. *Journal of Dairy Science*, 1994. **77**(10): p. 2907–2915.
26. Labuza, T. and B. Altunakar, Water activity prediction and moisture sorption isotherms. In *Water Activity in Foods: Fundamentals and Applications*, G.V. Barbosa-Cánovas, Editor. 2008, Blackwell Pub., Ames, IA. p. 109–154.
27. Labuza, T., et al., Storage stability of dry food systems: Influence of state changes during drying and storage. In 14th International Drying Symposium (IDS 2004). 2004: Sao Paulo, Brazil. p. 48–68.
28. Gordon, M. and J.S. Taylor, Ideal copolymer and the second-order transition of synthetic rubbers. I. Non-crystalline copolymers. *Journal of Applied Chemistry*, 1952. **2**(9): p. 493–500.
29. Couchman, P.R. and F.E. Karasz, A classical thermodynamic discussion of the effect of composition on glass-transition temperatures. *Macromolecules*, 1978. **11**(1): p. 117–119.
30. Williams, M.L., R.F. Landel, and J.D. Ferry, The temperature dependence of relaxation mechanisms in amorphous polymers and other glass-forming liquids. *Journal of the American Chemical Society*, 1955. **77**(14): p. 3701–3707.
31. Maroulis, Z.B., et al., Application of the GAB model to the moisture sorption isotherms for dried fruits. *Journal of Food Engineering*, 1988. **7**(1): p. 63–78.
32. Blahovec, J. and S. Yanniotis, GAB generalized equation for sorption phenomena. *Food and Bioprocess Technology*, 2008. **1**(1): p. 82–90.
33. Yazdanpanah, N. and T.A.G. Langrish, Mathematical modelling of the heat and moisture diffusion in a dairy powder particle with crystallisation phase change within the particle matrix. *International Journal of Heat and Mass Transfer*, 2013. **61**: p. 615–626.
34. Yazdanpanah, N. and T.A.G. Langrish, Comparative study of deteriorative changes in the ageing of milk powder. *Journal of Food Engineering*, 2013. **114**(1): p. 14–21.
35. Yazdanpanah, N. and T.A.G. Langrish, Releasing fat in whole milk powder during fluidized bed drying. *Drying Technology*, 2012. **30**(10): p. 1081–1087.

chapter 3

Powder properties and influencing factors

Yrjö H. Roos, Zahra Afrassiabian, Khashayar Saleh, Marie-Hélène Famelart, Alexia Audebert, Muhammad Gulzar, Thomas Croguennec, Jennifer Burgain, Tristan Fournaise, Claire Gaiani, Joël Scher, Jérémy Petit, Evandro Martins, Ramila Cristiane Rodrigues, Pierre Schuck, Ítalo Tuler Perrone, Solimar Gonçalves Machado and Antônio Fernandes de Carvalho

Contents

3.1 Glass transition and water activity

Yrjö H. Roos

3.1.1 Introduction

The amorphous state of lactose in dairy products, such as ice cream and dairy powders, is known as problematic particularly because of time-dependent lactose crystallization (Supplee, 1926; Troy and Sharp, 1930; Herrington, 1934; King, 1965; Berlin et al., 1968a,b; Saltmarch and Labuza, 1980). The manufacturing of dairy powders on the other hand has developed into an enormous industry, although lactose often tends to form amorphous states in dehydration (Roos and Drusch, 2015).

An amorphous solid state of a substance results from quench cooling to temperatures well below the equilibrium melting temperature of the crystalline substance or by rapid solvent removal. In dairy dehydration processes, water-soluble components, such as lactose, become saturated and often form noncrystalline (amorphous) solid structures (Jouppila et al., 1997). The noncrystalline state of food components is metastable and sensitive to water. Water in noncrystalline, hydrophilic food components acts as a plasticizer, i.e., water molecules distribute within the molecular assembly of the noncrystalline structures and increase the free volume. Water plasticization can be detected, for example, from a decrease of the glass transition temperature, T_g. Glass transition provides information on the physical state of noncrystalline solids (Roos and Drusch, 2015). The glass transition in dairy foods is often linked to that of lactose which has been discussed by numerous authors (Roos and Karel, 1991; Bhandari and Howes, 1999; Vuataz, 2002; Schuck et al., 2005; Shrestha et al., 2007; Carpin et al., 2016).

Noncrystalline hydrophilic food solids tend to be highly hygroscopic as their unorganized molecular structures support hydrogen bonding and strong interactions with water molecules. The quantity of water may be expressed as mass fraction, mole fraction or water content within a food or in an amorphous food component. The quantification of water in multicomponent dairy foods is often complex. That is because various food components vary in water sorption properties (Potes et al., 2012; Roos and Drusch, 2015). Conversely, water activity, a_w, defined as the ratio of vapor pressure of water within the material, p, and that of pure water, p_0, at the same temperature, applies throughout the material. Water activity can be assumed to be the same for various food components although there may be large differences in water distribution and quantities across components.

Understanding glass transition, water activity, water content and water plasticization relationships is fundamental for the successful dehydration of dairy foods as well as for the stability control of dried dairy foods. Such relationships of water and dairy solids also need to be taken into account in the manufacturing of foods and ingredients using dairy components. Here we discuss glass transitions and water relationships of dairy foods and particularly those of dehydrated dairy foods and ingredients.

3.1.2 *Glass transition in dairy solids*

The crystallization of materials from the noncrystalline state can be inhibited by various factors. Such inhibition is the requirement of glass formation (a glass is a transparent and brittle solid-like material). Glass formation may result from quench cooling as there is not sufficient time for molecular organization, and consequently nucleation and crystallization are inhibited. Likewise, the rapid removal of a solvent, such as water in spray drying, may occur more rapidly than nucleation and crystal growth in the supersaturated state. The resultant materials have disordered molecular arrangements and a solid but liquid-like structure. The glass-forming molecules can have a dense appearance typical of inorganic glasses or the glass-formers can remain as glassy membranes or walls around pores of an expanded structure. The latter is typical of freeze-dried materials and lactose in dairy powders (Roos and Drusch, 2015). The solid–liquid transformation of noncrystalline solids is known as the glass transition.

3.1.2.1 *Lactose and dairy powders*

Sugars often remain in a noncrystalline structure after dehydration which may be detected using X-ray diffraction measurements. An amorphous material shows no specific X-ray diffraction data typical of crystalline states such as those appearing in powder X-ray diffraction patterns of lactose (Figure 3.1.1). The glass transition occurs over a

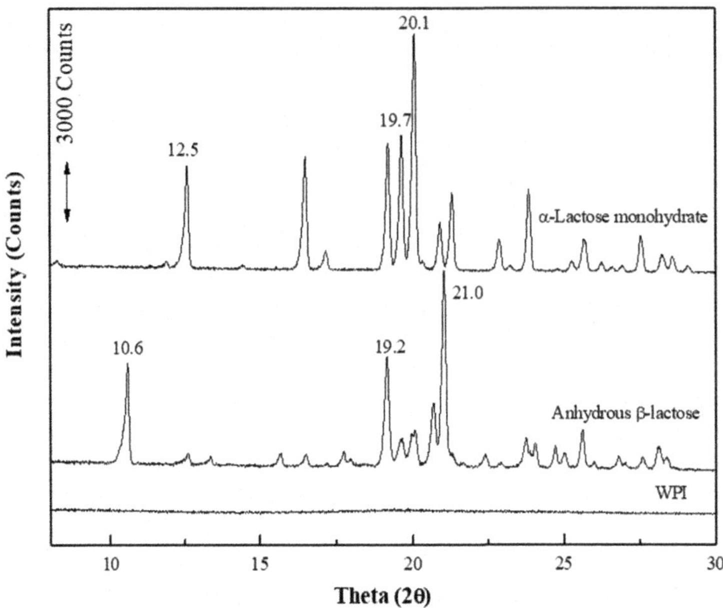

Figure 3.1.1 Powder x-ray diffraction patterns for alpha-Lactose monohydrate and anhydrous beta-lactose crystals (Fan and Roos, 2015).

specific temperature range for various sugars and is dependent on the molecular mass. Monosaccharides show glass transitions at lower temperatures than disaccharides and oligosaccharides (Roos, 1993). Lactose in an anhydrous state shows a heating induced calorimetric glass transition with onset at 105°C. The glass transition in a differential scanning calorimetry (DSC) study is a reversible thermal event occurring for lactose as an endothermic shift in heat capacity over approximately 20°C in heating. The lactose glass transition after spray drying is detectable in milk powders of various fat contents (Jouppila and Roos, 1994). It should be noted that a quite different glass transition behavior results from the hydrolysis of lactose to galactose and glucose components. That is because glass transitions of monosaccharides occur at substantially lower temperatures than those found for disaccharides (Roos, 1993).

3.1.2.2 *Water and dairy powders*

Dehydrated dairy powders have a low residual water content of 1 to 5% by mass. Such water is sorbed by hydrophilic protein components and lactose. In precrystallized dairy powders much of the lactose is likely to exist as monohydrate crystals with 5% water within the hydrate crystals. After spray drying, lactose in dairy powders, however, tends to exist in the noncrystalline, amorphous state. Amorphous lactose is plasticized by water which can be detected from the water content-dependent glass transition of the lactose. The glass transition of lactose, T_g, as measured during heating in a DSC scan, decreases with increasing water content. Figure 3.1.2 shows a typical DSC scan of amorphous lactose at a water content corresponding to 0.33 a_w. The DSC thermogram shows the glass transition with an enthalpy relaxation and the lactose crystallization endotherm. The physical state of lactose in dairy powders at various water contents has been described using state diagrams by several authors (e.g., Vuataz, 2002; Roos, 2008). A state diagram for amorphous

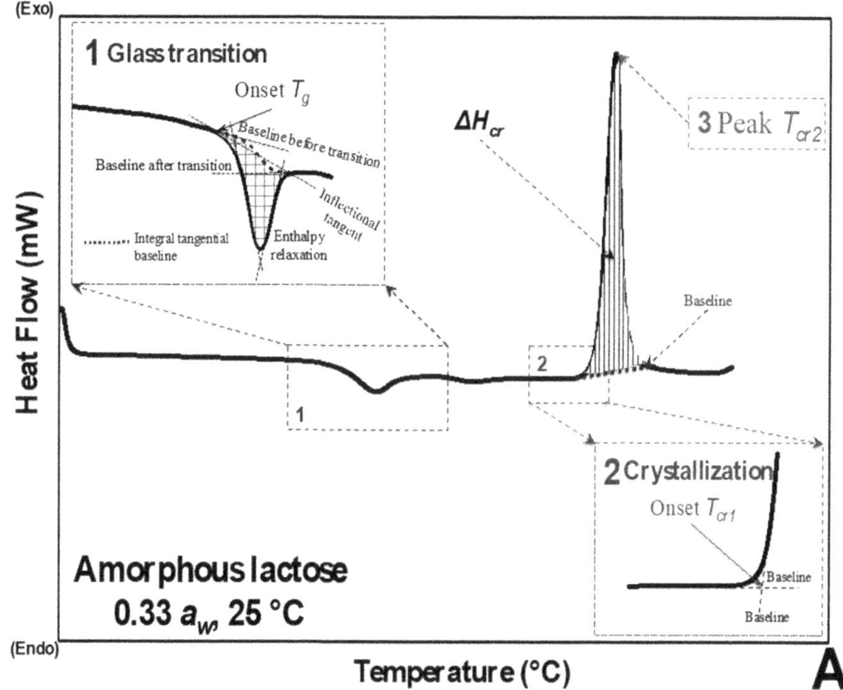

Figure 3.1.2 Typical differential scanning calorimetry curves obtained in heating of amorphous lactose. An enthalpy relaxation endotherm may be associated with the glass transition. Crystallization occurs above the glass transition (Fan and Roos, unpublished).

lactose, mapping its physical state at various water contents, is presented in Figure 3.1.3. The state diagram provides a means to locate temperature–water content relationships where lactose can exist as a metastable glass or highly time-dependent, unstable and crystallizing sugar. For example, water plasticization besides poor lactose solubility can cause a time-dependent lactose crystallization during the storage of dairy powders but also in ice cream (Roos and Drusch, 2015). Such crystallization occurs with increasing rates above the onset glass transition temperature, T_g. The time-to-crystallization followed an exponential Williams–Landel–Ferry relationship (Roos and Karel, 1992) and relaxation times above the T_g could be related to real-time rates using the Deborah number (Roos and Drusch, 2015). A maximum level of crystallinity at room temperature was obtained at a relative humidity (RH) of 70% (Jouppila et al., 1997).

3.1.3 Water and water activity

The water content of dairy liquids is high and generally above 90%. Dehydration in the manufacturing of dairy powders removes most of the water while all solids are concentrated to the final powder. Although Raoult's law provides a fairly accurate relationship of solutes and water vapor pressure only for dilute solutions, Raoult's law (Equation 3.1.1) is useful for the explanation of the effects of solutes composition on water vapor pressure, p_w, in dairy concentrates and powders.

$$p_w = xp_0 \tag{3.1.1}$$

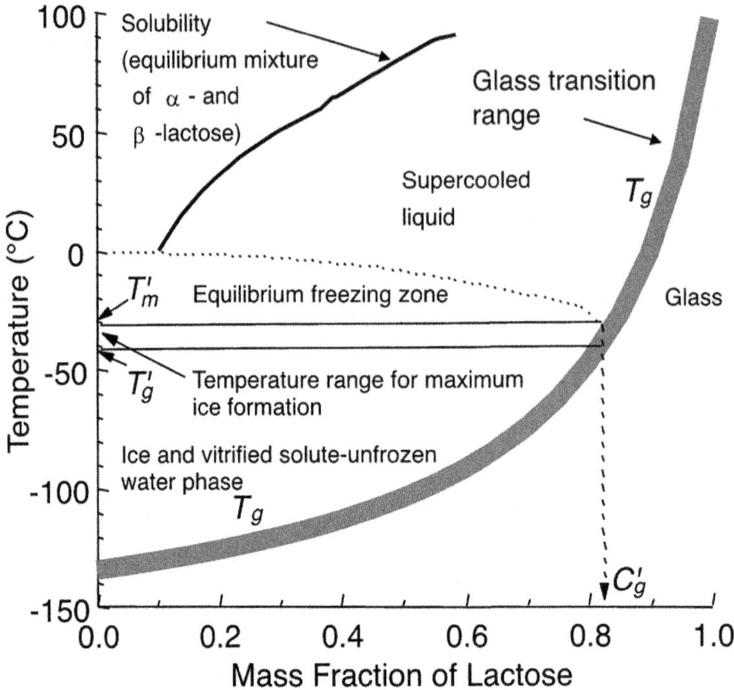

Figure 3.1.3 State diagram of lactose. The state diagram shows relationships of lactose content in water, temperature and the physical state. The T_g shows the water content-dependent glass transition of non-crystalline lactose. T_m' and T_g' are onset of ice melting in maximally freeze-concentrated lactose solution and onset temperature of maximally freeze-concentrated lactose, respectively.

where p_w is vapor pressure of water in solution, x is mole fraction of water, and p_0 is vapor pressure of pure water at the same temperature.

The mole fraction of water can be obtained from Equation 3.1.2 and it may also be described as the *effective concentration* of water.

$$x = \frac{[\text{water}]}{[\text{water}] + [\text{solutes}]} \tag{3.1.2}$$

Equation 3.1.2 is useful for understanding the importance of the effect of the size of solute molecules on the vapor pressure of water. In other words, the vapor pressure of water can be effectively reduced using solutes of low molecular mass, such as sugars and salts. That is, the molar concentration of solutes increases as the molecular mass of the solute decreases. In the dehydration of dairy liquids, the key components affecting water vapor pressure are lactose and minerals. As lactose makes up approximately 50% of skim milk solids it has a significant impact on the vapor pressure of water in dairy powders. Conversely, lactose hydrolysis doubles the molar concentration of sugars as each lactose molecule breaks into two monosaccharide molecules, one galactose and one glucose molecule. As a result of a high molar concentration of small molecules in dairy solids, the vapor pressure of the residual water is reduced to a low value compared to that of pure water.

3.1.3.1 *Water activity*

Water activity, a_w, is derived from the activity coefficient or fugacity, *f*, of water in a solution. For an ideal, dilute solution *water activity* is determined directly from its mole fraction (*effective concentration*) and Raoult's law simply gives the relationship of Equation 3.1.3. Such a relationship may also be derived from the chemical potential of water within a food and its vapor phase at equilibrium as that of the liquid and vapor phases are equal at equilibrium. Conversely, water vapor pressure surrounding a dairy powder is equal to that of its headspace, and it corresponds to the relative humidity of air (RH) surrounding the material (Equation 3.1.4) at equilibrium.

$$a_w = \frac{p_w}{p_0} \tag{3.1.3}$$

$$a_w = 0.01 * RH(\%) \tag{3.1.4}$$

Water activity as an intrinsic property of a material is an extremely useful measure of the effects of water on physicochemical and microbial deterioration (Labuza, 1977). In multicomponent dairy powders water is primarily associated with hydrophilic constituents, and dairy fat has little effect on measured a_w (Jouppila and Roos, 1994). In considering relationships of water activity and water content it is important to express water contents as quantities within non-fat solids.

3.1.3.2 *Water plasticization*

The noncrystalline state of hydrophilic food components attracts water molecules which results in rapid sorption and the formation of strong hydrogen bonding of water molecules within a dehydrated molecular assembly. The random molecular arrangement of amorphous sugars leaves unoccupied voids, i.e., free volume or "holes" within the structure. These holes can be investigated using positron annihilation lifetime spectroscopy, as was reported for trehalose glasses by Kilnburn et al. (2006). The holes increase in size with increasing water content (Kilnburn et al., 2006) which indicates the presence of highly mobile water molecules within the glassy solid. In other words, water mobility and solute mobility, i.e., water and glass-former mobility, glass are uncoupled. The increase in hole size corresponds to an increase in the free volume and a resultant decrease of the T_g of the glass. The decrease of T_g with increasing water content is known as *water plasticization*. Water plasticization at a level depressing the T_g to below the material temperature results in the glass transition and the decrease of structural relaxation times and flow corresponding to *thermal plasticization*. Both thermal plasticization and water plasticization above the T_g show a rapid decrease of structural relaxation times and softening of the material (Fan and Roos, 2017). The decrease of structural relaxation times can be quantified using direct measurements by dynamic mechanical analysis (DMA), dielectric spectroscopy or NMR, or it can be related to viscosity or other rheological measurements (Roos and Drusch, 2015).

3.1.3.3 *Water sorption*

The water sorption properties of dairy powders have been of much interest because of the importance of powder properties at various storage temperatures and relative humidity conditions. The time-dependent changes in water sorption and powder deterioration above a critical storage relative humidity have been well-documented. Berlin et al. (1968a) related a discontinuity of water sorption isotherm constructed for various dairy powders to water-induced crystallization of lactose, although such discontinuity occurred in whey powders around 0.35 a_w and at higher humidities in milk powders. The rate of crystallization of noncrystalline sugars is often dependent on composition, and crystallization typically is delayed by reduced water content and the presence of proteins and other carbohydrate species. Potes et al. (2012) also noted that small carbohydrates in blends with lactose were more effective in delaying lactose crystallization than larger carbohydrate molecules. Berlin et al. (1968b) showed that lactose crystallization was responsible for the unexpected loss of sorbed water in dairy powders. They also noted that at low humidities water sorption agreed with the sum of water sorbed by powder components. A similar additive water sorption or fractional water sorption model using water sorption by components of dairy powders was found accurate for the prediction of water sorption properties by Foster et al. (2005). We have also emphasized the importance of the additivity of component water sorption properties in understanding and extrapolating water sorption properties to higher humidity and temperature conditions (Potes et al., 2012).

The water sorption by dairy powders has been related to powder composition, and it appears that the water sorbed at various a_w conditions needs to be expressed as a quantity associated with the non-fat solids components (solids non-fat [SNF]) as described by Jouppila and Roos (1994a,b). Jouppila et al. (1997) investigated the effects of water sorption on the time-dependent lactose crystallization in skim milk powder. They related crystallization to lactose glass transition, and the maximum extent of crystallization was found to occur at 17% water content at 70% RH and a temperature of 61°C above the T_g. Since lactose crystallization in dairy powders occurs during water sorption there have been limited data available to quantify lactose water sorption at higher humidities. The study of Potes et al. (2012) used various maltodextrins co-dried with lactose which allowed the quantification of water sorption by lactose also at the higher relative humidities. Potes et al. (2012) noted that water sorption followed the fractional quantities sorbed by components and the total water sorption was additive based on that of the components. The indirect quantification of lactose water sorption allowed the modeling of water sorption isotherms over a wide range of water activities as well as the establishment of a more accurate state diagram for lactose and dairy powders (Figures 3.1.3 and 3.1.4).

3.1.3.4 *Critical water activity*

A lactose state diagram describes the water content dependence of the T_g which is fundamental to dairy powder manufacturing where given final water contents to ensure the solid state of amorphous lactose are required. The state diagram can be complemented using water sorption isotherms to present the relationships between a_w and water content. Most published water sorption isotherms for lactose are reliable only for the solid, noncrystalline state of lactose as time-dependent lactose crystallization takes place when water plasticization depresses the T_g to below ambient temperature.

Potes et al. (2012) reported water sorption data for amorphous lactose over a wide a_w range. They used lactose–maltodextrin blends where lactose crystallization was inhibited and assumed additivity of the component water contents in accordance with Foster et al. (2005). As maltodextrin sorption isotherms were available, lactose data could be derived from the water sorption data of the blends where lactose remained noncrystalline. The resultant water sorption isotherm for lactose is shown in Figure 3.1.4.

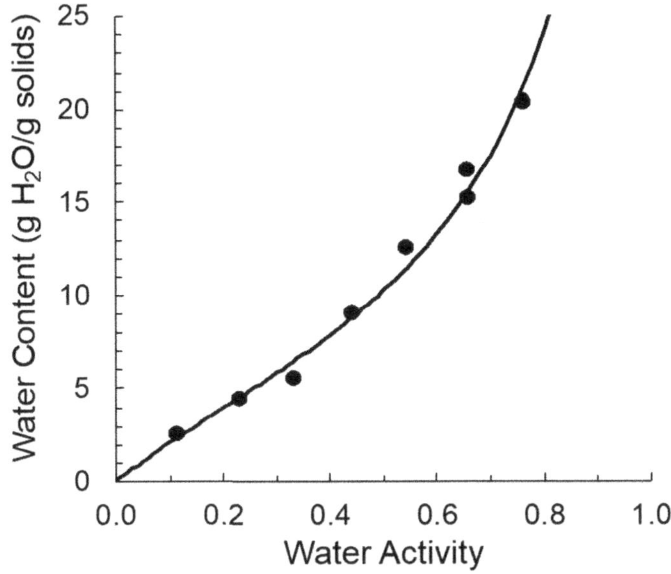

Figure 3.1.4 Water sorption isotherm of lactose. The experimental data are shown over a wide water activity range (Potes et al., 2012).

The water sorption isotherm, i.e., the water content plotted against a_w at a constant temperature, for a noncrystallizing material or when crystallization is absent can be modeled using various relationships (Jouppila and Roos, 1997). The Guggenheim–Anderson–De Boer (GAB) model is often reported as the choice as it successfully fits to the water content against a_w over a wide a_w range (Roos and Drusch, 2015). The GAB model may be used to establish critical a_w and water content values for dairy powders. Such critical values indicate a_w and corresponding water content limits where the T_g of lactose as a result of water plasticization is depressed to below ambient temperature as shown in Figure 3.1.5. It should be noted that the critical water activity of amorphous lactose at normal ambient temperature is 0.38 a_w. At a higher a_w lactose glass transition and crystallization from the amorphous state are observed. Such crystallization of amorphous lactose appears in sorption isotherms as a break and reduced water contents irrespective of a higher a_w which is typical of water sorption by dairy powders (Berlin et al., 1968a; Roos and Karel, 1991; Jouppila and Roos, 1994b; Jouppila et al., 1997; Roos and Drusch, 2015).

3.1.4 *Dehydration*

The dehydration of dairy liquids commonly involves evaporation and spray drying. The prerequisite for manufacturing powders from dairy solids using spray drying is that almost anhydrous solids are obtained from atomized particles which must exhibit characteristics of solid materials. For example, the constituents of skim milk solids are primarily lactose and milk proteins. Jouppila and Roos (1994a, 1994b) showed that the glass transition measured for milk solids followed that of lactose. On the other hand, proteins, as high-molecular-mass components, are not expected to show glass transitions at lower temperatures than lactose. Furthermore, it appears that the lactose and protein components in dairy solids are phase separated and poorly miscible (Silalai and Roos, 2010). Lactose glass

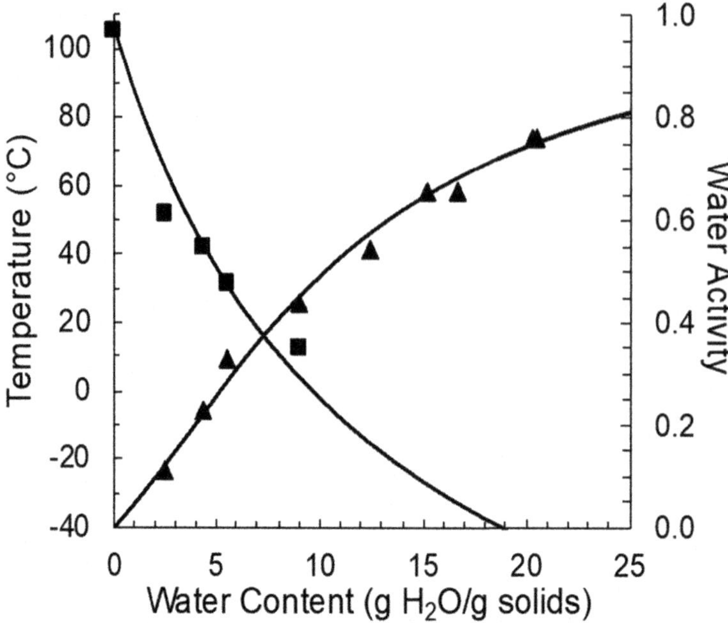

Figure 3.1.5 Water sorption and glass transition of amorphous lactose. A critical water activity can be taken as glass transition water activity at water sorption temperature corresponding to critical water content.

transition even above 5% water remains above typical ambient temperatures, and particles of dairy liquids can be converted by dehydration to a mix of noncrystalline, solid (glassy) lactose and milk proteins, i.e., milk powder.

Lactose in a dairy liquid must enter the glassy state on the surface of drying particles to avoid stickiness during spray drying as well as full vitrification prior to powder outlet and collection at the end of the dehydration process. The temperature–water content relationships for a successful spray drying can be schematically described using Figure 3.1.6. A similar approach for the control of temperature and water content relationships can be used in the control of other dehydration processes involving glass-forming carbohydrates and sugars in feed liquids. It is important to note that the success of a spray-drying process depends on the solidification properties of the drying particle components (Roos and Drusch, 2015). Some typical glass-formers in food liquids with respective glass transition temperature data are given in Table 3.1.1. The glass transition of fructose and glucose with even traces of water is lower than normal ambient temperatures. Consequently, liquids such as fruit juices cannot be spray dried without the use of drying aids, such as maltodextrins of higher glass transition temperature. The dehydration of lactose-hydrolyzed dairy liquids is also problematic because of the low glass transition temperatures of galactose and glucose.

The glass transition temperature of dairy powder components, particularly lactose, is often a significant determinant of powder stability. Well-known water content–dependent powder properties, such as stickiness and caking, can be related to the water plasticization of powder components. Stickiness of dried particles often occurs at 20°C above the onset temperature of glass transition (Roos and Drusch, 2015). Other time-dependent changes

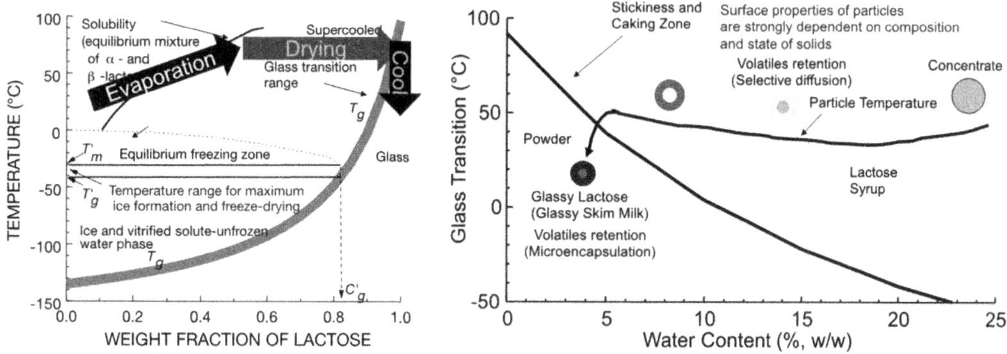

Figure 3.1.6 State diagram of amorphous lactose and water content–glass transition profile for lactose containing materials in spray drying.

resulting from an increase of temperature or water content include the crystallization of glass-forming components and browning reactions (Buera et al., 2011).

3.1.5 Structural relaxations and strength

Glass transition is a property of a metastable, nonequilibrium, noncrystalline materials. Such materials exhibit a thermodynamic driving force towards the equilibrium state. However, glass-forming molecules become frozen and immobilized in the glassy state which gives the materials solid-like properties. The glass transition is a reversible change in state: below glass transition translational mobility is inhibited and solid-like, brittle and often transparent glassy material properties dominate, while above glass transition an increasing thermal or water plasticization results in the exponential increase of translational molecular mobility and the loss of solid-like characteristics at the expense of the appearance of liquid-like viscous flow and rapidly increasing liquid characteristics.

Table 3.1.1 Glass transition temperatures, T_g, with onset of ice melting in a maximally freeze-concentrated solution, T_m', and glass transition temperature of maximally freeze-concentrated solute phase, T_g', for common glass-forming sugar and polyol components in foods (Roos, 1993)

Material	T_g (°C)	T_m' (°C)	T_g' (°C)
Fructose	5	–46	-57
Glucose	31	–46	-57
Lactose	105	–30	-41
Sucrose	70	–34	-46
Maltodextrin DE10	160	–13	-23
Starch	250	–6	-6
Sorbitol	–9	–49	-63
Xylitol	–29	–57	-72

Note: Glass transition temperatures are reported as onset temperatures during heating using differential scanning calorimetry.

Quite often one can assume stability and good flow properties of particles with components remaining in the glassy state. Structural relaxation times decrease rapidly over the glass transition temperature range and a critical decrease can be assumed at a four-decade reduction from 100 s in the glassy state to 0.01 s above the glass transition. Structural relaxation times less than 0.01 s result in stickiness, caking and significantly increasing rates of other physicochemical changes. Roos and Drusch (2015) described the use of the Williams–Landel–Ferry relationship of structural relaxation times and temperature to derive critical temperature–relaxation times relationships defined as strength, S, for food solids. Indeed, a number of studies have shown that S is dependent on the molecular size and components in blends, such as carbohydrates and proteins, besides the level of water plasticization (Maidannyk and Roos, 2016; Fan and Roos, 2017). Conversely, formulations using dairy solids and typical components of formulated dairy-based materials can be designed for dehydration and post-dehydration processes and stability control using relaxation times measurements and S determination at various water contents.

The significant differences in glass transition and strength measurements are related to rates of changes, particularly during the dehydration and storage of dried solids. Glass transition measurements provide a temperature, T_g, which is critical in the assessment of the suitability of materials for dehydration. Conversely, S values provide information on rates of changes around and above the glass transition. Such information is important in reducing stickiness in spray drying by using formulations increasing solid strength. S may also be used as a tool to manipulate the water plasticization of dried materials and thereby caking of powders or overall storage stability including physicochemical changes, such as nonenzymatic browning, of food solids.

3.1.6 Conclusions

The water content, water activity and glass transition of dairy solid components are fundamental properties affecting required dehydration and storage conditions as well as the selection of appropriate packaging materials. The glass transition of lactose and other carbohydrates often determines time–temperature–water content relationships of the successful manufacturing of dairy powders and storage stability. Furthermore, the crystallization of lactose depends on water content and the level of water plasticization at a specific storage temperature. Formulated materials can be designed for desired dehydration and storage characteristics using knowledge of structural relaxation times above glass transition based on glass transition and strength measurements.

References

Berlin, E., Anderson, B.A. and Pallansch, M.J. 1968a. Water vapor sorption properties of various dried milks and wheys. *J. Dairy Sci.* 51(9): 1339–1344.

Berlin, E., Anderson, B.A. and Pallansch, M.J. 1968b. Comparison of water vapor sorption by milk powder components. *J. Dairy Sci.* 51(12): 1912–1915.

Bhandari, B.R. and Howes, T. 1999. Implication of glass transition for the drying and stability of dried foods. *J. Food Eng.* 40(1–2): 71–79.

Buera, M.P., Roos, Y.H., Levine, H., Slade, L., Corti, H.R., Auffret, T. and Angell, C.A. 2011. State diagrams for improving processing and storage of foods, biological materials and pharmaceuticals. *Pure Appl. Chem.* 83(8): 1567–1617.

Carpin, M., Bertelsen, H., Bech, J.K., Jeantet, R., Risbo, J. and Schuck, P. 2016. Caking of lactose: A critical review. *Trends Food Sci. Technol.* 53: 1–12.

Fan, F. and Roos, Y.H. 2015. X-ray diffraction analysis of lactose crystallization in freeze-dried lactose–whey protein systems. *Food Res. Int.* 67: 1–11.

Fan, F. and Roos, Y.H. 2017. Glass transition-associated structural relaxations and applications of relaxation times in amorphous food solids: A review. *Food Eng. Rev.* 9(4): 257–270.

Foster, K.D., Bronlund, J.E. and Paterson, A.H.J. 2005. The prediction of moisture sorption isotherms for dairy powders. *Int. Dairy J.* 15(4): 411–418.

Herrington, B.L. 1934. Some physico-chemical properties of lactose. I. The spontaneous crystallization of supersaturated solutions of lactose. *J. Dairy Sci.* 17(7): 501–518.

Jouppila, K. and Roos, Y.H. 1994a. Water sorption and time-dependent phenomena of milk powders. *J. Dairy Sci.* 77(7): 1798–1808.

Jouppila, K. and Roos, Y.H. 1994b. Glass transitions and crystallization in milk powders. *J. Dairy Sci.* 77(10): 2907–2915.

Jouppila, K. and Roos, Y.H. 1997. Water sorption isotherms of dehydrated milk products: Applicability of linear and nonlinear regression analysis in modeling. *Int. J. Food Sci. Technol.* 32(6): 459–471.

Jouppila, K., Kansikas, J. and Roos, Y.H. 1997. Glass transition, water plasticization and lactose crystallization in skim milk powder. *J. Dairy Sci.* 80(12): 3152–3160.

Kilnburn, D., Townrow, S., Meunier, V., Richardson, R., Alam, A. and Ubbink, J. 2006. Organization and mobility of water in amorphous and crystalline trehalose. *Nat. Mater.* 5(8): 632–635.

King, N. 1965. The physical structure of dried milk. *Dairy Sci. Abstr.* 27: 91–104.

Labuza, T.P. 1977. The properties of water in relationship to water binding in foods: A review. *J. Food Process. Preserv.* 1: 167–190.

Maidannyk, V.A. and Roos, Y.H. 2016. Modification of the WLF model for characterization of the relaxation time-temperature relationship in trehalose-whey protein isolate systems. *J. Food Eng.* 188: 21–31.

Potes, N., Kerry, J.P. and Roos, Y.H. 2012. Additivity of water sorption, alpha-relaxations and crystallization inhibition in lactose–maltodextrin systems. *Carbohydr. Polym.* 89: 1050–1059.

Roos, Y.H. 2008. The glassy state. In *Food Materials Science*. Eds. Aguilera, J.M. and Lillford, P.J. New York: Springer.

Roos, Y. 1993. Melting and glass transitions of low molecular weight carbohydrates. *Carbohydr. Res.* 238: 39–48.

Roos, Y.H. and Drusch, S. 2015. *Phase Transitions in Foods*. San Diego: Academic Press.

Roos, Y. and Karel, M. 1991. Plasticizing effect of water on thermal behavior and crystallization of amorphous food models. *J. Food Sci.* 56(1): 38–43.

Roos, Y. and Karel, M. 1992. Crystallization of amorphous lactose. *J. Food Sci.* 57(3): 775–777.

Saltmarch, M. and Labuza, T.P. 1980. Influence of relative humidity on the physicochemical state of lactose in spray-dried sweet whey powders. *J. Food Sci.* 45(5): 1231–1236, 1242.

Schuck, P., Blanchard, E., Dolivet, A., Méjean, S., Onillon, E. and Jeantet, R. 2005. Water activity and glass transition in dairy ingredients. *Lait* 85(4–5): 295–304.

Shrestha, A.K., Howes, T., Adhikari, B.P., Wood, B.J. and Bhandari, B.R. 2007. Effect of protein concentration on the surface composition, water sorption and glass transition temperature of spray-dried skim milk powders. *Food Chem.* 104(4): 1436–1444.

Silalai, N. and Roos, Y.H. 2010. Roles of water and solids composition in the control of glass transition and stickiness of milk powders. *J. Food Sci.* 75(5): E285–E296.

Supplee, G.C. 1926. Humidity equilibria of milk powders. *J. Dairy Sci.* 9(1): 50–61.

Troy, H.C. and Sharp, P.F. 1930. α and β lactose in some milk products. *J. Dairy Sci.* 13(2): 140–157.

Vuataz, G. 2002. The phase diagram of milk: A new tool for optimising the drying process. *Le Lait* 82(4): 485–500.

3.2 Caking of dairy powders

Zahra Afrassiabian and Khashayar Saleh

3.2.1 Introduction

Dairy powders are products of the dehydration of dairy products (i.e., mammalian milk and its derivatives such as (ice) cream, cheese, yogurt or whey). They constitute a key ingredient in a number of foodstuffs, which are high in protein and essential vitamins. Some examples of dairy powders are given in Table 3.2.1 [1].

Dairy powders are widely used in everyday consumer products and in the food industry, including bakery, confectionery, ready meals and cooking aids, recombined milk and nutritional products. Their economic importance is undeniable and has been growing steadily for many years. For instance, regarding milk powder only (including whole milk, skimmed milk, buttermilk, fat-filled milk, whiteners and others), the global market was valued at about $28 billion in 2017 and is expected to grow at an annual rate of 4.4% to reach $38 billion by 2025 [2].

The industrial context of the dairy products is tightly connected to the key factors that govern the food industry in general. Indeed, faced with an increasingly competitive global context and, in order to maintain their leadership over emerging competitors, all market leaders are forced to integrate a quality approach based on customer satisfaction and the development of innovative products with higher added-value. Nowadays, these products must be more elaborated, more functional and, sometimes, more fanciful. For example, a dairy powder must have good physical, chemical and biological stability, good flowability, adequate organoleptic and nutritional properties, good reconstitution ability, etc.

Dairy powders are obtained by the dehydration of their original wet counterpart (e.g., milk, cream, cheese, whey) by evaporating the water using three main processes, namely, spray drying, roller drying or freeze-drying. They can be used as such, or can be formulated with other ingredients and additives (e.g., whiteners, which could contain, in addition to the milk powder, up to 20% of added sugars or other additives to reduce the acidity of the coffee). Powdering dairy products has several advantages including:

- Better preservation and far longer shelf life than wet dairy products due to low moisture content and subsequently low water activity of powders.
- Respond to the seasonality of production to better correspond to market demand, which is not periodic but constant.

However, despite the numerous benefits of powders, their handling and storage often cause some severe challenges. Among these problems, undesired and uncontrolled

Table 3.2.1 Some examples of dairy powders

Butter and buttermilk powder	Filled milk powder	Milk (skim and whole) powder
Caseinates	Infant formulae	Rennet casein
Cheese powder	Ice-cream powder	Whey powders

agglomeration of powders known as "caking" is undoubtedly one of the most important and can be detrimental to the quality of products.

This chapter deals with the caking of powders and presents the main mechanisms of caking, in relation to the properties of the products and the operating conditions, as well as the methods that can be used to assess and predict the risk of caking. Product–properties–process interactions in conjunction with the advent of caking are also discussed.

3.2.2 *Caking: definition, context, main issues*

Caking can be defined as the spontaneous and undesired formation of a coherent mass from individual grains. This unwanted agglomeration of the particles of a powder is often irreversible and difficult to predict.

Caking is a common root of many problems encountered when handling powders and can lead to unplanned shutdowns, client claims and product recycling/rejection due to the clogging of conveying lines, feeding devices and storage silos. From a product quality point of view, caked products have a penalizing appearance and a poor flowability. In addition to these problems related to the appearance and handling of powders, caking leads, in most cases, to a degradation of the end-use properties of powders. In this case, the powder no longer meets the product/process specifications, becomes unusable and is returned by customers, sometimes by whole cargoes.

The caking phenomenon occurs as a result of the formation of material bonds at the points of contact between particles. These links, whose fundamental mechanisms of appearance are still poorly understood, are reinforced by pressure, the migration of matter and variations in humidity and temperature. In addition to the problems caused by the recycling of degraded products (if this is even possible), the negative impacts of these incidents can be significant. The costs related to the caking can sometimes be important when the destruction of the product is unavoidable, and much more so if one also takes into account the costs relative to the recovery of unsatisfactory product.

However, the solutions proposed to avoid caking are far from satisfactory and are rather curative than preventive (sieving, crumbling, declogging, etc.). The rare preventive actions require the implementation of complex formulations based on anti-caking agents. In addition to the direct costs related to the purchase of these additives, these solutions require the implementation of additional operations (storage, transport, mixing, coating, etc.) as well as a preliminary development step to optimize the proposed formulation.

In this context, it is necessary to understand the theoretical aspects of the physicochemical phenomena occurring during the caking of the powders to bring reasoned, rational and sustainable solutions to this problem.

3.2.3 *Influential parameters and main mechanisms*

The caking of granular materials results from variations in environmental conditions, in particular those of temperature and relative humidity (the two parameters being linked elsewhere). These parameters vary with the weather (day/night, summer/winter) and the geographical location in which the product is stored and transported. These variations are sometimes significant and can largely exceed the limit conditions necessary for a good conservation of the product (e.g., the critical relative humidity of deliquescence, DRH, or the glass transition temperature, T_g).

Caking is usually induced by one or more of the following factors:

- Increase in pressure
- Increase in temperature beyond the melting temperature or the softening temperature of materials
- Presence of moisture in the granular medium

However, beyond these macroscopic and general aspects, the elementary mechanisms underlying the caking phenomenon differ substantially according to the physical properties (e.g., particle shape and size distribution), the molecular structure and, in particular, the solubility and the crystallinity of materials. In practice, the presence of vapor or liquid water considerably affects the caking process. Nevertheless, the underlying elementary mechanisms of caking may be substantially different (capillary condensation, dissolution/recrystallization, deliquescence/efflorescence, phase transition, etc.) according to the nature of product (crystalline or amorphous, soluble or insoluble, hydrophilic or hydrophobic, etc.) [3–14].

The caking phenomenon occurs when the interparticle adhesion forces are predominant forces within granular media. According to Rumpf [15], the presence or absence of material bridges can be considered as the overriding criterion for classifying these forces.

The non-material forces can be classified in four categories:

- The forces of entanglement and mechanical attachment which exist when using excipients with particular structures (filamentous or fibrous structure).
- The attractive van der Waals (vdW) forces. These forces of electromagnetic origin are omnipresent. Their field of action is restricted. The particles must be closely spaced (<50 nm) and very small (<100 μm), so that the intermolecular forces can prevail against the gravity.
- Electrostatic forces, which usually exist in systems where particles are in motion and appear when electrostatic charges are present. These forces decrease sharply with temperature or ambient humidity.
- Magnetic forces, which intervene only for metallic and magnetic powders.

The forces with material links are also divided into four categories:

- Adhesion forces in static or liquid bridges with restricted mobility. These forces are due to the presence of mono- or multi-molecular adsorbed layers strongly bonded to the surface of particles by van der Waals forces.
- Interfacial forces and capillary pressure on mobile liquid bridges. The resulting tensile strength of the granule depends on the quantity of liquid present in the interparticle space of the particle cluster.
- Adhesion forces due to solid bridges. Solid bridges are formed by crystallization or by the drying of dissolved substance present in liquid bridges. Little attention has been paid to the theoretical description of solid bridges, and in most cases the importance of these forces is estimated experimentally. Solid bridges have a much higher strength than liquid bridges.
- Adhesion forces due to sintered bridges. These bridges result from creep or partial melting of the solid handled on its surface during high-temperature operations. In general, a temperature above the glass transition temperature and/or 60% of the melting point of particles must be attained. This type of bond, which is considered to

be responsible for the caking of amorphous products, leads to the formation of very resistant agglomerates.

3.2.4 Classification and mechanisms of caking

The main caking mechanisms have been the subject of several bibliographical reviews. In his book entitled *Cake formation in particulate systems*, Griffith [16] gave a rather qualitative description of these mechanisms. He distinguished four main classes of caking based on the involved binding forces, namely: mechanical caking, plastic flow caking, chemical caking and electrical caking. Cleaver also provided an overview of the responsible mechanisms of powder caking [17]. More recently, Zafar et al. [18] established a review on the caking of powders, synthesizing the main works on this subject. More specifically, Hartmann and Palzer [19] focused on the caking of amorphous powders. They described the dynamics of the process based on a plastic creep of amorphous materials beyond their glass transition temperature. Finally, in a recent work, Carpin et al. [20] drew up a critical review on the caking of lactose.

A synthesis of these studies shows that it is difficult to establish a universal classification for the different types of caking. Indeed, the caking phenomenon is multi-dimensional and can be seen according to different criteria:

- The nature of the materials, which may be amorphous, polymorphous, semi-crystalline, crystalline, etc.
- The interparticle forces that cause the particles agglomeration.
- The phenomenology of the process and the different steps involved (capillary condensation, dissolution/drying, deliquescence/efflorescence, fusion/solidification, etc.).
- The initiating factors of caking (pressure, relative humidity, temperature). However, it is important to distinguish between the main factor and other influencing factors. For example, in the case of a hygroscopic product, the main factor initiating the caking is the relative humidity, and the temperature comes into play through its effect on RH. On the other hand, in the case of a plastic creep (as is the case of amorphous products), the main factor is the temperature and the humidity plays a role by its effect on the viscosity of the material.

Other criteria exist but they are of lesser importance. In what follows, we establish our classification based on the main factor triggering the caking. This choice is made because this classification best corresponds to the nature of dairy powders and their composition. In this regard, it is important to highlight the special role of water. Generally, whatever the main mechanism of caking is, the presence of water either in vapor, liquid, adsorbed or absorbed state could largely intensify the caking process. However, this effect arises from processes that can vary completely by nature. For example, in the case of insoluble materials, caking occurs as a result of water condensation at contact points between particles, whereas for amorphous materials, the presence of absorbed water decreases the glass transition temperature, T_g, to a level lower than the storage temperature, bringing the material to a rubbery state which is subject to caking [19, 21–24]. However, in this case, the process should not be considered as "wet" but as "thermal" caking. The water will then act as a plasticizer in the case of amorphous materials, as a binder for insoluble solids or as a solvent for crystalline powders.

3.2.5 Mechanical (pressure-induced) caking

This type of caking can be included in the broader category of caking in the absence of material bridges. In this case, the caking occurs following the rearrangement and consolidation of powder under pressure (weight, compression, etc.). The cohesion of the cake is ensured by van der Waals forces. This process is similar to the early stages of dry granulation in which the interparticle distance decreases and the number of contacts as well as the contact area between the particles increase. This results in a decrease in the void fraction and an increase in the cohesion of the powder. However, the pressures involved are much lower than those exerted during dry granulation and the compaction and consolidation levels obtained are significantly lower. Only rearrangement and dense packing stages are present during dry caking.

Generally, any variable that affects van der Waals forces (gap, contact surface area) or the number of contacts (particle shape and size distribution, packing, pressure, deformability, roughness, etc.) plays a role in dry caking. Furthermore, environmental conditions (RH, T) or additional forces (e.g., electrostatic or capillary) could accentuate the caking.

3.2.6 Wet (or RH-induced) caking

This category includes all cases where moisture is directly involved in the process. This may appear due to the formation of liquid bridges leading to adsorption and capillary condensation phenomena. Further dissolution and drying of the solid, if the latter is hydrosoluble, lead to the appearance of solid bridges.

3.2.6.1 Origin and emergence of water within granular media

The common basis of wet caking is the presence of liquid water within the granular medium. Initially, the water can be contained either in the particles themselves (incomplete drying) or in the ambient air. Liquid water could then appear from the ambient humidity as a consequence of water sorption and capillary condensation phenomena.

3.2.6.1.1 Water sorption The water vapor from the surrounding air could be fixed on the particles' surface by adsorption due to intermolecular interactions. The amount of adsorbed water in equilibrium with the relative humidity of air is given by the sorption isotherm, where the water content of a solid is plotted against the water activity (i.e., relative humidity at equilibrium). A typical example of a sorption isotherm is shown in Figure 3.2.1.

At low water activities (i.e., at early stages of adsorption – generally between 5% and 40%) the solid's surface is covered by a few layers of water molecules. This phase is characterized by a linear and slight increase in the water sorption isotherm and can be modeled by classical BET or GAB models. Indeed, the amount of adsorbed water in this zone is low, but even these small amounts could reinforce vdW forces because of strong interactions between water and solid molecules. However, as long as the adsorption takes place in this zone, the main caking mechanism remains similar to dry caking except that the extent of the vdW forces is greater than for dry powder.

At higher water activities, water molecules become mobile enough to form liquid bridges. In addition, a capillary condensation phenomenon occurs, leading to the filling of small gaps at contact points between particles. In this zone, all spaces that could be

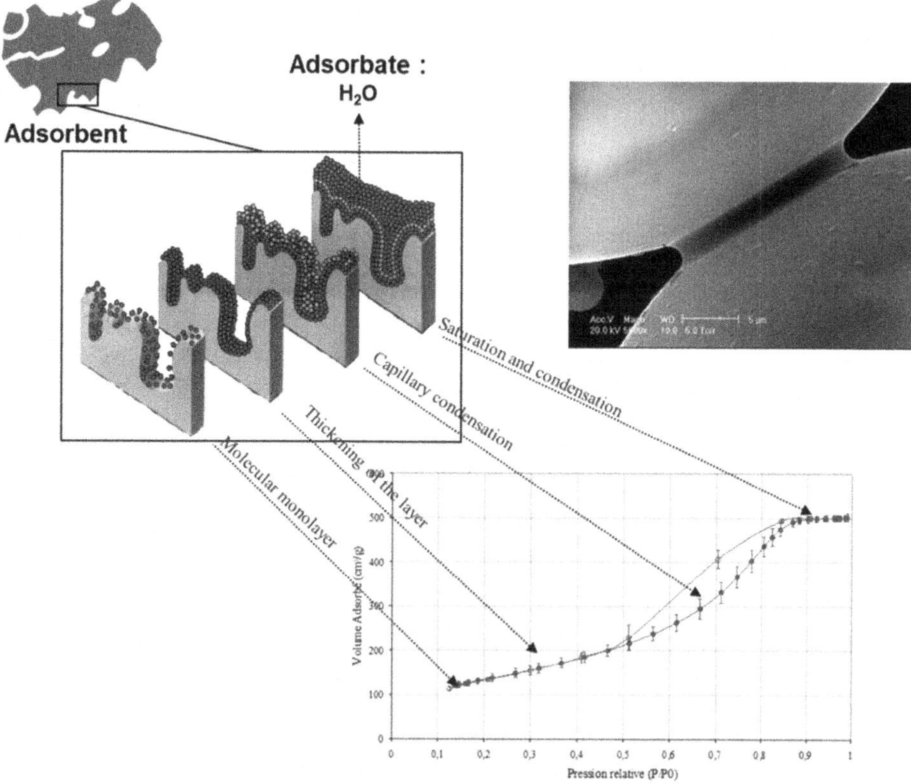

Figure 3.2.1 Example of a water sorption isotherm and liquid bridge formation due to capillary condensation (ESEM image: SAPC/UTC).

considered as mesopores (pore size < 50 nanometers), such as particle porosity and asperities, are filled (Figure 3.2.1).

3.2.6.1.2 Capillary condensation The capillary condensation phenomenon occurs in very confined spaces of nanometric size, such as internal porosities (mesopores), contact points between particles or surface roughnesses. It is well-established that at a given temperature, a vapor phase present in such a space can condensate at a partial pressure lower than its normal vapor pressure. In other words, the water activity (the ratio between the equilibrium vapor pressure in the pores and the normal vapor tension) is less than unity. This decrease in vapor tension can be related to the equivalent pore radius, r_k, by the Kelvin model:

$$\ln\frac{p_v}{p_v^*} = \frac{-2\upsilon_L \sigma_{LV}\cos\theta}{r_k RT} \tag{3.2.1}$$

where:

- p_v is the equilibrium vapor pressure.
- p_v^* is the normal vapor pressure at saturation.
- r_k is the mean equivalent capillary size.

- σ_{LV} is the liquid/vapor surface tension.
- υ_L is the liquid molar volume.
- R is the ideal gas constant.
- T is the absolute temperature.

According to this model, for a wetting system which is characterized by $\cos\theta > 0$, in a confined space, a vapor could condensate at vapor pressures lower than its normal vapor pressure and then at water activities smaller than unity. This relation shows that the water activity inside pores decreases exponentially with the reduction of the equivalent pore radius and with the increase of the wettability of the solid. The water activity is very affected within a part of mesopores (2 to 20 nm) [25]. So, in a large space the water vapor begins to condensate at a relative humidity of 100%, but in a pore of 10 nm in a completely hydrophilic solid ($\theta = 0$), it condensates from a relative humidity as low as 80%.

Furthermore, the presence of liquid water bridges between particles results in capillary forces, which are attractive for hydrophilic systems. These forces are amenable to a theoretical description for simple geometries by applying the Laplace theory for capillary forces and have been extensively described in the literature [14, 26–29].

3.2.6.2 *Water–solid interactions*

Once liquid water is present, water–solid interactions of different types come into play. Their extent depends on the amount of water and the chemical nature as well as the physico-chemical properties of the solids that define their affinity with respect to water. In fact, three categories of solids in the ascending order of the intensity of their interactions with water are distinguished, namely insoluble, soluble (but not deliquescent) and hygroscopic (deliquescent) solids.

3.2.6.2.1 Insoluble powders The main cause of caking for insoluble solids, including a large variety of mineral powders (talc, calcite, mica, alumina, etc.), is the formation of liquid bridges due to capillary condensation. The ESEM images presented in Figure 3.2.2 illustrate an example of capillary condensation in a sample of hydrophilic particles. This phenomenon leads to the formation of liquid bridges between particles. The intensity and

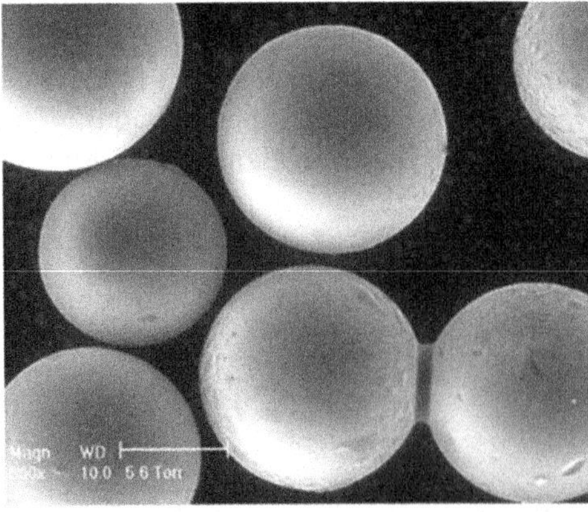

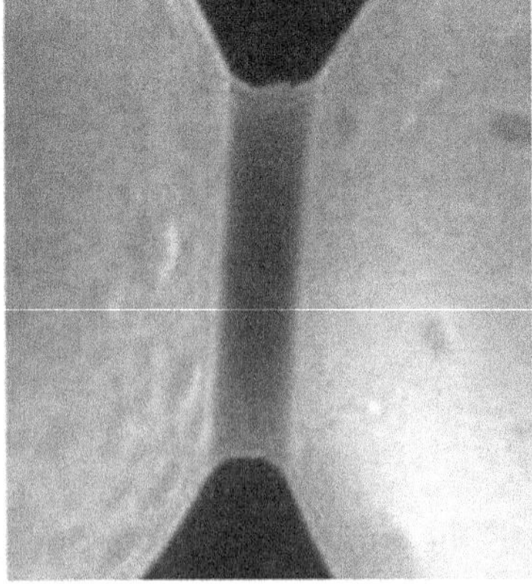

Figure 3.2.2 Capillary condensation in contact points between particles.

range of the connection forces induced by these bridges being several orders of magnitude greater than those of the van der Waals forces, this leads to a wet clogging of the powder.

3.2.6.2.2 Non-deliquescent water-soluble powders For hydrosoluble solids, local liquid bridges appearing after capillary condensation could dissolve a part of solids and form solid bridges after solidification by drying or recrystallization. An example of this kind of caking has been reported by Cleaver et al. [30] for boric acid which is a fairly soluble hygroscopic solid.

3.2.6.2.3 Deliquescent powders Problems are accentuated for soluble hygroscopic solids. In fact, an important but not universal feature of crystalline solids is the existence of a critical (deliquescence) relative humidity (DRH), also called the point of deliquescence. This point corresponds to the relative humidity at which the solid absorbs substantial amounts of water vapor. Generally, deliquescence refers to the conversion of a solid substance into a liquid as a result of the absorption of moisture or water vapor from the air. Typical examples of this kind of solid are sodium chloride, NaCl, and sucrose (which can be part of the formulation of some dairy powders).

Figure 3.2.3 shows a superposition of a sequence of ESEM images taken during a humidification/drying cycle of NaCl crystals on the water sorption isotherm of NaCl established by a dynamic vapor sorption (DVS) instrument. Three segments are distinguished during the sorption:

- Until DRH (76% RH), there is no significant water uptake, and only multilayer molecular adsorption and capillary condensation in very narrow spaces can take place in this zone.

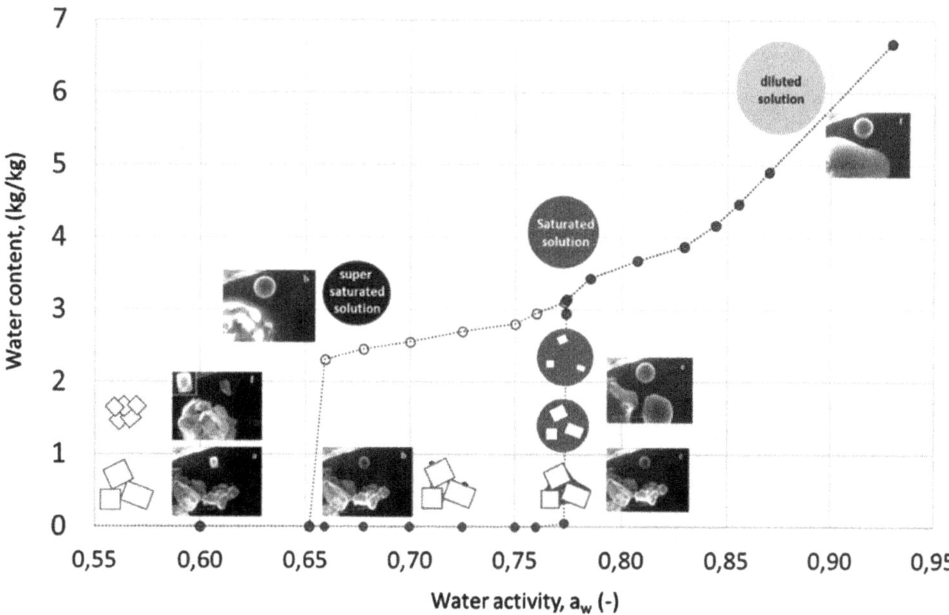

Figure 3.2.3 Sorption isotherm of NaCl at 25°C, and the sequence of deliquescence/efflorescence, Langlet et al. [31].

- From 76% to 77% RH, deliquescence happens and a sharp increase of water uptake is observed. The equilibrium point at 77% RH is a saturated aqueous solution with a concentration of 340 g NaCl/kg water.
- Above DRH, the aqueous solution of NaCl is diluted while increasing RH.

Three particular branches are also distinguished during desorption:

- From 93% to 77% RH, desorption points overlap the dilution curve until reaching saturation. Below 77% RH, the evaporation of water continues and the solution becomes supersaturated. The decrease is smooth and no sharp variation in water uptake is observed, suggesting that no recrystallization occurs. In fact, for crystallization to occur, the solution must reach a state of supersaturation.
- Between 66% and 65% RH, the desorption curve undergoes a sharp decrease due to the efflorescence of NaCl. The water is completely eliminated, and there is no residual water in the solids.

Consequently, starting from a dispersed powder, this procedure leads to a mass of particles linked by the appearance of solid bridges. It should be noted that in practice, these hydration/dehydration cycles could come from variations in atmospheric conditions.

It is also important to note that some solids could form amorphous phases during the drying. This is especially the case for viscous solutions. A good example of this type of solid is sucrose, which is also largely present in formulated dairy products. Figure 3.2.4 shows that the deliquescence of sucrose proceeds in the same manner as NaCl [11]. However, its behavior is very different during the drying phase because the solution becomes viscous and water is entrapped within the solid matrix. Therefore, in this case, departing from a crystalline and well-dispersed state, the product becomes a cake, which is partly amorphous.

The situation becomes more complicated when mixtures of two or more deliquescent components are involved. Several works [10, 31–34] have pointed out the existence of a

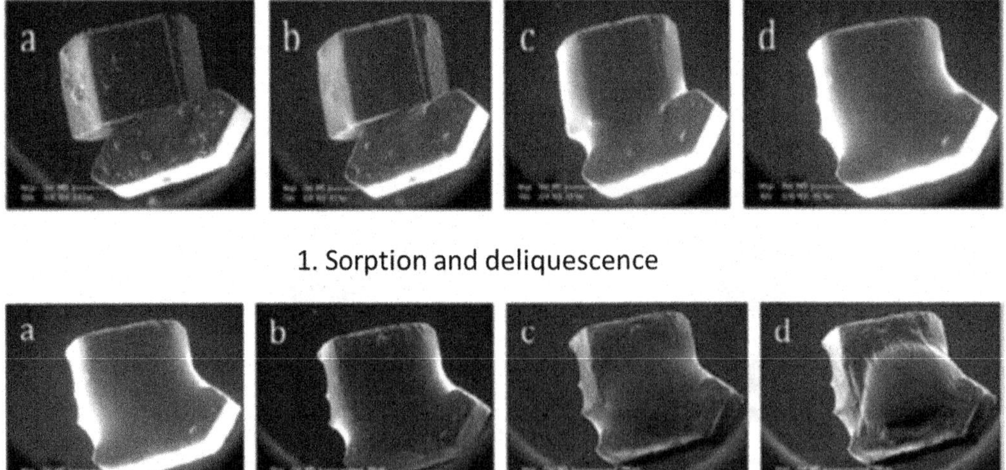

Figure 3.2.4 Time sequence of ESEM images of sucrose deliquescence (1) and efflorescence (2) [11].

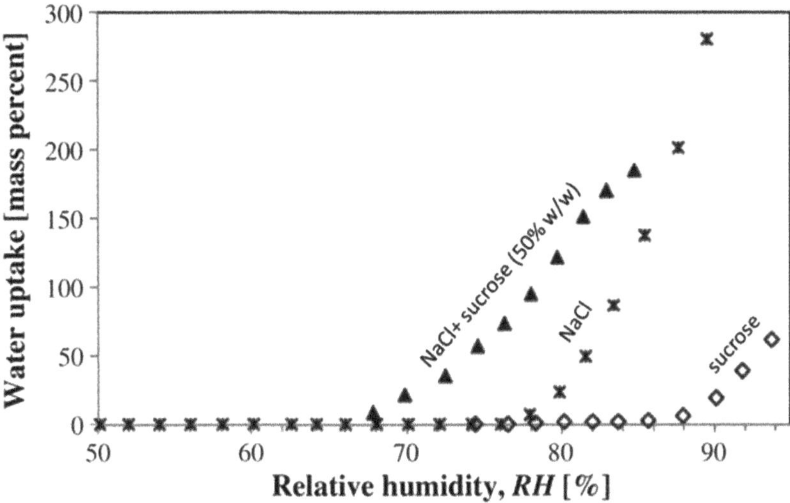

Figure 3.2.5 Water uptake of pure sucrose (◇), pure NaCl (x) and a mixture of NaCl-sucrose $f_{NaCl/sucrose} = 50$ (w/w) (▲) at varying relative humidity (from [32]).

mutual deliquescence relative humidity (MDRH). For example, Figure 3.2.5 shows the mutual deliquescence of a binary mixture of NaCl and sucrose (50% w/w). It can be seen that the mutual deliquescence takes place at an RH lower than the DRH of both salt and sucrose, which was equal to 76% for salt and 85% for sucrose.

This singular phenomenon, which is known as "deliquescence lowering," can be explained by some thermodynamic concepts to describe the water activity (a_w) in aqueous solutions of electrolytes or of organic substances.

In general, the storage of water-soluble crystals at relative humidity below their critical relative humidity, DRH, does not present major problems. On the other hand, an oscillation of the relative humidity around the DRH can lead to caking by the following elementary steps:

- Capillary condensation followed by the dissolution of the solid during humidification which can go as far as the deliquescence.
- Evaporation and recrystallization (efflorescence) during drying.

Regarding the mutual deliquescence, the main point is that the critical RH of the mixture is lower than that for the major component and even both components. For example, small amounts of salt in a sucrose solution could significantly lower the critical RH. The maximum effect (DRH = 64%) is obtained at the eutonic point which is reached at 16% NaCl [10].

3.2.7 *Thermal (temperature-induced) caking*

Another common caking mechanism is thermal caking. Involved mechanisms are:

- Solid-phase diffusion (the same as that involved in sandstone formation) which accelerates with temperature
- Phase transition mechanisms like melting, softening, change of crystalline phase

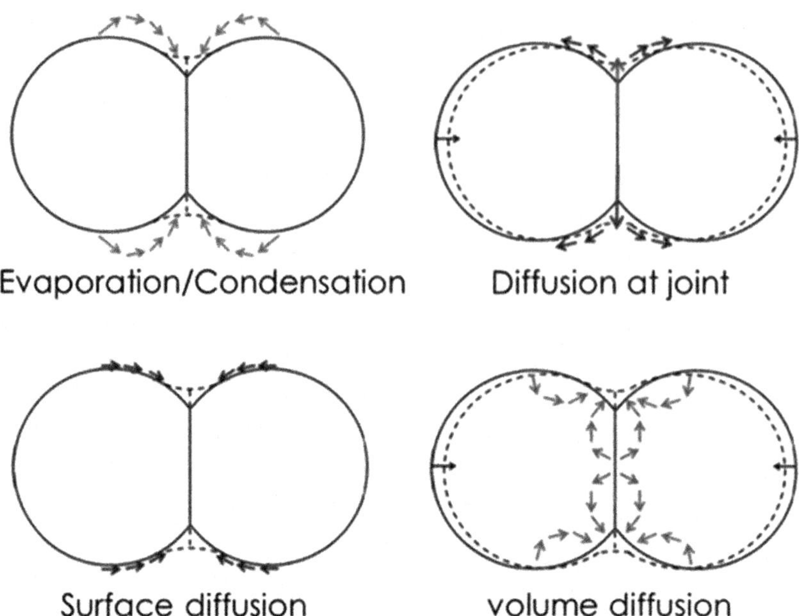

Figure 3.2.6 Mechanisms of solid sintering.

The common thermodynamic base for this kind of caking is the lowering of surface energy. Indeed, it is well-known that if two objects, which are free to move, are brought into contact, the system will tend towards minimum energy. This is, for example, what happens during the coalescence of two liquid droplets. Therefore, any factor that could make a solid flow tends toward this evolution.

One of these mechanisms is solid phase diffusion. This type of caking is very similar to solid phase sintering. The main mechanisms are presented in Figure 3.2.6 [35, 36].

All these mechanisms are limited by the molecular diffusivity which is highly temperature-dependent. Using a simple rule of thumb, if the time necessary for an atom to travel 1 μm is 3 months at 25°C, it is only a few hours at 90°C. For solids with a low melting point, even at ordinary temperatures, a partial melting of the solid could occur which increases the diffusivity and then accelerates considerably the caking. This is for example the case for cocoa powder whose melting temperature is low enough and which could cake under thermal stress at around 30°C.

Another important class of thermal caking concerns amorphous powders. Unlike crystalline materials, amorphous materials do not have critical RH. Their water content increases gradually with moisture but at a much lower level than that encountered for crystals. Although the possibility of wet caking for water-soluble amorphous powders cannot be completely ruled out, in their case, it is actually the temperature that is the key factor. A remarkable feature of amorphous materials is the existence of a so-called glass transition temperature, T_g, corresponding to the transition between the vitreous (glass) state and the rubbery state. An amorphous solid above its T_g is a soft solid (or hard liquid). In this so-called rubbery state, volume and surface diffusion mechanisms are accelerated. This process is analogous to the solid sintering process in the fields of metallurgy and plastics. This is an agglomeration process under the effect of heat at temperatures below the melting temperature of the material. A simple model to describe the kinetics of caking for two equal-sized particles was established by Rumpf's model, which considers the surface tension and the external pressure as the driving forces of caking and the viscosity as the opposing force [19, 23, 24]. According to this model, the extent of sintering, characterized by the ratio of the diameter of the bridge to the particle diameter, is:

$$\left(\frac{x}{d}\right)^2 = \left(\frac{4}{5}\cdot\frac{\gamma}{d}+\frac{2F_t}{5\pi d^2}\right)\frac{t}{\mu} \tag{3.2.2}$$

where x is the diameter of the bridge, d is the particle diameter, γ is the surface tension of rubbery material, μ is the viscosity, F_t is the external force and t is time.

Note that more accurate models have been also proposed considering unequal-sized or multiparticle systems, but their resolution requires the use of sophisticated numerical methods [37, 38].

Regarding the caking of amorphous powders, several crucial consecutive or competitive phenomena should be remembered:

- The glass transition temperature decreases as the water content of the solid increases. This evolution is generally represented by the Gordon–Taylor (G–T) model:

$$T_g = \frac{x_1 T_{g1} + k x_2 T_{g2}}{x_1 + k x_2} \tag{3.2.3}$$

  where x_1 and x_2 are the mole fractions of the amorphous material and the water, T_{g1} and T_{g2}, their respective T_g and k is a constant.
- The viscosity of a material in a rubbery state decreases sensibly with the difference between T and T_g. This decrease in viscosity results in an increase in the mobility of the molecules that, in turn, gives rise to two caking mechanisms: plastic creep and solid bridge formation due to crystallization. Note that these mechanisms (and therefore, the sintering extent through the effect of viscosity according to Equation 3.2.2) accelerate exponentially by $T - T_g$. Indeed, according to the literature, the evolution of both viscosity and characteristic crystallization time, t_{cr}, with temperature can be conveniently expressed by the William–Lendel–Ferry (WLF) model [39]:

$$\log_{10}\left(\frac{\eta \text{ or } t_{cr}}{\eta_g \text{ or } t_{cr,g}}\right) = \frac{-C_1\left(T-T_g\right)}{C_2+\left(T-T_g\right)} \tag{3.2.4}$$

  Regarding the viscosity, Williams et al. [39] reported so-called "universal" constant values of 17.44 and 51.60 for C_1 and C_2, respectively.

  At their T_g, amorphous materials have a very high viscosity (≈ 10^{14} Pa.s) so that they appear as solids. As can be seen from the WLF model, the viscosity decreases rapidly to a critical value of about 10^7 Pa.s where they become sticky. According to the literature [40, 41], this critical viscosity is reached when $T - T_g$ reaches 10 to 20°C.

Therefore, on the one hand, an amorphous solid which is exposed to the relative humidity of surrounding air could absorb water until equilibrium (the amount of which is given by its sorption isotherm). On the other hand, an increase in the water content of the solid could decrease the glass transition temperature. Therefore, even at constant storage temperature, the T_g could change due to RH and fall to the storage temperature, and the caking could take place because of creep and/or crystallization phenomena. Note, in addition, that crystallization is accompanied by a release of excess water, which in turn enhances the caking.

Lactose, which is the main component of dairy powders, is a good example of this category of materials. Figure 3.2.7 shows the water sorption isotherm (inside box), as well as

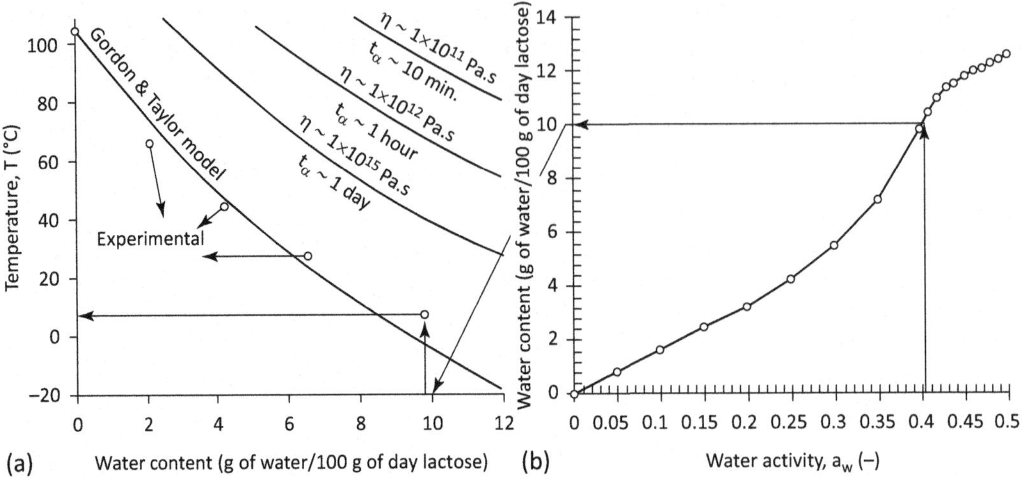

Figure 3.2.7 (a) Glass transition vs water content [21] and (b) the sorption isotherm [42] of pure amorphous lactose.

the variations of the T_g as a function of the water content for an amorphous lactose sample produced by freeze-drying. The indicative iso-η and iso-t_{cr} curves based on the data from the literature are also presented [42].

As can be seen, the presence of water, even in small quantities, leads to a sustained decay in T_g. For example, if an initially dry sample of lactose is exposed to moist air at 40% RH, its equilibrium water content would be about 10% which would lead to a sharp decrease in T_g: from 106°C (dry solid) to about 10°C (see the path of the arrows in Figure 3.2.7). Storage above this temperature would be detrimental to the quality of the product.

3.2.8 *Chemical (reaction-induced) caking (phase transition)*

Caking can also occur due to chemical bonding at contact points between particles. The chemical bond could result from a chemical reaction with oxygen or water (e.g., cement) or from a change of crystallinity of the solid. It should be noted that most polymorphic products may present caking risks due to phase changes. Indeed, during these changes, links can be created at the contact surfaces between particles. In addition, water can be released during amorphous to crystalline transition or from the dehydration reaction leading to the formation of liquid bridges. Lactose, which is the major component of dairy powders, is a very good example of this latter mechanism. Except the unstable anhydrous α-lactose, other lactose polymorphs are not hygroscopic and have a low solubility compared to other sugars. However, lactose could undergo several phase transitions at room temperature (Figure 3.2.8) [42].

More precisely, the transition of amorphous and anhydrous α- and β-lactose to α-monohydrate form, which occurs at room temperature and RH values exceeding 50%, could cause caking [42].

3.2.9 *Caking testing methods*

A multitude of methods has been reported in the literature to characterize the propensity of powders to cake. The test methods reported in the literature range from very basic tests (drop test) to much more sophisticated tests such as shear tests.

The common point of all these methods is that they include two stages:

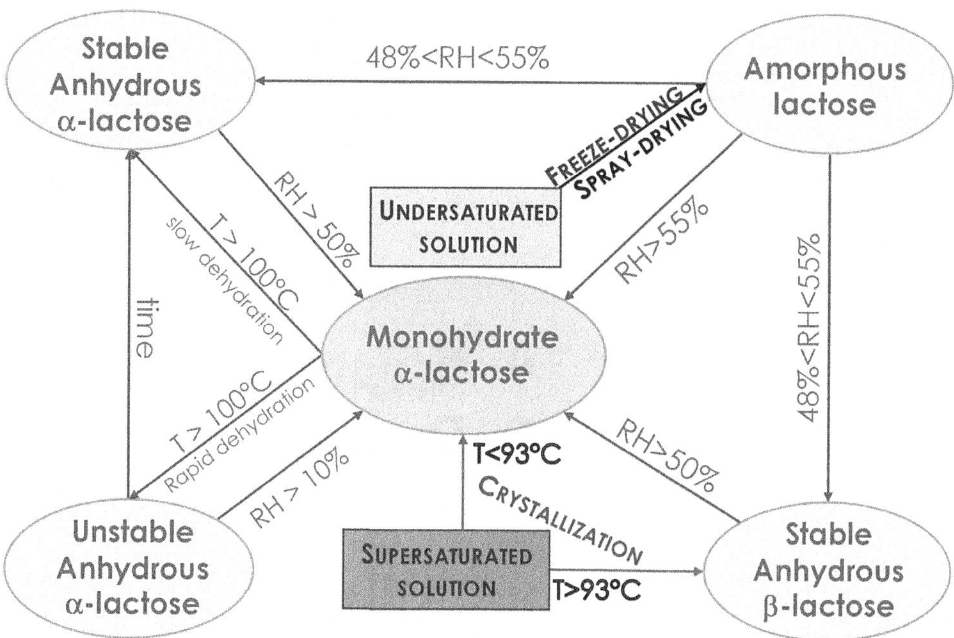

Figure 3.2.8 Solid-state transitions of lactose [42].

- The first step consists in placing the powder under conditions such that bonds will form between the particles. This can be done by applying a stress (formation of van der Waals bonds or solid bonds), by humidifying the powder (liquid bridges) or by subjecting the powder to a moisture and/or temperature cycle (formation of solid bridges).
- Once the powder has been caked, the sample can then be characterized mechanically by methods such as shear, compression, traction and indentation, which will be detailed below.

Samain [43] provided an extensive review of the caking tests by crossing the methods to prepare and those used to characterize the cakes. Zafar et al. [18] and Carpin et al. [20] reviewed the different tests reported in the literature. These authors made a classification of tests based on the method used to measure the mechanical resistance of cakes. They distinguished mechanical test methods including shear cell, uniaxial compression, tensile, ICI, penetration and powder rheometer tests as well as alternative tests (sensory, sieving, sticky point measurement, blow tester, microscopic observations, etc.). Because the caking is a slow process, the preparation of cakes is generally accelerated by prior wetting of powders or by using an air current instead of stagnant air.

3.2.10 *Special case of dairy and dairy-like emulsion powders*

Dairy powders are generally complex formulations including components involving almost all caking mechanisms. The main constituents of milk (i.e., caseins, lactose, whey proteins and fat) as well as additional components such as sugar, salt and oils are commonly present in dairy powders. The caking behavior of dairy powders does not depend as much on the overall composition of the particles but on the composition on their surface. The surface composition depends on operating conditions during the homogenizing of the emulsion and its drying. Vega and Roos [41] provided a review of spray-dried dairy

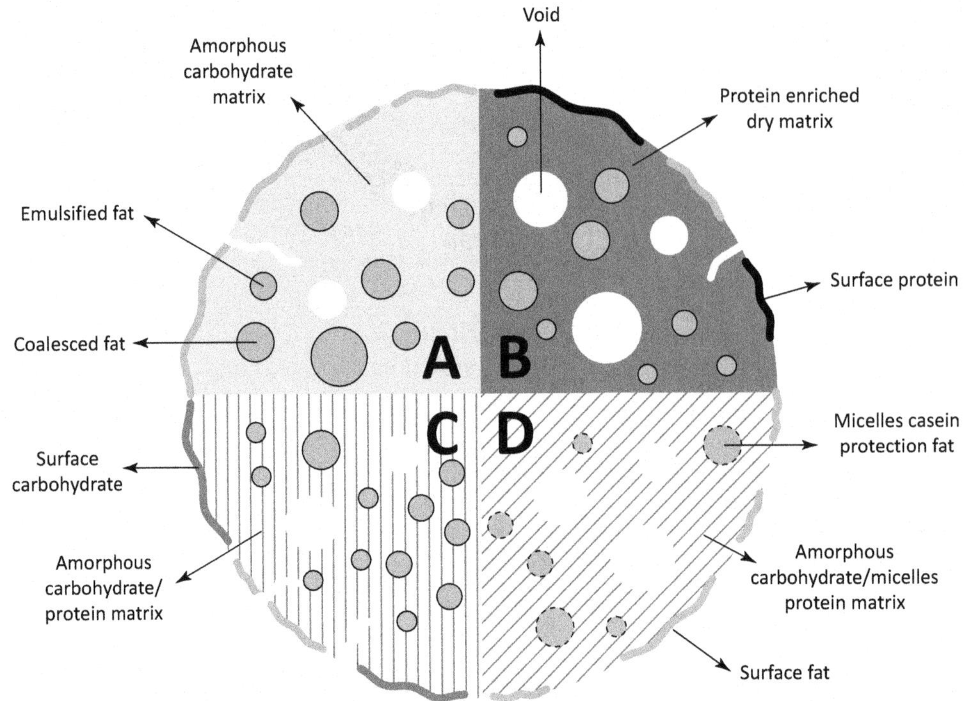

Figure 3.2.9 Surface and bulk composition of spray-dried dairy and dairy-like emulsions [41].

and dairy-like emulsions and highlighted four different compositions of dried particles as shown in Figure 3.2.9. According to this synthesis, the surface composition, and in particular the surface fat, could vary significantly: very high (>80%) for case A, low (~15%) for case B, medium to high (>50%) for case C and medium (~15%) for case D.

Regarding the caking of dairy powders, it is well-established that it is the presence of fat at the surface that predominates over other factors such as hygrometry, amorphous content and particle size [44]. Indeed, the surface fat can melt at low temperatures, ranging from 5 to 25°C, and form liquid bridges between particles.

For low fat contents, the plastic creep and crystallization of amorphous lactose or other carbohydrates come into play and besides temperature, RH should also be considered.

Finally, for sugar- and salt-containing formulations, (mutual) deliquescence could also take place.

3.2.11 *Conclusion*

The caking of powders is a very complex process bringing into play a multitude of interactions and mechanisms. To prevent caking, it is essential to better understand the underlying phenomena. Regarding dairy powders, the main phenomena are the melting of surface fat, the creep and/or crystallization of amorphous carbohydrates (lactose, maltodextrins, etc.) and the deliquescence of hygroscopic crystals (sugar, salt, etc.). The abundance of studies carried out on this subject has allowed considerable progress in understanding these phenomena. However, this knowledge, although very useful, is not sufficient to fully describe the caking phenomenon and prevent it. Indeed, for the moment, most works only give a static view of the phenomenon because they are based solely on thermodynamic equilibria. In reality, caking is a dynamic process that progresses slowly out of equilibrium. Unfortunately, the current state of knowledge does not allow predictions of the time that a powder would take to cake once the environmental conditions were favorable. Similarly, the role of the mechanical stresses on the powder is not yet well-established. To answer these questions, a detailed knowledge of the dynamics of the mechanisms involved at different scales (particles, agglomerates, heaps) with a view to their coupling in a dynamic model is required.

References

1. Kelly, A.L. and P.F. Fox, Manufacture and properties of dairy powders. In *Advanced dairy chemistry: Volume 1B: Proteins: Applied aspects*, P.L.H. McSweeney and J.A. O'Mahony, Editors. 2016, Springer New York: New York, NY. p. 1–33.
2. Bhandalkar, S. and D. Das, Milk powder market. 2019; Available from: https://www.alliedmarketresearch.com/milk-powder-market.
3. Rondeau, X., *Processus Physicochimiques intervenant dans le phénomène de mottage de solides divisés*. 2000, Université de Technologie de Compiègne. p. 259.
4. Affolter, C., *Etude des modes d'action d'additifs sur les phénomènes de dissolution et de cristallisation de sels hygroscopiques: Application au mottage du nitrate d'ammonium*. 2003, Université de Technologie de Compiègne.
5. Rondeau, X., C. Affolter, L. Komunjer, D. Clausse and P. Guigon, Experimental determination of capillary forces by crushing strength measurements. *Powder Technology*, 2003. **130**(1): p. 124–131.
6. Komunjer, L. and C. Affolter, Absorption–evaporation kinetics of water vapour on highly hygroscopic powder: Case of ammonium nitrate. *Powder Technology*, 2005. **157**(1): p. 67–71.
7. Dupas-Langlet, M., *De la déliquescence au mottage des poudres cristallines: Cas du chlorure de sodium*. 2013, Université de Technologie de Compiègne. p. 226.
8. Dupas-Langlet, M., M. Benali, I. Pezron, K. Saleh and L. Metlas-Komunjer, Deliquescence lowering in mixtures of NaCl and sucrose powders elucidated by modeling the water activity of corresponding solutions. *Journal of Food Engineering*, 2013. **115**(3): p. 391–397.
9. Dupas-Langlet, M., M. Benali, I. Pezron, K. Saleh and L. Metlas-Komunjer, The impact of deliquescence lowering on the caking of powder mixtures. *Powder Technology*, 2015. **270**(Part B): p. 502–509.
10. Dupas-Langlet, M., M. Benali, I. Pezron and K. Saleh, Characterization of saturated solutions and establishment of "aw-phase diagram" of ternary aqueous inorganic-organic and organic-organic systems. *Journal of Food Engineering*, 2017. **201**: p. 42–48.
11. Samain, S., M. Dupas-Langlet, M. Leturia, M. Benali and K. Saleh, Caking of sucrose: Elucidation of the drying kinetics according to the relative humidity by considering external and internal mass transfer. *Journal of Food Engineering*, 2017. **212**: p. 298–308.
12. Samain, S., Caractérisation multi-échelle de l'efflorescence et du mottage du saccharose. In *Génie des procédés industriels*. 2018, Université de Technologie de Compiègne, Compiègne, France.
13. Samain, S., M. Leturia, S. Mottelet, M. Benali and K. Saleh, Characterization of caking for crystalline materials: Comparison and statistical analysis of three mechanical tests. *Chemical Engineering Science*, 2019. **195**(23): 218–229.
14. Afrassiabian, Z., M. Leturia, M. Benali, M. Guessasma and K. Saleh, An overview of the role of capillary condensation in wet caking of powders. *Chemical Engineering Research and Design*, 2016. **110**: p. 245–254.
15. Rumpf, H., The strength of granules and agglomerates. In *Agglomeration*, W.A. Knepper, Editor. 1962, John Wiley, New York. p. 379–418.
16. Griffith, E.J., *Cake formation in particulate systems*. 1991, VCH Publishers Inc., New York.
17. Cleaver, J. *Powder Caking – An Overview of the Responsible Mechanisms*, 2007. **79**(9): p. 1387.
18. Zafar, U., V. Vivacqua, G. Calvert, M. Ghadiri and J.A.S. Cleaver, A review of bulk powder caking. *Powder Technology*, 2017. **313**: p. 389–401.
19. Hartmann, M. and S. Palzer, Caking of amorphous powders—Material aspects, modelling and applications. *Powder Technology*, 2011. **206**(1–2): p. 112–121.
20. Carpin, M., H. Bertelsen, J.K. Bech, R. Jeantet, J. Risbo and P. Schuck, Caking of lactose: A critical review. *Trends in Food Science and Technology*, 2016. **53**: p. 1–12.
21. Descamps, N., S. Palzer, Y.H. Roos and J.J. Fitzpatrick, Glass transition and flowability/caking behaviour of maltodextrin DE 21. *Journal of Food Engineering*, 2013. **119**(4): p. 809–813.

22. Descamps, N., S. Palzer and U. Zuercher, The amorphous state of spray-dried maltodextrin: Sub-sub-Tg enthalpy relaxation and impact of temperature and water annealing. *Carbohydrate Research*, 2009. **344**(1): p. 85–90.
23. Descamps, N., E. Schreyer and S. Palzer, Modeling the sintering of water soluble amorphous particles. In *PARTEC*. 2007, Nuremburg, Germany.
24. Palzer, S., The effect of glass transition on the desired and undesired agglomeration of amorphous food powders. *Chemical Engineering Science*, 2005. **60**(14): p. 3959–3968.
25. Gomez, F. and K. Saleh, Mise en forme des poudres Séchage par atomisation. *Principles, Techniques de l'ingénieur Opérations Unitaires: Évaporation et Séchage*, 2012. Base documentaire: TIB316DUO (ref. article: j2256).
26. Rumpf, H., Particle technology. In *Powder technology series*, 1990, Chapman and Hall, London.
27. Guessasma, M., H. Silva Tavares, Z. Afrassiabian and K. Saleh, Numerical modelling of powder caking at REV scale by using DEM. In *EPJ Web of Conferences*, Volume 140, 2017. *Powders and Grains 2017 – 8th International Conference on Micromechanics on Granular Media*, Montpellier, France.
28. Khashayar, S. and G. Pierre, Mise en œuvre des poudres Granulation humide: Bases et théorie. *Techniques de l'ingénieur Cosmétiques*, 2009. Base documentaire: TIB634DUO (ref. article: j2253).
29. Khashayar, S. and G. Pierre, Mise en œuvre des poudres Techniques de granulation humide et liants. *Techniques de l'ingénieur Mise en Forme des Médicaments*, 2009. Base documentaire: TIB611DUO (ref. article: j2254).
30. Cleaver, J.A.S. and G. Karatzas, S. Louis and I. Hayati, *Moisture-induced caking of boric acid powder. Powder Technology*, 2004. **146**(1): p. 93–101.
31. Langlet, M., F. Nadaud, M. Benali, I. Pezron, K. Saleh, P. Guigon and L. Metlas-Komunjer, Kinetics of dissolution and recrystallization of sodium chloride at controlled relative humidity. *KONA Powder and Particle Journal*, 2011. **29**: p. 168–179.
32. Langlet, M., M. Benali, I. Pezron, K. Saleh, P. Guigon and L. Metlas-Komunjer, Caking of sodium chloride: Role of ambient relative humidity in dissolution and recrystallization process. *Chemical Engineering Science*, 2013. **86**: p. 78–86.
33. Mauer, L.J. and L.S. Taylor, Deliquescence of pharmaceutical systems. *Pharmaceutical Development and Technology*, 2010. **15**(6): p. 582–594.
34. Mauer, L.J. and L.S. Taylor, Water-solids interactions: Deliquescence. *Annual Review of Food Science and Technology*, 2010. **1**(1): p. 41–63.
35. Martin, S., M. Guessasma, J. Léchelle, J. Fortin, K. Saleh and F. Adenot, Simulation of sintering using a Non Smooth Discrete Element Method. Application to the study of rearrangement. *Computational Materials Science*, 2014. **84**: p. 31–39.
36. Martin, S., R. Parekh, M. Guessasma, J. Léchelle, J. Fortin and K. Saleh, Study of the sintering kinetics of bimodal powders. A parametric DEM study. *Powder Technology*, 2015. **270**(Part B): p. 637–645.
37. Kamyabi, M., K. Saleh, R. Sotudeh-Gharebagh and R. Zarghami, Simulation of viscous-flow agglomerate sintering process: Effect of number of particles and coordination number. *Particuology*, 2019. **43**: 76–83.
38. Kamyabi, M., R. Sotudeh-Gharebagh, R. Zarghami and K. Saleh, Principles of viscous sintering in amorphous powders: A critical review. *Chemical Engineering Research and Design*, 2017. **125**: p. 328–347.
39. Williams, M.L., R.F. Landel and J.D. Ferry, The temperature dependence of relaxation mechanisms in amorphous polymers and other glass-forming liquids. *Journal of the American Chemical Society*, 1955. **77**(14): p. 3701–3707.
40. Roos, Y. and M. Karel, Phase transitions of mixtures of amorphous polysaccharides and sugars. *Biotechnology Progress*, 1991. **7**(1): p. 49–53.
41. Vega, C. and Y.H. Roos, Invited review: Spray-dried dairy and dairy-like emulsions—Compositional considerations. *Journal of Dairy Science*, 2006. **89**(2): p. 383–401.
42. Afrassiabian, Z., *Etude multi-échelles du phénomène de mottage des poudres amorphes : De la physicochimie des matériaux aux applications industrielles*. 2019, Université de Technologie de Compiègne.
43. Samain, S., *Caractérisation multi-échelle de l'efflorescence et du mottage du saccharose*. 2017, Université de Technologie de Compiègne.
44. Vignolles, M.-L., R. Jeantet, C. Lopez and P. Schuck, Free fat, surface fat and dairy powders: Interactions between process and product: A review. *Lait*, 2007. **87**(3): p. 187–236.

3.3 Whey proteins pre-texturized by heating in dry state

Marie-Hélène Famelart, Alexia Audebert, Muhammad Gulzar and Thomas Croguennec

3.3.1 Introduction

Dry heating as a means to pre-texturize protein ingredients is a process involving the incubation of a protein powder at a set temperature (typically in the range 50–130°C) and at a set relative humidity (RH, typically in the range 30–80%) for a pre-determined time (from min to days). Dry heating was used first in the pharmaceutical and food industries for the viral and microbial decontamination of heat-sensitive proteins, because the moisture reduction induces an increase in the denaturation temperature of proteins. Beside this function, dry heating is used to pre-texturize proteins through chemical reactions (glycation, phosphorylation) in order to improve their functional properties in food products. These reactions are conducted in the presence of reducing sugars or a phosphate derivative under soft dry heating conditions (Enomoto et al., 2007, 2008, 2009; Fenaille et al., 2003; Gauthier et al., 2001; Li et al., 2005; Liu and Zhong, 2014; Morgan et al., 1999b, 1999a; O'Mahony et al., 2017; Schong and Famelart, 2017).

Protein glycation by dry heating is the first step of the Maillard reaction when a covalent bond forms between a reducing sugar and one amino group of a protein (O'Mahony et al., 2017). Glycation by dry heating (50°C up to 2 days, pH 7.2 and under 65% RH) is more efficient than glycation by heat treatment in solution, and it preserves most of the native structure of whey proteins (Morgan et al., 1999a). Dry heating at higher temperatures (80–120°C) and for prolonged heating times (up to days) in the presence of a reducing sugar also produces very large insoluble supramolecular structures (Famelart et al., 2018; Gulzar et al., 2012).

Whey protein phosphorylation by dry heating consists of the addition of phosphate derivatives to the hydroxyl group of serine, threonine, tyrosine side chain of a protein (Donato et al., 2009; Enomoto et al., 2007, 2009; Li et al., 2005; Schmitt et al., 2007). The protein phosphorylation efficiency is enhanced if the proteins are glycated (Li et al., 2005). Phosphorylation by dry heating has a very slight effect on the secondary structure of whey proteins; however significant modifications in the tertiary structure have been observed (Enomoto et al., 2007).

In addition, prolonged dry heating also induces modifications in protein structures (in the absence of added reducing sugars and phosphate derivatives) leading to modifications in their functional properties (water holding capacity, gelling, foaming and emulsifying). Dry heating has been extensively used for pre-texturizing egg-white proteins but much less work has been conducted on milk proteins, especially whey proteins. However, dry heating for pre-texturizing food proteins constitutes an alternative to heating in solution for which a huge literature is available. In this chapter, we aim to fill the gap by addressing in the first part the chemical modifications occurring at the molecular level during protein dry heating. The supramolecular structures formed by dry heating and the parameters affecting their size, shape and reactivity are given in the second part. Finally, some functional properties of the pre-texturized whey proteins are presented in the third part.

3.3.2 Whey protein modifications at the molecular level

The temperature of dry heating is usually below the denaturation temperature of whey proteins, which is considerably increased at low water content (Figure 3.3.1, Zhou and

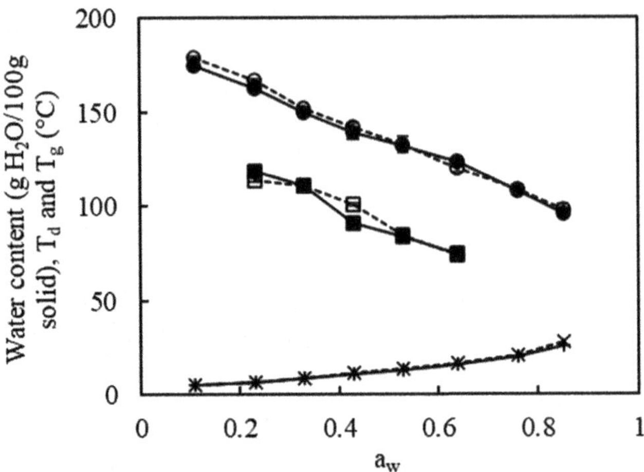

Figure 3.3.1 The changes in water content, denaturation and glass transition temperatures (T_d and T_g) of whey proteins and β-Lg with the water activity: water content (+ ×), T_d (● ○), T_g (■ □); whey proteins: full symbols; β-Lg: open symbols (adapted from Zhou and Labuza, 2007).

Labuza, 2007). This could explain specific behaviors of whey proteins during dry heating such as a greater preservation of the whey protein 3-dimensional structure and some specific chemical modifications (dehydration and/or deamidation, peptide bond cleavage).

The studies tend to agree that the secondary structure of whey proteins is only very slightly modified during dry heating in the absence or presence of added reducing sugars or phosphate derivatives (Augustin and Udabage, 2007; Enomoto et al., 2007; Gulzar et al., 2013; Morgan et al., 1999a). Some modifications in the tertiary structure of whey proteins, resulting in an increased accessibility of aromatic residues and hydrophobic patches on the surface of proteins, were evidenced by fluorescence spectroscopy (Gulzar, 2011; Ibrahim et al., 1993). These modifications in secondary and tertiary structures were more pronounced for α-La as compared to β-Lg and in samples dry heated at acidic pH as compared to at neutral pH (Gulzar et al., 2013).

Dry heating also induces chemical modifications in the primary structure of the whey proteins. The prevalent modification concerns the cyclization of the N-terminal glutamic acid of α-La leading to the formation of N-terminal pyroglutamic acid and the release of one water molecule (Figure 3.3.2a) (Gulzar et al., 2013). This reaction occurs spontaneously at room temperature (Dick et al., 2007), but the temperature increase accelerates its kinetics (Park et al., 2001). The reaction is favored in the dry state due to favorable mass balance and shows a large majority of the α-La molecules were modified after dry heating. Other dehydration reactions were identified in whey proteins leading to the formation of internal cyclic imide. This results from the reaction between the side chain carbonyl group of a glutamic acid or aspartic acid and the α-amino group of the n+1 residue in the polypeptide chain (Figure 3.3.2b). This reaction is favored when the residues are located in an unstructured domain with a glycine at position n+1 (Geiger and Clarke, 1987) even if other facilitating amino acids were identified (Desfougères et al., 2011). These conditions explain that the dehydration reaction concerns only a limited amount of glutamic acid or aspartic acid (Gulzar et al., 2013). The formed internal cyclic imide is stable in solution below pH 5 but when the pH of the solution is increased above

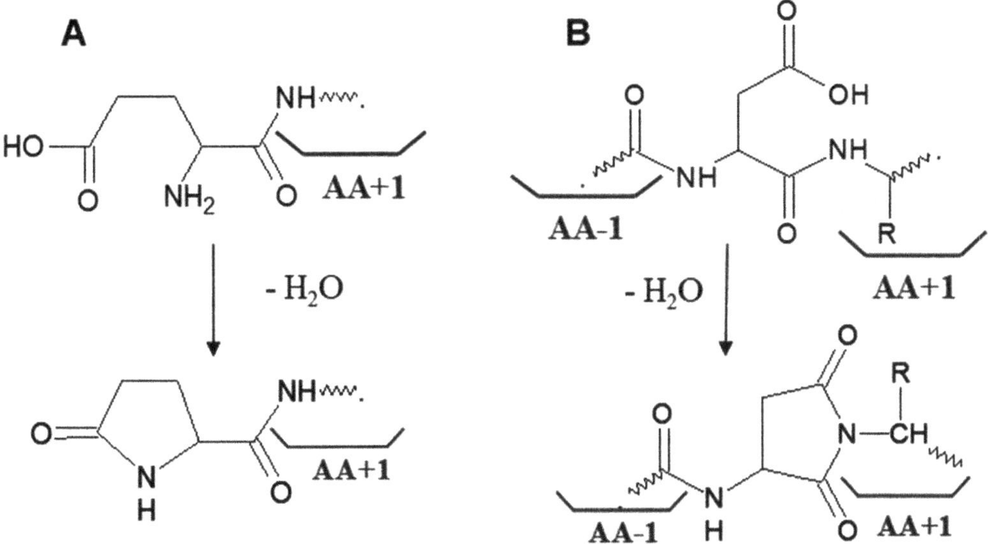

Figure 3.3.2 Chemical structure of (A) the N-terminal pyroglutamic acid resulting from the cyclization of the N-terminal glutamic acid of α-lactalbumin, (B) a cyclic imide resulting from the cyclization of an internal aspartyl residue (adapted from Gulzar et al., 2013).

pH 5 it is hydrolyzed, resulting in the isomerization of aspartyl residue into isoaspartyl (Desfougères et al., 2011). This structural change affects the intrinsic protein stability (Desfougères et al., 2011).

Dry heating also induced the hydrolysis of whey proteins into fragments as shown by SDS-PAGE combined with mass spectrometry. Nevertheless, their identification is tricky because protein hydrolysis occurs concomitantly with protein aggregation through inter-polypeptide covalent bonds (Gerrard, 2002) other than disulfide bonds (see next section).

3.3.3 *Aggregation and polymerization reactions*

Dry heating at temperatures >50°C in the presence of different carbohydrates leads to whey protein polymerization (Famelart et al., 2018; Gulzar et al., 2011a, 2012, 2013; Guyomarc'h et al., 2015; Hiller and Lorenzen, 2010; Liu and Zhong, 2013, 2014; Liu et al., 2014; Schmitt et al., 2005). Even without carbohydrates, β-Lg forms dimers and oligomers by dry heating at 100°C, a_w 0.23 for 24 h (Gulzar et al., 2011b) or soluble aggregates at 80°C for 5 d at 7.5% moisture content (Ibrahim et al., 1993). Soluble and insoluble aggregates can both be formed (Gulzar et al., 2011a, 2012). Moreover, the reticulation of whey proteins in the presence of a small amount of lactose by dry heating at 100°C, $a_w = 0.23$, at an alkaline pH value allows the production of stable large microparticles (Famelart et al., 2018).

The size of the aggregates formed by the dry heating of whey proteins increases: (i) with the time of dry heating; (ii) with the pH increase of the concentrate used to prepare the powder; (iii) with the water activity (a_w) of powders or their RH (Famelart et al., 2018; Gulzar et al., 2011a, 2012; Hiller and Lorenzen, 2010; Ibrahim et al., 1993; Liu et al., 2014;

Norwood et al., 2017). The fraction of polymerized proteins also increases with increases in temperature and lactose content (Norwood et al., 2017).

The powder moisture content is a major parameter of the aggregation of whey proteins, because water molecules increase the mobility of protein molecules. But for very high moisture contents, the content in polymers decreases, due to the dilution and to a lower rate of aggregation (Zhou et al., 2008).

The effect of pH (2.5, 4.5, 6.5) has been studied on the dry heating of a mixture of whey proteins in the presence of a small amount of lactose (P/L = 100 g/g) at 100°C for up to 24 h under $a_w = 0.23$ (Gulzar et al., 2011a, 2012). Only soluble polymers were identified at pH 2.5, while soluble and insoluble aggregates form at pH 4.5 and 6.5. This effect is due to an increase in the thiolate content at alkaline pH values that enhances SS/SH interchanges.

Whey protein polymerization depends on the nature of the carbohydrate present during the dry heating. Covalent crosslinked polymers of M_w around 50 kDa form after dry heating (80°C, RH 80%) whey proteins at pH 7 with glucose, lactose and maltodextrin, but none with sucrose, a non-reducing disaccharide (Liu and Zhong, 2013). Another study deals with the effect of saccharides on the polymerization of whey proteins by dry heating at 70°C and 65% RH (Hiller and Lorenzen, 2010). The lowest polymerization was observed with dextran (1–2% of total mass with $M_w > 1{\cdot}10^5$ g/ml), then glucose (11%), lactose (19%) and pectin (27%). Schmitt et al. (2005) dry heated β-Lg alone or with acacia gum at 60°C under $a_w = 0.79$ for 14 d. After a 2-day incubation without gum, some polymers of $M_w = 43–67$ kDa and even >94 kDa form by covalent bonds other than disulfide ones or by hydrophobic bridging. The polymerization was reduced by the presence of the gum.

Polymerization also depends on proteins. During the dry heating at 80°C of pure α-Lac or β-Lg at 7.5% moisture content without carbohydrate, a large part of β-Lg formed soluble aggregates after 5 d of incubation, while α-Lac did not (Ibrahim et al., 1993). In contrast, pure β-Lg or α-Lac both formed 30% of soluble dimers and small oligomers during dry heating for 24 h at 100°C, pH 6.5 and $a_w = 0.23$ without carbohydrate (Gulzar et al., 2013). β-Lg polymerization also occurred during dry heating for 4 d at 60°C of a whey protein mixture in the presence of glucose (Liu et al., 2014). Anyway, these authors showed that α-Lac, β-Lg and BSA all disappeared into crosslinked polymers of sizes around 200 kDa after dry heating.

Many authors agree that disulfide bonds play a major role, either in the absence of carbohydrate for pure BSA (Liu et al., 1991), β-Lg (Gulzar et al., 2011b, 2013) and α-Lac (Gulzar et al., 2013) or in the mixture of whey proteins in the presence of lactose (Gulzar et al., 2011a; Norwood et al., 2017). According to Gulzar et al. (2011a, 2011b), disulfide bonds are the only links in polymers formed at pH 2.5, while covalent bonds other than disulfide bonds are formed at pH > 4.5. Regarding the formation of disulfide bonds, the sulfhydryl/disulfide interchange reactions seem more likely than the oxidation of sulfhydryl residues (Liu et al., 1991), but the latter was also suggested (Gulzar et al., 2011a).

Some authors also reported on other covalent bonds, resistant to reduction (Gulzar et al., 2011b, 2013; Guyomarc'h et al., 2015; Norwood et al., 2017) and on non-covalent interactions (Zhou et al., 2008). The other covalent bonds could originate from isopeptidic links (Gulzar et al., 2011b, 2013), or via the dehydroalanine pathway (Gulzar et al., 2013; Norwood et al., 2017) or lysino-alanine (Gulzar et al., 2011b) or finally from Maillard intermediates such as α-dicarbonyl (Guyomarc'h et al., 2015; Norwood et al., 2017). Such Maillard intermediates are produced by the degradation of carbohydrates via dehydration, oxidation and/or fragmentation during the dry heating (Okitani et al., 1984). Once produced, they are grafted to arginine, tryptophan or lysine residues of proteins during dry heating. As these Maillard compounds form preferentially at pH 6.5, it could explain why the addition of lactose to a pure mixture of β-Lg and α-Lac at pH 2.5 does not increase the amount of polymers formed by dry heating, while it does at pH 6.5 (Guyomarc'h et al., 2015).

Non-covalent bonds between whey proteins were observed in small amounts between pH 5 and 6 (Zhou et al., 2008). These authors assumed that non-covalent bonds could first form, and then, that sulfhydryl/disulfide interchanges could be initiated.

3.3.4 *Functional properties of spray-dried whey powders*

3.3.4.1 *Viscosity and water-holding capacity*

Whey proteins dry heated at 70°C and 65% RH for 4 h in the presence of glucose or 96 h in the presence of lactose show an increase of 10.6% and 18.7% in polymers having a M_w > $2{\cdot}10^5$ g/mol. Their suspensions (5%, w/w) have increased apparent viscosity (164% and 152%, respectively, as compared to ~20% for unheated controls). The increase in the *Mw*, steric hindrance, and water-binding properties of Maillard products can explain the viscosity increases (Hiller and Lorenzen, 2010).

Famelart et al. (2018) dry heated whey protein powder containing lactose at pH 9.5 at 100°C and $a_w = 0.23$ for up to 36 h. This leads to a 3–4-fold increase of powder browning indicative of a severe Maillard reaction, when advanced glycation end-products are formed. Over the dry heating time, increasing amounts of suspended stable microparticles with heterogeneous sizes (50–2000 μm) are observed. The size of these microparticles is close to the size of the powder particles. The authors concluded that whey proteins are irreversibly crosslinked in the powder grain during dry heating at pH 9.5 and that the dry-heated powder is no longer soluble. Maillard-induced microparticles dispersed in water at pH 6.5 are able to swell due to electrostatic repulsions between whey proteins. Increasing the dry heating time reduces their swelling due to an increased number of crosslinks inside the particles. Surprisingly, the microparticles formed by dry heating for 36 h retain 30 times their weight of water, because powder particles have a low density due to large amounts of occluded air and, as a result, insoluble microparticles retain a large amount of water when dispersed in solution (Schong and Famelart, 2018) (Table 3.3.1). Moreover, these microparticles are able to shrink when placed at acidic pH values and then to re-swell when brought back to neutral pH values, due to changes with pH of the electrostatic repulsions between whey proteins inside the crosslinked particles. Due to their large water content, these macroparticles may have significant food potential as dairy additives.

3.3.4.2 *Interfacial, foaming and emulsifying properties*

At hydrophobic interfaces (gas, oil) whey proteins form a concentrated protein layer inside which the proteins have the capacity to self-assemble. The resulting viscoelastic interfacial film limits gas bubble and oil droplet deformation and consequently retards the dispersion destabilization. In addition, bulk aggregates by confinement inside the foam plateau border may also improve foam stability, especially against liquid drainage (Fameau and Salonen, 2014). They increase the viscosity of the continuous phase, and they prevent the interfacial layers of adjacent bubbles from coming too close together. The formation of

Table 3.3.1 The apparent viscosity of 2.3% (w/w) suspensions of microparticulated whey proteins (mean size 10 μm) (Simplesse 100, CPQuelco) and of microparticles (mean size 16 μm) produced by dry heating (100°C for 24, 36, 48 and 72 h, under $a_w = 0.23$) whey protein powder at pH 9.5 (Schong and Famelart, 2018)

Samples at 2.3% (w/w)	η_{app} at 1 s^{-1} (mPa·s)	η_{app} at 100 s^{-1} (mPa·s)
Simplesse 100	2	1
Dry-heated powder (24 h)	3288	100
Dry-heated powder (36 h)	2319	84
Dry-heated powder (48 h)	1335	118
Dry-heated powder (72 h)	46	9

soluble unfolded proteins and aggregates by dry heating was identified as a means to improve the interfacial, foaming and emulsifying properties of the proteins, such as egg-white proteins. Table 3.3.2 lists the interfacial, emulsifying and foaming properties of dry-heated whey protein powders. Ibrahim et al. (1993), Norwood et al. (2016a), Medrano et al. (2009a) and Audebert et al. (2019a) have demonstrated a foam stability improvement after whey protein dry heating. However, this improvement is weak in comparison to the one observed for dry-heated egg-white proteins (Talansier et al., 2009) (three-fold improvement). The rigidity of whey protein foam is also improved with dry heating (Audebert et al., 2019b).

Foaming properties depend on the intrinsic molecular properties of the proteins such as their flexibility, surface hydrophobicity and the ability to establish intermolecular interactions when concentrated at the hydrophobic interface. The higher the surface hydrophobicity, the faster the surface tension decreases (Nakai, 1983). Because of their compact structure stabilized by several intra-molecular disulfide bonds, whey proteins require relatively high energy input in order to unfold (Horiuchi et al., 1978). This could explain the different interfacial behavior of whey proteins in comparison to egg-white proteins after dry heating. Apo-α-Lac, α-Lac depleted of its bound calcium, has an increased flexibility and could reach the molten globule state required to develop interesting foaming properties (Kinsella and Whitehead, 1989). In addition, dry heating increases α-Lac surface hydrophobicity (Ibrahim et al., 1993; Van der Plancken et al., 2007). In contrast, β-Lg exhibits lower surface hydrophobicity after dry heating (Ibrahim et al., 1993). In the native state, β-Lg has higher surface hydrophobicity than major egg-white proteins (ovalbumin, lysozyme) (Nakai, 1983), but these hydrophobic sites could be hidden during the aggregation reactions occurring during dry heating. Although whey protein aggregates could improve foam stability (Fameau and Salonen, 2014), the limited change in the flexibility and surface hydrophobicity of whey proteins during dry heating could represent a clue of the non-exacerbated interfacial properties.

The presence of sugar in whey protein powders during dry heating affects whey protein interfacial properties (Table 3.3.2). Protein glycation increases the protein's hydrophilic character and changes the protein's hydrodynamic properties. Thus, protein solubility and viscosity can be increased and foaming properties are enhanced. Lactose promotes Maillard-induced aggregation reactions (Guyomarc'h et al., 2015), which, in excess of a threshold, may also reduce foaming properties and justify the de-sugaring stage before egg-white dry heating (Campbell et al., 2003). Thus, discrepancies observed between

Table 3.3.2 Interfacial, emulsifying and foaming properties of spray-dried whey powders

Spray-dried whey powders	Dry-heating treatment	Powder conditions (pH prior to dry heating, a_w or moisture content)	Interfacial properties (pH 7.0)	Foaming properties (pH 7.0)	Emulsifying properties (pH 7.0)	Literature
β-Lg α-Lac	80°C; 0–10 days	Moisture content 7,5%	–	Foamability and foam stability increased.	Emulsifying activity was improved. Emulsion stability increased for α-Lac and decreased for β-lg.	Ibrahim et al., 1993
β-Lg + sugars (galactose, mannose, glucose and lactose, ribose and glyceraldehyde)	60°C; 48 h	pH 6.5	–	–	Emulsifying properties of dry-heated β-Lg in the presence of sugars were better than in the absence of sugar.	Nacka et al., 1998
β-Lg β-Lg + sugars (glucose, lactose)	50°C; 51 or 96 h	pH 7.0; a_w 0.65	Slower adsorption of glycated β-Lg than untreated β-lg at the air/water interface.	Liquid drainage in foam made with glycated proteins decreases in comparison to untreated β-lg (especially with lactose). Higher foam density in the following conditions: lactose, 51 h.	–	Medrano et al., 2009

(Continued)

Table 3.3.2 (Continued) Interfacial, emulsifying and foaming properties of spray-dried whey powders

Spray-dried whey powders	Dry-heating treatment	Powder conditions (pH prior to dry heating, a_w or moisture content)	Interfacial properties (pH 7.0)	Foaming properties (pH 7.0)	Emulsifying properties (pH 7.0)	Literature
β-Lg β-Lg + galactose (GAL)	40 or 50°C; 24 or 48 h	pH 7.0; a_w 0.44	Except for dry-heated β-Lg + GAL (48 h; 50°C), the diffusion to air/water interface of dry-heated proteins is faster than for untreated proteins. The interfacial layer dilatational modulus (E) of dry-heated β-Lg increased slightly except for dry-heated β-Lg + GAL (48 h; 50°C) which decreased it. Native β-Lg showed higher interfacial elasticity than dry-heated β-Lg in the absence and presence of GAL regarding shift angle (dilatational surface rheology).	β-Lg and β-Lg + GAL dry heating had no effect on foamability and foam density except dry- heated β-Lg + GAL (48 h; 50°C) which decreased them. Dry-heated β-Lg and dry-heated β-Lg + GAL decreased foam stability in comparison to native β-Lg.	–	Corzo-Martínez et al., 2012

(*Continued*)

Table 3.3.2 (Continued) Interfacial, emulsifying and foaming properties of spray-dried whey powders

Spray-dried whey powders	Dry-heating treatment	Powder conditions (pH prior to dry heating, a_w or moisture content)	Interfacial properties (pH 7.0)	Foaming properties (pH 7.0)	Emulsifying properties (pH 7.0)	Literature
Whey protein isolate (WPI) containing trace of lactose	4, 20, 40, 60°C; up to 12 months	a_w 0.23 or 0.36	–	6 months' storage at 60°C (a_w 0.36) slightly improved foam stability and foam density in comparison to untreated WPI.	–	Norwood et al., 2016
WPI (2% lactose) WPI (0.2% lactose)	70°C; 0 or 125 h	a_w 0.12, 0.23 or 0.52 pH 3.5, 5.0 or 6.5	–	Dry heating (125 h) at pH 5.0 and pH 6.5 decreased foam stability and the effect got worse in the presence of lactose. Whatever the lactose content, dry heating (125 h) at pH 3.5 slightly improved WPI foam stability, especially against drainage.	–	Audebert et al., 2019a
WPI (2% lactose) β-Lg	70°C; 0 or 125 h	a_w 0.23 pH 3.5 or 6.5	Whatever the pH, WPI and β-Lg dry heating decreased the interfacial layer dilatational storage modulus (E') and the surface tension.	Whatever the pH, WPI dry heating decreased disproportionation rates and increased foam visco-elasticity at low deformation strain (1%). WPI dry heating at pH 3.5 increased foam rigidity (yield strain). β-Lg dry heating decreased foam visco-elasticity at low deformation (1%).	–	

the studies reported in Table 3.3.2 are probably due to the broad variety of dry-heating parameters (mild or extensive dry-heating treatments, purified proteins or mixture of proteins, absence or presence of sugars, etc.). Those parameters may act at different levels in the foaming properties (interfacial or bulk properties, steric repulsion, etc.), which may explain controversial results.

3.3.4.3 Heat-set gelling properties

Gelation is an aggregation process of proteins, in which protein–protein and protein–solvent interactions are so balanced that a self-supporting tertiary network or matrix which has liquid and other elements entrapped in it is formed (Brodkorb et al., 2016; O'Mahony et al., 2017). Protein heat gelation is a three-step mechanism. The first step involves the unfolding of the proteins that aggregate and form soluble aggregates in a second step. If the protein concentration is sufficient then a gel network is formed in a third step. The ability of proteins to form heat-set gels depends upon their structure, interactions with other components and processing conditions (heating temperature, protein concentration, pH and ionic strength, etc.). The heating of concentrated solutions of whey proteins results in strong gels with high water holding capacity.

Dry-heated whey proteins demonstrated better heat-set gelation properties as compared to native proteins (Gulzar et al., 2012). These authors related the heat-set gelation properties with the level of whey protein aggregation; the water holding capacity and the strength of heat-set gels are increased as long as the amount of soluble aggregates increases; however, the presence of insoluble aggregates in the dry-heated powder drastically affected the water holding capacity and the strength of the heat-set gels. Heat intensity, water activity, pH and lactose contents, etc., are shown to increase the dry heat–induced aggregation of whey proteins (Gulzar et al., 2012; Norwood et al., 2017).

The rheological properties and strength of a gel vary with the number of interactions and types of interactions between proteins (O'Mahony et al., 2017). The strain properties of heat-set whey protein gels are mainly related to the number of crosslinks (Errington and Foegeding, 1998). A lower number of protein crosslinks in heat-set gels makes the gel more brittle, while increasing crosslinks makes the gel more elastic. The occurrence of protein crosslinks also depends on the surface properties of the proteins, which might increase or decrease the access to the reactive parts. The gelation properties of proteins can be increased if the surface properties promote crosslinking during heat-set gelation (O'Mahony et al., 2017) by exposing the reactive parts. Dry heating might have generated the characteristics on the surface of the whey proteins, which are more prone to crosslinking during heating in solution. A two-fold increase in gelling properties was observed in the case of dry-heated whey proteins (Gulzar et al., 2012).

3.3.5 Conclusion

Dry heating is an original way to modify the structure and the functionalities of whey proteins. At the moment, dry heating is not widespread in the dairy industry mainly because of the discontinuity of the process, the energy and immobilization costs linked to the time and temperature required to modify milk protein functionalities and the lack of reproducibility between batches. Indeed, protein modification by dry heating is very sensitive to small changes in the powder composition (amount of lactose, pH) and conditions of dry heating (a_w, temperature/time of heating in the dry state). However, some of these drawbacks could be reduced by heating the dairy powders at controlled a_w and temperature in a fluidized bed or heated roller fed directly by the powder coming from the spray-drier. It is also possible to ensure the protein structural changes by including physiochemical treatments conducted on the protein concentrate before the dry heating treatment. Dry heating remains a unique process to obtain stable microparticles by reticulation of the powder particles without the use of chemicals. These microparticles have exceptional water retention properties, and they could pave the way for new applications such as the encapsulation of a large diversity of sensitive ingredients and use as a fat replacer.

References

Audebert, A., Beaufils, S., Lechevalier, V., Le Floch-Fouéré, C., Saint-Jalmes, A., and Pezennec, S. (2019a). How film stability against drainage is affected by conditions of prior whey protein powder storage and dry-heating: a multidimensional experimental approach. *Journal of food Engineering. 242*, 153–162.

Audebert, A., Saint-Jalmes, A., Beaufils, S., Lechevalier, V., Le Floch-Fouéré, C., Cox, S., Leconte, N., and Pezennec, S. (2019b). Interfacial properties, film dynamics and bulk rheology: A multi-scale approach to dairy protein foams. *Journal of Colloid and Interface Science 542*(15), 222–232.

Augustin, M.A., and Udabage, P. (2007). Influence of processing on functionality of milk and dairy proteins. *Advances in Food and Nutrition Research 53*, 1–38.

Brodkorb, A., Croguennec, T., Bouhallab, S., and Kehoe, J.J. (2016). Heat-induced denaturation, aggregation and gelation of whey proteins. In *Advanced Dairy Chemistry*, Paul L. H. McSweeney, James A. O'Mahony Editors (Springer, New York, NY), pp. 155–178.

Campbell, L., Raikos, V., and Euston, S.R. (2003). Modification of functional properties of egg-white proteins. *Food / Nahrung 47*(6), 369–376.

Corzo-Martínez, M., Carrera Sánchez, C., Moreno, F.J., Rodríguez Patino, J.M., and Villamiel, M. (2012). Interfacial and foaming properties of bovine β-lactoglobulin: Galactose Maillard conjugates. *Food Hydrocolloids 27*(2), 438–447.

Desfougères, Y., Jardin, J., Lechevalier, V., Pezennec, S., and Nau, F. (2011). Succinimidyl residue formation in hen egg-white lysozyme favors the formation of intermolecular covalent bonds without affecting its tertiary structure. *Biomacromolecules 12*(1), 156–166.

Dick, L.W., Kim, C., Qiu, D., and Cheng, K.-C. (2007). Determination of the origin of the N-terminal pyro-glutamate variation in monoclonal antibodies using model peptides. *Biotechnology and Bioengineering 97*(3), 544–553.

Donato, L., Schmitt, C., Bovetto, L., and Rouvet, M. (2009). Mechanism of formation of stable heat-induced β-lactoglobulin microgels. *International Dairy Journal 19*(5), 295–306.

Enomoto, H., Li, C.P., Morizane, K., Ibrahim, H.R., Sugimoto, Y., Ohki, S., Ohtomo, H., and Aoki, T. (2007). Glycation and phosphorylation of β-Lactoglobulin by dry-heating: Effect on protein structure and some properties. *Journal of Agriculture and Food Chemistry 55*(6), 2392–2398.

Enomoto, H., Li, C.-P., Morizane, K., Ibrahim, H.R., Sugimoto, Y., Ohki, S., Ohtomo, H., and Aoki, T. (2008). Improvement of functional properties of bovine serum albumin through phosphorylation by dry-heating in the presence of pyrophosphate. *Journal of Food Science 73*(2), C84–91.

Enomoto, H., Hayashi, Y., Li, C.P., Ohki, S., Ohtomo, H., Shiokawa, M., and Aoki, T. (2009). Glycation and phosphorylation of α-lactalbumin by dry heating: Effect on protein structure and physiological functions. *Journal of Dairy Science 92*(7), 3057–3068.

Errington, A.D., and Foegeding, E.A. (1998). Factors determining fracture stress and strain of fine-stranded whey protein gels. *Journal of Agriculture and Food Chemistry 46*(8), 2963–2967.

Fameau, A., and Salonen, A. (2014). Effect of particles and aggregated structures on the foam stability and aging. *Comptes Rendus Physique 15*(8–9), 748–760.

Famelart, M.-H., Schong, E., and Croguennec, T. (2018). Dry heating a freeze-dried whey protein powder: Formation of microparticles at pH 9.5. *Journal of Food Engineering 224*, 112–120.

Fenaille, F., Campos-Giménez, E., Guy, P.A., Schmitt, C., and Morgan, F. (2003). Monitoring of β-lactoglobulin dry-state glycation using various analytical techniques. *Analytical Biochemistry 320*(1), 144–148.

Gauthier, F., Bouhallab, S., and Renault, A. (2001). Modification of bovine beta-lactoglobulin by glycation in a powdered state or in aqueous solution: Adsorption at the air-water interface. *Colloids and Surfaces, Part B: Biointerfaces 21*(1–3), 37–45.

Geiger, T., and Clarke, S. (1987). Deamidation, isomerization, and racemization at asparaginyl and aspartyl residues in peptides. Succinimide-linked reactions that contribute to protein degradation. *Journal of Biological Chemistry 262*(2), 785–794.

Gerrard, J.A. (2002). Protein–protein crosslinking in food: Methods, consequences, applications. *Trends in Food Science and Technology 13*(12), 391–399.

Gulzar, M. (2011). Dry heating of whey proteins under controlled physicochemical conditions: Structures, interactions and functionalities *207*, Rennes1.

Gulzar, M. (2011). Dry heating of whey proteins under controlled physicochemical conditions: Structures, interactions and functionalities. PhD thesis n°2011-31 Food Engineering. Université de bretagne occidentale - AGROCAMPUS OUEST.

Gulzar, M., Bouhallab, S., and Croguennec, T. (2011b). Structural consequences of dry heating on Beta-Lactoglobulin under controlled pH. *Procedia Food Science 1*, 391–398.

Gulzar, M., Lechevalier, V., Bouhallab, S., and Croguennec, T. (2012). The physicochemical parameters during dry heating strongly influence the gelling properties of whey proteins. *Journal of Foods Engineering 112*(4), 296–303.

Gulzar, M., Bouhallab, S., Jardin, J., Briard-Bion, V., and Croguennec, T. (2013). Structural consequences of dry heating on α-lactalbumin and beta-lactoglobulin at pH 6.5. *Food Research International 51*(2), 899–906.

Guyomarc'h, F., Famelart, M.-H., Henry, G., Gulzar, M., Leonil, J., Hamon, P., Bouhallab, S., and Croguennec, T. (2015). Current ways to modify the structure of whey proteins for specific functionalities—a review. *Dairy Science and Technology 95*(6), 795–814.

Hiller, B., and Lorenzen, P.C. (2010). Functional properties of milk proteins as affected by Maillard reaction induced oligomerisation. *Food Research International 43*(4), 1155–1166.

Horiuchi, T., Fukushima, D., Sugimoto, H., and Hattori, T. (1978). Studies on enzyme-modified proteins as foaming agents: Effect of structure on foam stability. *Food Chemistry 3*(1), 35–42.

Ibrahim, H.R., Kobayashi, K., and Kato, A. (1993a). Improvement of the surface functional-properties of β-lactoglobulin and α-lactalbumin by heating in a dry state. *Bioscience, Biotechnology and Biochemistry 57*, 1549–1552.

Kinsella, J.E., and Whitehead, D.M. (1989). Proteins in whey: Chemical, physical, and functional properties. In *Advances in Food and Nutrition Research*, (Elsevier), pp. 343–438.

Li, C.P., Enomoto, H., Ohki, S., Ohtomo, H., and Aoki, T. (2005). Improvement of functional properties of whey protein isolate through glycation and phosphorylation by dry heating. *Journal of Dairy Science 88*(12), 4137–4145.

Liu, G., and Zhong, Q. (2013). Thermal aggregation properties of whey protein glycated with various saccharides. *Food Hydrocolloids 32*(1), 87–96.

Liu, G., and Zhong, Q. (2014). Removal of milk fat globules from whey protein concentrate 34% to prepare clear and heat-stable protein dispersions. *Journal of Dairy Science 97*(10), 6097–6106.

Liu, Q., Kong, B., Han, J., Sun, C., and Li, P. (2014). Structure and antioxidant activity of whey protein isolate conjugated with glucose via the Maillard reaction under dry-heating conditions. *Food Structure 1*(2), 145–154.

Liu, W.R., Langer, R., and Klibanov, A.M. (1991). Moisture-induced aggregation of lyophilized proteins in the solid state. *Biotechnology and Bioengineering 37*(2), 177–184.
Medrano, A., Abirached, C., Panizzolo, L., Moyna, P., and Añón, M.C. (2009). The effect of glycation on foam and structural properties of β-lactoglobulin. *Food Chemistry 113*(1), 127–133.
Morgan, F., Léonil, J., Mollé, D., and Bouhallab, S. (1999b). Modification of bovine β-lactoglobulin by glycation in a powdered state or in an aqueous solution: Effect on association behavior and protein conformation. *Journal of Agriculture and Food Chemistry 47*(1), 83–91.
Morgan, F., Venien, A., Bouhallab, S., Molle, D., Leonil, J., Peltre, G., and Levieux, D. (1999a). Modification of bovine β-lactoglobulin by glycation in a powdered state or in an aqueous solution: Immunochemical characterization. *Journal of Agriculture and Food Chemistry 47*(11), 4543–4548.
Nacka, F., Chobert, J.-M., Burova, T., Léonil, J., and Haertlé, T. (1998). Induction of new physicochemical and functional properties by the glycosylation of whey proteins. *Journal of Protein Chemistry 17*(5), 495–503.
Nakai, S. (1983). Structure-function relationships of food proteins: With an emphasis on the importance of protein hydrophobicity. *Journal of Agriculture and Food Chemistry 31*(4), 676–683.
Norwood, E.-A., Le Floch-Fouéré, C., Briard-Bion, V., Schuck, P., Croguennec, T., and Jeantet, R. (2016). Structural markers of the evolution of whey protein isolate powder during aging and effects on foaming properties. *Journal of Dairy Science 99*(7), 5265–5272.
Norwood, E.-A., Pezennec, S., Burgain, J., Briard-Bion, V., Schuck, P., Croguennec, T., Jeantet, R., and Le Floch-Fouéré, C. (2017). Crucial role of remaining lactose in whey protein isolate powders during storage. *Journal of Food Engineering 195*, 206–216.
Okitani, A., Cho, R.K., and Kato, H. (1984). Polymerization of lysosyme and impairment of its amino acid residues caused by reaction with glucose. *Agricultural and Biological Chemistry 48*, 1801–1808.
O'Mahony, J.A., Drapala, K.P., Mulcahy, E.M., and Mulvihill, D.M. (2017). Controlled glycation of milk proteins and peptides: Functional properties. *International Dairy Journal 67*, 16–34.
Park, C.B., Lee, S.B., and Ryu, D.D. (2001). L-pyroglutamate spontaneously formed from L-glutamate inhibits growth of the hyperthermophilic archaeon *Sulfolobus solfataricus*. *Applied and Environment Microbiology 67*(8), 3650–3654.
Schmitt, C., Bovay, C., and Frossard, P. (2005). Kinetics of formation and functional properties of conjugates prepared by dry-state incubation of β-lactoglobulin/acacia gum electrostatic complexes. *Journal of Agriculture and Food Chemistry 53*(23), 9089–9099.
Schmitt, C., Bovay, C., Rouvet, M., Shojaei-Rami, S., and Kolodziejczyk, E. (2007). Whey protein soluble aggregates from heating with NaCl: Physicochemical, interfacial, and foaming properties. *Langmuir 23*(8), 4155–4166.
Schong, E., and Famelart, M.-H. (2017). Dry heating of whey proteins. *Food Research International 100*(2), 31–44.
Schong, E., and Famelart, M.-H. (2018). Dry heating of whey proteins leads to formation of microspheres with useful functional properties. *Food Research International 113*, 210–220.
Talansier, E., Loisel, C., Dellavalle, D., Desrumaux, A., Lechevalier, V., and Legrand, J. (2009). Optimization of dry heat treatment of egg white in relation to foam and interfacial properties. *LWT - Food Science and Technology 42*(2), 496–503.
Van der Plancken, I., Van Loey, A., and Hendrickx, M. (2007). Effect of moisture content during dry-heating on selected physicochemical and functional properties of dried egg white. *Journal of Agricultural and Food Chemistry 55*(1), 127–135.
Zhou, P., and Labuza, T.P. (2007). Effect of water content on glass transition and protein aggregation of whey protein powders during short-term storage. *Food Biophysics 2*(2–3), 108–116.
Zhou, P., Liu, X., and Labuza, T.P. (2008). Moisture-induced aggregation of whey proteins in a protein/buffer model system. *Journal of Agriculture and Food Chemistry 56*(6), 2048–2054.

3.4 *Physical properties of spray-dried dairy powders in relation with their flowability and rehydration capacity*

Jennifer Burgain, Tristan Fournaise, Claire Gaiani, Joël Scher and Jérémy Petit

3.4.1 *Introduction*

In most cases, dairy powders are produced by spray-drying a concentrate. This rapid dehydration method allows the production of high-quality products. The resulting powder is composed of a matrix of proteins, lipids and carbohydrates, in which vitamins, minerals and air vacuoles are embedded. The physicochemical properties (physical properties, such as particle size, shape, surface structure and chemical properties, like surface and core proximate compositions) of dairy powders depend on the bulk composition (Sadek et al., 2015a), concentrate pretreatment (homogenization, heat treatment such as pasteurization) and physicochemical properties (pH, solids content, viscosity) of the concentrate. Spray-drying process conditions (nozzle design, spraying conditions, drying air inlet and outlet temperatures) have also a great influence on the physicochemical properties of dairy powders (Schuck et al., 2016; Toikkanen et al., 2018). Powder physical properties are crucial as they directly impact powder functionalities (e.g., flowability, rehydration capacity). Depending on storage conditions, powder physical properties can be altered, which impairs powder functionalities, leading to economic losses (Toikkanen et al., 2018; Sharma et al., 2012). Indeed, milk reconstituted from powder should ideally reflect the original properties of fresh milk (organoleptic, nutritional, physicochemical and structural properties). In the present chapter, the physical properties of dairy powders will be addressed with a particular focus on their relation to functional properties.

3.4.2 *Powder physical properties*

3.4.2.1 *Particle size, shape and density*

Particle morphology can be described in terms of particle size, shape, internal structure and surface properties (Vehring, 2008). During spray-drying, atomization influences the shape, structure, velocity and size distribution of droplets to be dried, which is in turn determining for particle size distribution and shape, as well as the surface structure and composition of the resulting powder. The physical properties of the concentrate (e.g., viscosity, surface tension and density) are likely to influence the breakup of the liquid jet, and hence, the droplet size distribution in the spray (Petit et al., 2015; Mandato et al., 2012). Droplet drying occurs immediately after atomization and a particle is formed when the solids concentration at the droplet surface reaches a certain critical level depending on components' solubility (Elversson et al., 2003). Droplet drying kinetics (e.g., drying time, shell flexibility, evaporation rate) are central to the understanding of particle morphology (Vehring, 2008). Evaporation rate is controlled by the drying air temperature, while shell flexibility is strongly linked to solids concentration. As the droplet surface dries, component saturation leads to the formation of a skin, either rigid or smooth and flexible (Sadek et al., 2015b). A rigid skin is known to result in hollow, dense or broken particles. In contrast, a flexible skin allows inflation and expansion cycles of the droplet during drying and results in folded particles. As a general rule, particles tend to be spherical in rapid drying conditions (high drying air temperature or flow-rate) and high solids content, whereas they tend to be shriveled in slow drying conditions and at low solids concentration.

The main methods currently used to measure the particle size distribution of dairy powders are sieve analysis, laser-light scattering and image analysis. Sieve analysis is performed by mixing the powder with a free-flowing agent of known particle size distribution, putting the mixture on a stack of sieves and shaking it at defined amplitude during a sufficient time. The size range for sieve analysis is about from 5 μm to 125 mm. One of the major advantages of the sieving technique is that equipment and consumables are relatively cheap. Typically, measurements at industrial scale are performed off-line.

Laser-light scattering involves powder dispersion by air and passing through a monochromatic laser beam, which causes light scattering by reflection, refraction and diffraction. According to the Mie theory, the angle of diffraction is directly dependent on the particle size. This method has many advantages, e.g., short analysis time and the possibility to implement it on process lines to perform quality control analysis.

During the last decade, dynamic image analysis has become widely used for determining particle size and shape distributions. It is a digital image-processing technique used to characterize the shape of particles in motion with a high-speed digital camera. The powder is dispersed by air and a series of images is taken. Using mathematical algorithms, particle size and shape factors, such as aspect ratio and sphericity, are extracted. Image analysis may be used for particles from 1 μm to 10 mm (Yu and Hancock, 2008).

Density and porosity, which can be determined from measurements of particle size distribution, particle shape and surface area, are also important physical properties of powders. Different types of powder densities can be defined, mainly true, bulk and tapped densities. The true density stands for the density of the particle material; it can be determined using a gas pycnometer. The bulk density, which takes into account interparticular space, is defined as powder mass divided by the volume of the powder bulk that has not been consolidated. Powder bulk density primarily depends on particle size and shape distributions along with powder cohesion (the tendency of particles to adhere to each other). Tapped density is measured after mechanical tapping in given conditions (i.e., tap number and amplitude), which are chosen to reach a consolidated powder state where almost no volume change occurs upon further tapping. Hence, the tapped density is always higher than the bulk density. Bulk and tapped densities can be used to calculate the compressibility index (CI) and Hausner ratio, which are measures of the propensity of a powder to be compressed and to flow, respectively (see Section 3.4.3, Flow properties). As for porosity, two main definitions exist: either interparticular porosity which corresponds to the porosity of the powder bed (i.e., the volume of empty spaces between particles divided by the volume of the powder bed) or intraparticular porosity which corresponds to the porosity of particles themselves (i.e., the volume of internal pores divided by the volume of the particle).

3.4.2.2 *Particle microstructure*

The microstructure of a dairy powder designates the location of components inside particles (core vs. surface) and their possible interactions. Spray-dried particles are generally spherical with diameters ranging from 10 to 250 μm. Occluded air is often entrapped within particles in the form of either a large central vacuole or small vacuoles distributed throughout the particle matrix (Kosasih et al., 2016). Cracks and pores can also be present (Figure 3.4.1).

The surface structure of dairy powders is mainly controlled by chemical composition and spray-drying conditions. For example, skim milk and semi-skim milk particles are usually wrinkled, whereas high-protein powders are relatively smooth (Figure 3.4.2).

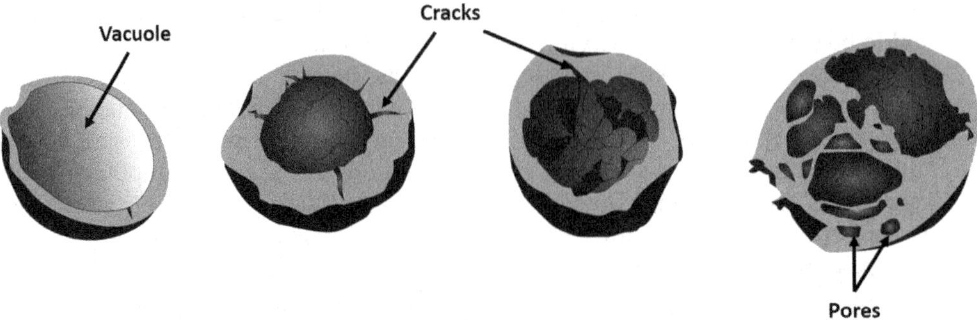

Figure 3.4.1 Internal microstructure of particles presenting vacuole, cracks and pores.

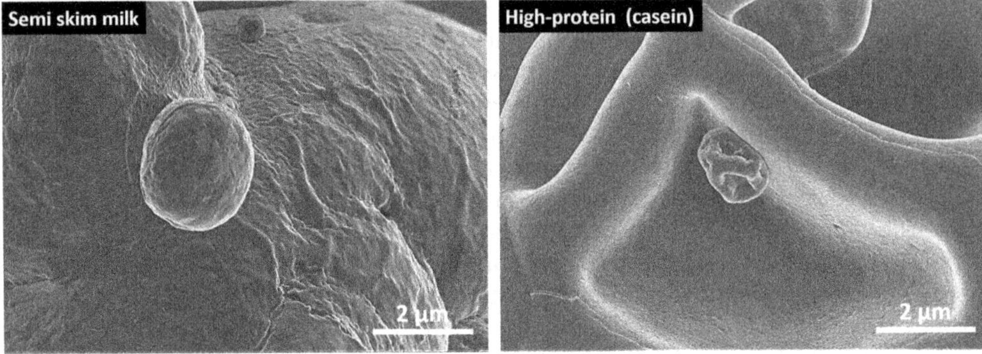

Figure 3.4.2 Analysis of surface structure of semi-skim milk and casein particles by scanning electron microscopy.

The core and surface microstructures of dairy powders are often studied using scanning electron microscopy (SEM) and confocal laser scanning microscopy (CLSM). In SEM, a focused beam of electrons scans the surface and the interactions between incident electrons and the sample surface generate secondary electrons of low energy from which topographical information can be recovered. Thus, SEM allows the investigation of the particle surface microstructure (e.g., cracked, smooth, porous) as well as the degree of particle agglomeration (Burgain et al., 2017). SEM has been widely used to study the particle size, shape and surface characteristics (presence of pores, wrinkles, lactose crystals, etc.) of dairy powders. The internal structure can also be observed with SEM after the particles have been cut (with an ultramicrotome for example). CLSM is an optical imaging technique permitting the capture of multiple two-dimensional images at different depths of a sample (optical sectioning) in order to rebuild three-dimensional images of particle structure. Using CLSM, the distribution of fat and proteins in dairy powders can be evidenced by using appropriate fluorescent probes and solvents (Kosasih et al., 2016). For instance, CLSM showed that the surface of whole milk powder is largely covered by free fat, either in the form of irregular patches or a homogeneous fat layer (Vignolles et al., 2009).

3.4.2.3 Mechanical properties

Dairy powders may exhibit elastic, plastic or viscoelastic behaviors, and they can sometimes be hard, tough or brittle. These mechanical properties play an important role in powder flowability and behavior upon compaction. In general, when applying a stress to a powder, particles begin to undergo elastic deformation. In this case, the deformation is completely reversible: particles return to their original shapes on the release of applied stress. In contrast, plastic deformation represents a permanent change in material shape caused by applied stress. The plastic properties of a material can be determined by an indentation test. At elevated values of applied stress, materials may fail by brittle or ductile fractures. The viscoelastic properties of powders are also important, as they reflect the time-dependent nature of stress–strain response. The surface mechanical properties of dairy powders are controlled by the phenomenon of skin formation during spray-drying (as previously indicated in Section 3.4.2.1, Particle size, shape and density). It was evidenced that the droplet surface becomes elastic when a critical solids concentration is approached, owing to the solidification of the protein skin (Sadek et al., 2015). The elastic modulus has been shown to be higher for caseins than for whey, meaning that the casein powder surface is stiffer.

Atomic force microscopy (AFM), especially the AFM nanoindentation technique, can be performed on individual particles to access mechanical properties at the nanometer scale. Indentation tests consist in touching the sample surface with a probe of known mechanical properties. Performing nanoindentation with an atomic force microscope allows the mapping of the elastic modulus of a defined region of interest of the particle surface, giving access to local information about surface hardness. This can be achieved by acquiring a force-distance curve and fitting the latter with theoretical models (e.g., Hertz, JKR models) to determine sample mechanical properties such as stiffness and Young's modulus (Masterson and Cao, 2008). For example, nanoindentation was employed for lactose-rich pharmaceutical powders in order to correlate particle hardness with powder compaction performance. Also, a link between the surface mechanical properties of individual particles and the rehydration capacity of spray-dried casein powder was established (Burgain et al., 2016); see Section 3.4.4, Reconstitution properties: links with physical properties, for more details.

3.4.3 *Powder flow properties: links with physical properties and characterization methods*

Powder flow properties aim to describe the way by which powders behave when they are put in motion. These functional properties of powders are crucial from an industrial point of view, as powders need to be transported, packed into containers or discharged from silos and hoppers. Flowability not only depends on powder physical properties (size, shape, density, surface area, etc.), but also on handling conditions. Consequently, various indices have been proposed to define flowability, including the Jenike flow function, the Hausner ratio, the compressibility index and the angle of repose.

3.4.3.1 *Powder flow properties: links with physical properties*

Powder flowability is the ability of a powder to flow as individual particles. It is mainly influenced by powder cohesion and compressibility, which are strongly related to powder physical properties such as particle size and shape distributions. Particle size has a major influence on powder flowability (Fu et al., 2012). Particles larger than 200–250 μm usually flow freely, while fine particles smaller than 100 μm are generally prone to cohesion, leading to poor flowability (Juliano and Barbosa-Canovas, 2010). A powder composed of small particles tends to be cohesive, and its flow through an orifice is restricted, because the cohesive forces between the particles are of the same order of magnitude as gravitational forces. Indeed, particles of lower size almost systematically have higher specific surface area, allowing more interparticular contact points, making them more cohesive and impairing their flowability.

The width of the particle size distribution may also have a significant impact on powder flowability. Indeed, the presence of several particle populations differing in size within a powder is generally detrimental to its flow properties, as the presence of fine particles between large ones increases the number of interparticular contact points. Hence, monomodal powders or powders with a thin particle size distribution are expected to exhibit better flow properties than polydisperse ones. This has been evidenced by the authors of the present chapter in a recently published study (Fournaise et al., 2020), where an homogenization pretreatment of spray-dried skim milk concentrate led to smaller particles and a lower span (width of particle size distribution) but improved the flow properties compared to skim milk powder obtained without pretreatment.

Particle shape is also of great importance for powder flowability: in fact, irregular particles are able to establish more contact points than spherical ones (Fu et al., 2012). Indeed, the formation of liquid bridges and the action of capillary forces at high moisture content, two phenomena that can trigger powder caking (see Section 3.2, Caking), are known to reduce powder flowability (Fitzpatrick, 2013). It has been shown in an unpublished study by the authors of this chapter that an increase in casein proportion in dairy powders spray-dried at pilot scale led to less spherical powders with a more irregular surface, which caused a slight decrease in flowability despite their higher mean particle size (data not shown).

Stickiness is a major problem encountered during the drying and handling of dried dairy products. Sticky powder is very cohesive, leading to powder adhesion to surfaces and powder loss. Eventually, powder caking can also occur (see Section 3.2, Caking). Dairy powders, initially free- or easy-flowing, may become sticky during handling and storage in adverse moisture and temperature conditions. In particular, fatty powders are extremely sensitive to temperature, as, above the fat-melting temperature, fat migration toward the powder surface can occur, making it sticky and dramatically altering flowability. Also, the powder composition and environmental conditions are crucial as regards the glass transition phenomenon (see Section 3.1, Water activity/glass transition). Indeed, most dairy powders produced by spray-drying are amorphous, and they often contain a significant proportion of amorphous lactose, which is likely to crystallize when the temperature exceeds the glass transition temperature of lactose (see Section 2.3, Lactose crystallization). Above the glass transition temperature of lactose, dairy powders become sticky, which increases cohesion, impairs flowability and may induce caking (Ebrahimi and Langrish, 2015; Fernández et al., 2003; Mercan et al., 2018). It should be kept in mind that the glass transition temperature of classical dairy powders, containing about 3–5% (w/w) moisture (Schuck et al., 2012), is often near ambient (Fernández et al., 2003); thus the impairment of flow properties is relatively common during handling or storage (Fitzpatrick et al., 2007).

3.4.3.2 *Characterization of powder flow properties*

Various techniques have been developed to measure the flow properties of powders. Pioneering measurements concerned the application of shear cell techniques (Jenike, 1967). The shear test measures the shear stress needed to obtain the failure of a preconsolidated powder bed, i.e., the putting in motion of particles relative to one another, as a function of applied normal stress (Figure 3.4.3; Table 3.4.1). Linear regression of the failure shear stress (τ) vs. applied normal stress (σ) curve is called the yield locus and allows the calculation of various powder flow parameters (Freeman et al., 2009).

The flow function is very useful for sorting powders according to their flowability. According to Jenike, powders can be classified as not flowing for $ff_c < 1$, very cohesive for $1 < ff_c < 2$, cohesive for $2 < ff_c < 4$, easy-flowing for $4 < ff_c < 10$ and free-flowing for $ff_c > 10$. For instance, it was shown in an unpublished study by the authors of the present chapter that

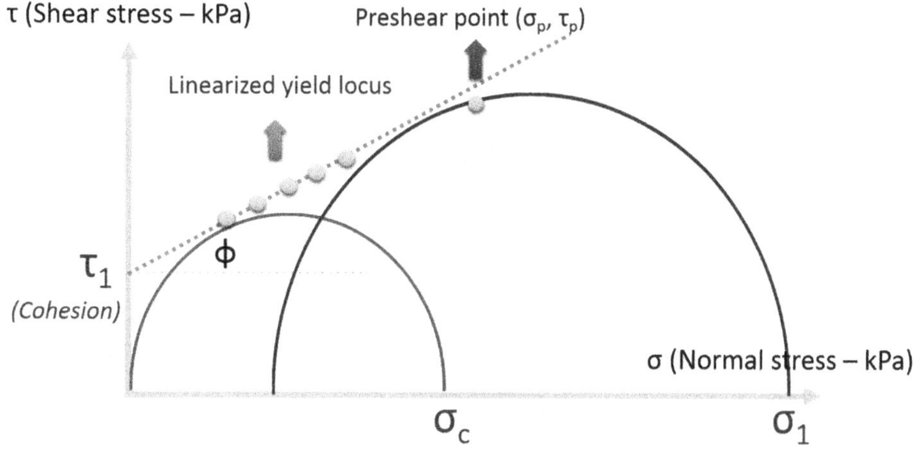

Figure 3.4.3 Yield locus and flow properties deduced from the shear-stress curve.

Table 3.4.1 Flow parameters obtained in a shear cell test

Parameter	Symbol (unit)	Definition
Unconfined yield strength (UYS)	σ_c(Pa)	The stress causing the failure of the consolidated powder bed in unconfined conditions
Major principal stress (MPS)	σ_1(Pa)	The largest of all normal stresses acting during steady-state flow in all possible cutting planes of the powder bed
Flow function	ff_c(-)	Jenike flow function, defined as: $ff_c = \dfrac{\sigma_1}{\sigma_c}$
Angle of internal friction	φ(-)	The angle between the axis of normal stress (abscissa) and the linearized yield locus

the flow behavior of milk powders spray-dried at pilot scale was markedly affected by their total fat content: $ff_c = 5.59$ for skim milk powder (2% (w/w) fat), 2.30 for semi-skim milk powder (14% (w/w) fat) and 2.28 for whole milk powder (26% (w/w) fat).

Currently, sophisticated equipments are available for measuring powder flowability: for example, the FT4 Powder Rheometer (Freeman Technology, Worcestershire, UK) and the Powder Flow Analyser (Stable Micro Systems Ltd, Surrey, UK). These powder rheometers are composed of a rotating blade that moves up and down through a vessel filled with the sample powder. The torque applied to rotate the blade and the normal force on the blade can be measured to assess powder flowability (Freeman, 2007; Freeman et al., 2009).

3.4.3.2.1 Compressibility index The compressibility index (CI) is a measure of the propensity of a powder to consolidate. As such, it is a measure of the degree of interparticular interactions. CI can be calculated according to Equation 3.4.1:

$$\mathrm{CI}(\%) = 100 \times \left(1 - \frac{d_{\mathrm{bulk}}}{d_{\mathrm{tapped}}}\right) \tag{3.4.1}$$

Table 3.4.2 Classifications of powder flow behavior according to compressibility index and Hausner ratio

Flow behavior	Compressibility index (%)	Hausner ratio (-)
Excellent	≤ 10	1.00–1.11
Good	11–15	1.12–1.18
Fair	16–20	1.19–1.25
Passable	21–25	1.26–1.34
Poor	26–31	1.35–1.45
Very poor	32–37	1.46–1.59
Extremely poor	>38	>1.60

where d_{bulk} and d_{tapped} respectively correspond to bulk and tapped densities, expressed with the same unit. A general scale of powder flowability based on CI values is given in Table 3.4.2.

Typically, a free-flowing powder should have a low CI, i.e., very close bulk and tapped densities, because the interparticular forces are lower than for a poorly flowing powder. As a general rule of thumb, a CI higher than 25% indicates poor powder flowability. A study of the physical characteristics and flow properties of a series of milk protein concentrate (MPC) powders obtained by spray-drying revealed that the increase in protein content led to smaller particles and lower bulk and tapped densities, leading to CI values up to 91% (Crowley et al., 2014).

3.4.3.2.2 Hausner ratio The Hausner ratio can also be determined from the measurements of bulk and tapped densities: it is calculated as the ratio between tapped and bulk densities. A higher Hausner ratio generally implies that the powder is more cohesive and has poorer flow properties (cf. Table 3.4.2). Indeed, powder cohesion should be overcome (i.e., interactions between particles should be disrupted) before powder is put in motion, thus requiring more energy to make it flow. The correlation between the Hausner ratio and flow behavior has been empirically established, based on the fact that a cohesive powder tends to form a powder bed with higher interparticular porosity, thus leading to a great difference between bulk and tapped densities. For example, spray-dried skim milk powders generally present better flow properties than whole milk ones, which can be explained by the much higher surface fat content of the latter (Murrieta-Pazos et al., 2011), making them more sticky. This difference in flow behavior is well-correlated with Hausner ratio: Schuck et al. (2012) obtained Hausner ratios of 1.22 and 1.76 for skim milk and whole milk, classifying them as fairly flowing and extremely poorly flowing, respectively.

3.4.3.2.3 Angle of repose The angle of repose is defined as the slope angle of the free surface of a powder gently poured from a funnel under gravity. A smaller angle of repose is generally indicative of better flowability: for example, Schuck et al. (2012) evidenced that the angle of repose is lower for spray-dried skim milk powders (41°) than for whole milk ones (57°), in agreement with the better flow properties of the former. The angle of repose is an easy and direct empirical method to assess flow properties, but it is only applicable for powders presenting low or intermediate cohesion. Several pieces of equipment have been developed to standardize this analysis, such as the GranuHeap apparatus (GranuTools, Awans, Belgium).

3.4.4 *Reconstitution properties: links with physical properties*

The ability to rehydrate readily in aqueous liquids is an essential quality attribute of food powders, especially for dairy powders that are rehydrated before use in many food preparations (Gaiani et al., 2007). The rehydration phenomenon can be divided into the following steps (Felix da Silva et al., 2018):

- Wetting: absorption of water by the powder when poured at water surface
- Swelling: increase in particle size following water absorption
- Sinking: powder immersion in the liquid
- Dispersion: separation of the powder into single particles
- Dissolution: disappearance of the particle structure by the solubilization of powder constituents

Nowadays, numerous analytical methods are available to monitor the global rehydration process (e.g., low-field nuclear magnetic resonance, focused-beam reflectance and ultrasound techniques) or focus on some peculiar steps (e.g., turbidimetry, light scattering, microscopy, rheology) (Felix da Silva et al., 2018).

It has been shown that large particles tend to facilitate rehydration (Ji et al., 2016). The production of large particles can be achieved by selecting appropriate spray-drying conditions or employing an additional agglomeration technique (Ji et al., 2015). Granulation is the most employed process for instant milk production (Chever et al., 2017). Particle agglomerates present an open structure and interparticular pores that ease water capillary rise during the wetting step (Forny et al., 2011). In contrast, small particles tend to disperse very poorly, especially in cold water, and often lead to lump formation, which dramatically slows down the reconstitution process.

Casein powders are particularly difficult to rehydrate (Felix da Silva et al., 2018). Their low lactose content and the presence of cross-linked casein micelles have been pointed out as responsible for their poor rehydration ability (Mimouni et al., 2010a, 2010b). It was also reported that the decrease in solubility during powder storage, leading to increased reconstitution times, was related to the formation of a crust at the particle surface. SEM analyses revealed that the crust was composed of fused casein micelles and its formation was favored during storage at high temperatures (Nasser et al., 2017; Burgain et al., 2016). In order to more deeply understand crust formation, the surface mechanical properties of spray-dried casein powders subjected to long-term storage at high temperatures were investigated by AFM topography and nanoindentation (Burgain et al., 2016, 2017). It was shown that casein micelles were spatially separated for the native casein powder, whereas storage at high temperatures induced the aggregation of casein micelles, leading to a heterogeneous surface formed of hard bumps, where caseins were tightly packed, and soft hollows. More details about the impairment of powder reconstitution after storage will be given in Section 3.6, Rehydration versus storage.

3.4.5 *Conclusion*

The physical properties of dairy powders obtained by spray-drying are strongly dependent on feed concentrate composition and processing conditions. Moreover, the functional properties of powders (flowability, rehydration capacity) reflect the product quality and are directly linked to powder physical properties. In the present chapter, chemical properties were not addressed but it is evident that they should also be considered for controlling powder functional properties. Therefore, it appears obvious that producing dairy powders with targeted functionalities requires the perfect understanding and mastering of the relationship between process conditions, powder structure and functionalities, which explains the tremendous interest of scientists and companies in these research topics.

Acknowledgments

The authors wish to thank Faustine GOMAND for drawing Figure 3.4.1.

References

Burgain J., J. Petit, J. Scher, R. Rasch, B. Bhandari and C. Gaiani (2017). Surface chemistry and microscopy of food powders. *Progress in Surface Science*, 92(4), 409–29.

Burgain J., J. Scher, J. Petit, G. Francius and C. Gaiani (2016). Links between particle surface hardening and rehydration impairment during micellar casein powder storage. *Food Hydrocolloids*, 61, 277–85.

Chever S., S. Méjean, A. Dolivet, F. Mei, C. M. Den Boer, G. Le Barzic, R. Jeantet and P. Schuck (2017). Agglomeration during spray drying: Physical and rehydration properties of whole milk/sugar mixture powders. *LWT - Food Science and Technology*, 83, 33–41.

Crowley S. V., I. Gazi, A. L. Kelly, T. Huppertz and J. A. O'Mahony (2014). Influence of protein concentration on the physical characteristics and flow properties of milk protein concentrate powders. *Journal of Food Engineering*, 135, 31–8.

Ebrahimi A. and T. A. G. Langrish (2015). Spray drying and crystallization of lactose with humid air in a straight-through system. *Drying Technology*, 33(7), 808–16.

Elversson J., A. Millqvist-Fureby, G. Alderborn and U. Elofsson (2003). Droplet and particle size relationship and shell thickness of inhalable lactose particles during spray drying. *Journal of Pharmaceutical Sciences*, 92(4), 900–10.

Felix da Silva D., L. Ahrné, R. Ipsen and A. B. Hougaard (2018). Casein-based powders: Characteristics and rehydration properties. *Comprehensive Reviews in Food Science and Food Safety*, 17(1), 240–54.

Fernández E., C. Schebor and J. Chirife (2003). Glass transition temperature of regular and lactose hydrolyzed milk powders. *LWT - Food Science and Technology*, 36(5), 547–51.

Fitzpatrick J. (2013). 12 – Powder properties in food production systems. In *Handbook of Food Powders*, edited by B. Bhandari, N. Bansal, M. Zhang and P. Schuck, 285–308. Woodhead Publishing Series in Food Science, Technology and Nutrition, Cambridge, UK

Fitzpatrick J., K. Barry, P. S. M. Cerqueira, T. Iqbal, J. O'Neill and Y. H. Roos (2007). Effect of composition and storage conditions on the flowability of dairy powders. *International Dairy Journal*, 17(4), 383–92.

Forny L., A. Marabi and S. Palzer (2011). Wetting, disintegration and dissolution of agglomerated water soluble powders. *Powder Technology*, 206(1–2), 72–8.

Fournaise T., J. Burgain, C. Perroud, J. Scher, C. Gaiani and J. Petit (2020). Impact of formulation on reconstitution and flowability of spray-dried milk powders. *Powder Technology*, 372, 107–16.

Freeman R. (2007). Measuring the flow properties of consolidated, conditioned and aerated powders—A comparative study using a powder rheometer and a rotational shear cell. *Powder Technology*, 174(1–2), 25–33.

Freeman R. E., J. R. Cooke and L. C. R. Schneider (2009). Measuring shear properties and normal stresses generated within a rotational shear cell for consolidated and non-consolidated powders. *Powder Technology*, 190(1–2), 65–9.

Fu X., D. Huck, L. Makein, B. Armstrong, U. Willen and T. Freeman (2012). Effect of particle shape and size on flow properties of lactose powders. *Particuology*, 10(2), 203–8.

Gaiani C., P. Schuck, J. Scher, S. Desobry and S. Banon (2007). Dairy powder rehydration: Influence of protein state, incorporation mode, and agglomeration. *Journal of Dairy Science*, 90(2), 570–81.

Jenike A. W. (1967). Quantitative design of mass-flow bins. *Powder Technology*, 1(4), 237–44.

Ji J., K. Cronin, J. Fitzpatrick, M. Fenelon and S. Miao (2015). Effects of fluid bed agglomeration on the structure modification and reconstitution behaviour of milk protein isolate powders. *Journal of Food Engineering*, 167, 175–82.

Ji J., J. Fitzpatrick, K. Cronin, P. Maguire, H. Zhang and S. Miao (2016). Rehydration behaviours of high protein dairy powders: The influence of agglomeration on wettability, dispersibility and solubility. *Food Hydrocolloids*, 58, 194–203.

Juliano P. and V. Barbosa-Canovas (2010). Food powders flowability characterization: Theory, methods, and applications. *Annual Review of Food Science and Technology*, 1, 211–39.

Kosasih L., B. Bhandari, S. Prakash, N. Bansal and C. Gaiani (2016). Physical and functional properties of whole milk powders prepared from concentrate partially acidified with CO2 at two temperatures. *International Dairy Journal*, 56, 4–12.

Mandato S., E. Rondet, G. Delaplace, A. Barkouti, L. Galet, P. Accart, T. Ruiz and B. Cuq (2012). Liquids' atomization with two different nozzles: Modeling of the effects of some processing and formulation conditions by dimensional analysis. *Powder Technology*, 224, 323–30.

Masterson V. M. and X. Cao (2008). Evaluating particle hardness of pharmaceutical solids using AFM nanoindentation. *International Journal of Pharmaceutics*, 362(1–2), 163–71.

Mercan E., D. Sert and N. Akın (2018). Determination of powder flow properties of skim milk powder produced from high-pressure homogenization treated milk concentrates during storage. *LWT*, 97, 279–88.

Mimouni A., H. C. Deeth, A. K. Whittaker, M. J. Gidley and B. Bhandari (2010a). Investigation of the microstructure of milk protein concentrate powders during rehydration: Alterations during storage. *Journal of Dairy Science*, 93(2), 463–72.

Mimouni A., H. C. Deeth, A. K. Whittaker, M. J. Gidley and B. Bhandari (2010b). Rehydration of high-protein-containing dairy powder: Slow- and fast-dissolving components and storage effects. *Dairy Science and Technology*, 90(2–3), 335–44.

Murrieta-Pazos I., C. Gaiani, L. Galet, B. Cuq, S. Desobry and J. Scher (2011). Comparative study of particle structure evolution during water sorption: Skim and whole milk powders. *Colloids and Surfaces, Part B: Biointerfaces*, 87(1), 1–10.

Nasser S., R. Jeantet, P. De-Sa-Peixoto, G. Ronse, N. Nuns, F. Pourpoint, J. Burgain, C. Gaiani, A. Hédoux and G. Delaplace (2017). Microstructure evolution of micellar casein powder upon ageing: Consequences on rehydration dynamics. *Journal of Food Engineering*, 206, 57–66.

Petit J., S. Méjean, P. Accart, L. Galet, P. Schuck, C. Le Floch-Fouéré, G. Delaplace and R. Jeantet (2015). A dimensional analysis approach for modelling the size of droplets formed by bi-fluid atomisation. *Journal of Food Engineering*, 149, 237–47.

Sadek C., L. Pauchard, P. Schuck, Y. Fallourd, N. Pradeau, C. Le Floch-Fouéré and R. Jeantet (2015a). Mechanical properties of milk protein skin layers after drying: Understanding the mechanisms of particle formation from whey protein isolate and native phosphocaseinate. *Food Hydrocolloids*, 48, 8–16.

Sadek C., P. Schuck, Y. Fallourd, N. Pradeau, C. Le Floch-Fouéré and R. Jeantet (2015b). Drying of a single droplet to investigate process–structure–function relationships: A review. *Dairy Science and Technology*, 95(6), 771–94.

Schuck P., R. Jeantet, B. Bhandari, X. D. Chen, I. T. Perrone, A. Fernandes de Carvalho, M. Fenelon and P. Kelly (2016). Recent advances in spray drying relevant to the dairy industry: A comprehensive critical review. *Drying Technology*, 34(15), 1773–90.

Schuck P., R. Jeantet and A. Dolivet (2012). *Analytical Methods for Food and Dairy Powders*. John Wiley & Sons, Chichester, West Sussex, UK.

Sharma A., A. H. Jana and R. S. Chavan (2012). Functionality of milk powders and milk-based powders for end use applications—A review. *Comprehensive Reviews in Food Science and Food Safety*, 11(5), 518–28.

Toikkanen O., M. Outinen, L. Malafronte and O. J. Rojas (2018). Formation and structure of insoluble particles in reconstituted model infant formula powders. *International Dairy Journal*, 82, 19–27.

Vehring R. (2008). Pharmaceutical particle engineering via spray drying. *Pharmaceutical Research*, 25(5), 999–1022.

Vignolles M. L., C. Lopez, J. J. Ehrhardt, J. Lambert, S. Méjean, R. Jeantet and P. Schuck (2009). Methods' combination to investigate the suprastructure, composition and properties of fat in fat-filled dairy powders. *Journal of Food Engineering*, 94(2), 154–62.

Yu W. and B. C. Hancock (2008). Evaluation of dynamic image analysis for characterizing pharmaceutical excipient particles. *International Journal of Pharmaceutics*, 361(1–2), 150–57.

3.5 *The microbiology of milk powder processing*

Evandro Martins, Ramila Cristiane Rodrigues, Pierre Schuck, Ítalo Tuler Perrone, Solimar Gonçalves Machado and Antônio Fernandes de Carvalho

3.5.1 *Introduction*

Liquid milk is a perishable product that may be preserved by concentration operations and eliminating water using heat or membrane filtration, then drying. The advantages of removing water from fresh milk include the reduction of storage and transportation costs, convenience, production seasonality management, industrial process optimization, final product diversification and, in some cases, a longer shelf life.

The microbiology of dehydrated dairy products is determined by the quality of the used raw product, how it is processed and what the contamination conditions are post-processing. Certain sporulated and thermoduric microorganisms are able to survive thermal treatments and can compromise the quality of dehydrated products. The presence of thermostable extracellular enzymes produced primarily by psychrotrophic microorganisms in milk may also cause undesirable effects in the final products, which is why raw milk must be tested for microbiological quality before being used.

The equipment, utensils, facilities and health of handlers must be monitored throughout the manufacturing process, from the collection time of raw ingredients to the packaging of the final product, in order to guarantee the quality and microbiological safety of these foods.

3.5.2 *Microbiota of raw milk*

The milk leaving the mammary glands of healthy animals is considered a sterile product. Subsequently, several factors can alter its microbiota between milking and storage (Figure 3.5.1). Initial milk contamination may occur due to intramammary infections and the presence of microorganisms on the external udder surface; also, poorly sanitized milking equipment may be a source of contamination (Figure 3.5.1).

Milk can be contaminated with pathogenic microorganisms from sick animals, including *Mycobacterium bovis, Brucella abortus, Coxiella burnetii, Listeria monocytogenes* and *Salmonella* spp. In animals with mastitis, a disease which causes the mammary gland to become infected and inflamed, milk can contain different species of microorganisms, most commonly *Staphylococcus aureus, Streptococcus agalactiae, Streptococcus uberis* and *Escherichia coli* (Vissers and Driehuis, 2009).

The external udder surface can also be an important source of microbial contamination of raw milk. Pathogenic and deteriorating bacteria, such as *Salmonella* spp., *Campylobacter* spp., *L. monocytogenes, Clostridium* spp., psychrotrophic and bacterial spores, can be transferred from udder to milk during milking (Vissers and Driehuis, 2009). These microorganisms have different origins, but generally come from the animal breeding environment (Table 3.5.1). Enteric bacteria may also be transferred to the udder when handlers have poor hygiene habits (Figure 3.5.1).

The microbiota composition on the external surface of the udder varies qualitatively and quantitatively from one rural location to another (Verdier-Metz et al., 2012) due to factors that include animal breed, feeding, confinement, milking systems and the

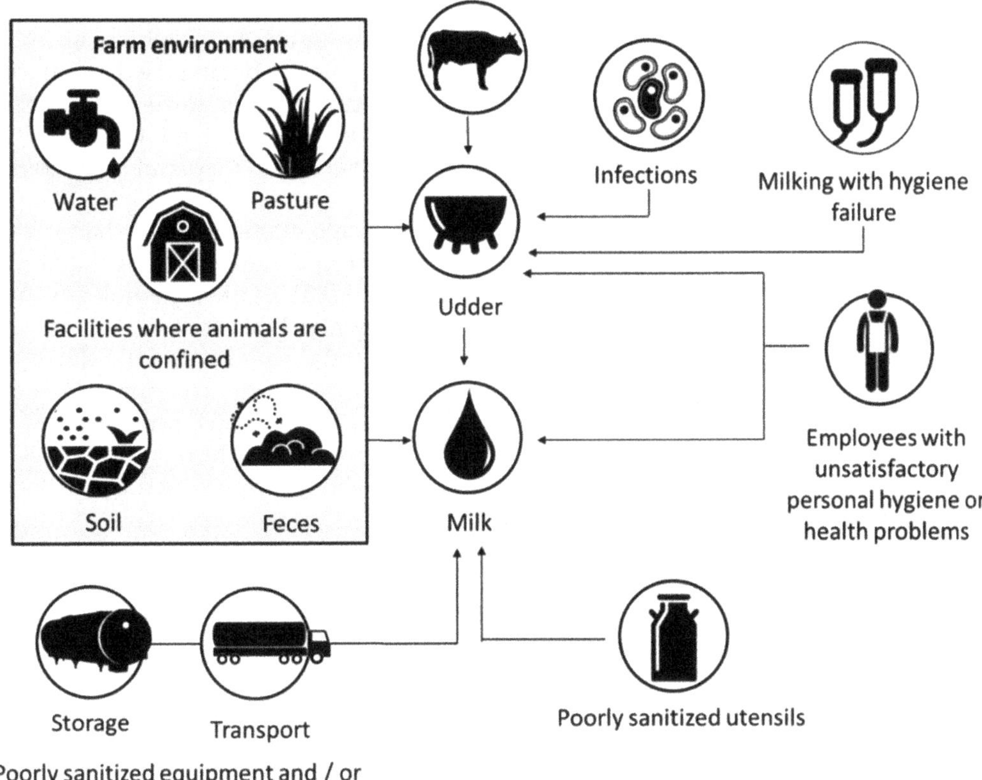

Figure 3.5.1 Factors affecting the microbiological quality of raw milk.

hygienic-sanitary quality of milking (Monsallier et al., 2012). These factors are mainly responsible for altering microbiota composition within technologically relevant groups (Mallet et al., 2012).

Milking equipment and utensils may expose the milk to Gram-negative microorganisms, such as *Pseudomonas, Flavobacterium, Enterobacter, Cronobacter, Klebsiella, Acinetobacter, Aeromonas, Alcaligenes* and *Achromobacter*, Gram-positive microorganisms, as *Corynebacterium, Microbacterium* and *Micrococcus* and Gram-positive spore-forming microorganisms, such as *Bacillus* and *Clostridium*. Most of these bacteria are able to form biofilms in equipment cracks, rubbers and joints as well as on utensils containing milk residues. During milking or processing, part of the biofilm they form may detach and come into direct contact with raw milk (Vissers and Driehuis, 2009).

Surfaces and equipment contamination levels may be influenced by the quality of the water used for cleaning. The bacterial population of a milk storage tank may increase by 12% when the equipment is sanitized with non-chlorinated water (Vilar et al., 2008).

Raw milk temperatures during transportation and storage may also influence the milk's microbiota composition. When milk leaves the udder, its temperature is ideal for mesophilic bacteria whose optimal multiplication temperatures are between 25 and 35°C. At these temperatures, contaminating microbiota have a high multiplication rate, which can lead to rapid food spoilage. Lactic acid bacteria are the main spoilage microorganisms found in raw milk; they produce organic acids from lactose fermentation which can lead to destabilization, protein precipitation and casein acid coagulation during the thermal treatment processes. To slow down lactic acid microbiota and pathogen development (Table 3.5.2), milk should be cooled to 4 to 7°C right after milking. However, a low temperature does not prevent the multiplication of psychrotrophic bacteria, and these can form biofilms that remain in the refrigeration tanks.

Pseudomonas spp. and *L. monocytogenes* are psychrotrophic contaminants that produce proteases and lipases which degrade milk proteins and lipids during refrigerated milk storage (Barclay et al., 1989; Shumi et al., 2004; Côrrea et al., 2011). Other examples of bacteria producing proteolytic and lipolytic enzymes are shown in Table 3.5.3.

These protease and lipase activities may cause sensorial changes in milk due to the presence of amino acids and free fatty acids. Moreover, casein degradation can lead to reduced cheese yield and to the formation of low-molar-mass nitrogen compounds which can constitute nutrients for other contaminating microorganisms (Brito and Brito, 2001). Proteases also alter casein heat stability and lead to protein precipitation during concentrated and UHT milk production.

To inhibit spoilage microbiota development, milk should be stored on farms for as short a time as possible and be shipped to processing plants in isothermal transport vehicles.

During the refrigeration, storage and transport stages, lapses in hygienic-sanitary conditions can lead to milk contamination from microorganisms of the genera *Pseudomonas, Micrococcus, Bacillus, Clostridium, Achromobacter, Lactobacillus* and *Flavobacterium* (Brito and Brito, 2001).

Table 3.5.1 Primary sources of microorganisms associated with dairy product spoilage and safety

Microbial species	Associated problem	Source of contamination	Possible multiplication in storage tanks
Bacillus cereus (spores)	Spoilage of pasteurized dairy products	Food, feces, soil and equipment	Yes
Bacillus sporothermodurans (spores)	Spoilage of ultra-pasteurized dairy products	Food and feces	No
Campylobacter jejuni	Food safety	Feces	No
Escherichia coli	Spoilage and food safety	Feces	Yes
Listeria monocytogenes	Food safety	Food and feces	Yes
Mycobacterium paratuberculosis	Food safety	Feces	No
Pseudomonas spp.	Spoilage	Soil and equipment	Yes
Salmonella spp.	Food safety	Feces	Yes
Streptococcus thermophilus	Spoilage	Feces, soil and equipment	Yes
Staphylococcus aureus	Food safety	Udder interior	Yes

Source: Vissers and Driehuis, 2009.

Table 3.5.2 Potential pathogens in raw milk

Microorganism	Growth at <6°C
Bacillus cereus	Yes[a]
Campylobacter jejuni	No
Clostridium spp.	No[b]
Escherichia coli	–
Listeria monocytogenes	Yes
Mycobacterium paratuberculosis	–
Salmonella spp.	No
Staphylococcus aureus	No
Yersinia enterecolitica	Yes

[a] Only a few species.
[b] Certain proteolytic species can grow at low temperatures.
Source: Schuck, 2014.

Table 3.5.3 Raw milk spoilage bacteria and associated extracellular enzyme activity

Proportion of isolates with enzymatic activity (%)	*Pseudomonas*		Other Gram-negative bacteria[a]
	Fluorescent	Non-fluorescent	
Only lipases	5	32	0–25
Only proteases	2	1	0–9
Lipases and proteases	71	11	24–92

[a] Including bacteria classified as *Enterobacteriaceae, Aeromonas, Pasteurella* or *Vibrio, Acinetobacter, Moraxella* or *Brucella; Flavobacterium, Chromobacterium, Alcaligenes.*
Source: Schuck, 2014.

Collection, storage and transportation conditions must be monitored to ensure the milk microbiological quality and ultimately to reduce the risk of contamination from pathogenic microorganisms.

However, certain microorganisms among raw milk microbiota can survive the later stages of dehydrated milk processing. In this case, the quality of the final product is compromised, which is why raw milk quality is crucial for the dairy industry.

3.5.3 *Effects of processing on milk microbiota*

Once raw milk reaches the processing plant for dehydration, it undergoes a series of processes including cooling, heat treatments, evaporative procedures and membrane filtration (Figure 3.5.2). These operations inhibit microbial growth and contribute to remove or destroy microorganisms (Figure 3.5.3).

Following thermal treatments and membrane filtration, caution should be taken to ensure the hygienic-sanitary conditions of equipment and utensils to avoid product recontamination (Figure 3.5.2). In this section, the effect of major operations on dairy product microbiota will be discussed in detail.

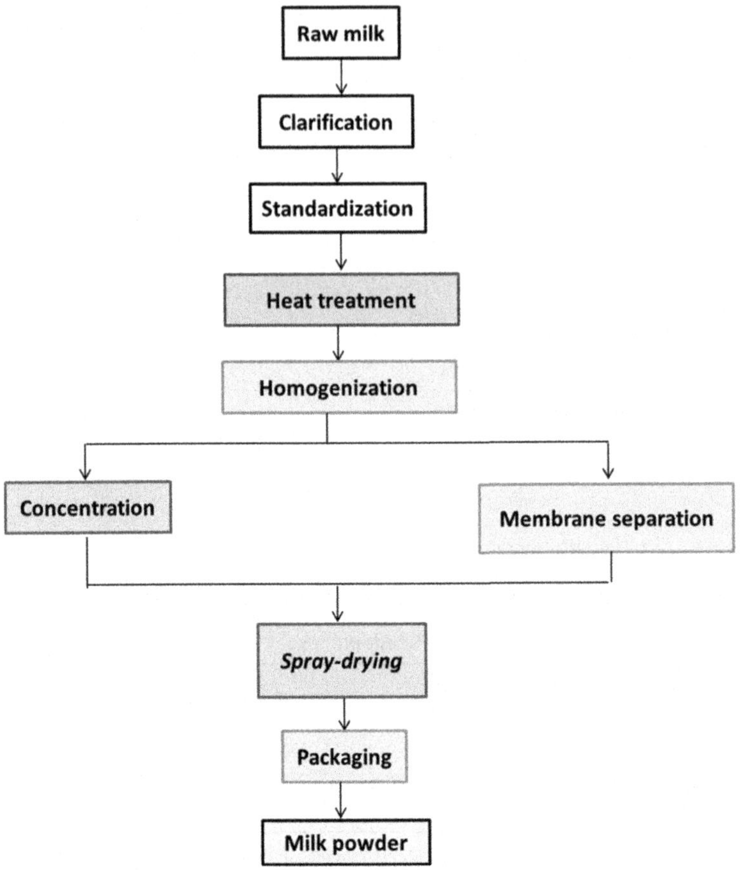

Figure 3.5.2 Dehydrated dairy product production stages. Clarification, concentration, membrane separation, spray-drying, packaging: operations that slow microbial multiplication in raw milk or in the final product; Heat treatment: processes that eliminate spoilage and pathogen microbiota; Homogenization, concentration, membrane separation, packaging: steps where recontamination from pathogenic and spoilage microorganisms is possible.

3.5.3.1 *Centrifugation: clarification and fat standardization*

At the processing plant, contaminating and deteriorating microbiota can be separated from milk using a process called clarification or bactofugation. The milk is placed in a high-performance centrifuge that separates the particulate material (viable cells and spores) from the fluid milk due to density difference. Bacterial spores found in raw milk have a higher density than the milk. When a modern ultracentrifugation process is used, a single passage through the clarifier can result in a 90% reduction in spore amount (Schuck, 2014). Although the process does not effectively eliminate viable bacteria, it is particularly significant for eliminating *Bacillus* spp. and *Clostridium* spp. spores which are not inactivated by heat treatments like pasteurization. Consequently, the clarification can be an excellent opportunity to reduce the contamination, and it helps prevent the propagation of pathogens in dehydrated dairy products.

The clarification is followed by standardization, a process that adjusts the proportions of milk fat and total solids required in the final product. While clarification aims to separate the raw milk into particulate material and whole milk, standardization is a centrifugal process that aims to separate the whole milk into cream (fat) and skim milk.

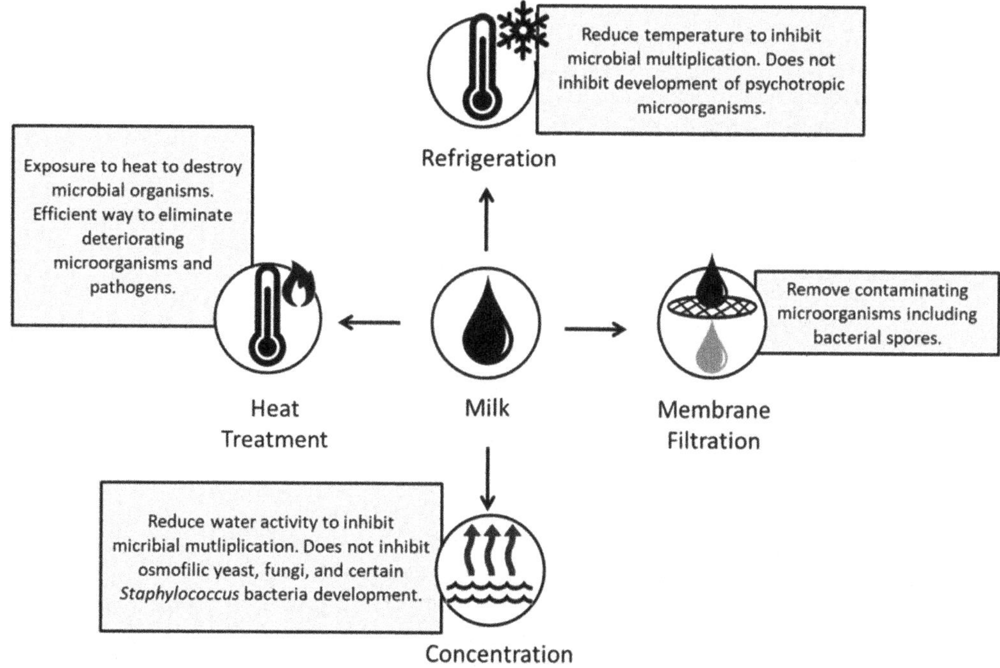

Figure 3.5.3 Effect of refrigeration, heat treatment, membrane filtration and concentration on contaminating microbiota in raw milk.

During standardization, skim milk and cream are recombined in specific proportions with no major change to the microbial population in the milk. Particles that are denser than milk plasma such as soil, somatic cells and even certain microorganisms can be removed during the centrifugation phase. The rate of removal of these particles depends on the temperature of the milk, as plasma density decreases with an increase in temperature.

In refrigerated raw milk and thermized milk, somatic and microbial cells may participate in the agglutination of fat globules and thus be removed with the cream when the milk is at temperatures below 35°C (Walstra et al., 2006).

3.5.3.2 *Heat treatment*

Heat treatment is commonly applied indirectly using heat exchangers in order to inactivate enzymes, spoilage and pathogenic microorganisms and to minimize the chemical reactions of milk (Lewis and Deeth, 2009). In the case of milk powder production, heat treatment is applied to denature milk proteins and thus to reduce lipid oxidation during storage.

Thermization is a mild heat treatment (68°C for 10 s or 65°C for 20 s) that decreases the milk's microbial load to a factor of 10^3 or 10^4 colony-forming units per milliliter of milk (UFC•mL^{-1}) (Jong, 2008). Thermization is usually employed prior to storage in refrigerated silos, and it is intended to reduce the microbial load of psychrotrophic bacteria which produce lipases (Lewis and Deeth, 2009).

Pasteurization (63°C × 30 min^{-1} or 72°C × 15 s^{-1}) is a more rigorous treatment compared to thermization, and it eliminates most of the pathogens (Table 3.5.4) and Gram-negative psychrotrophic bacteria present in milk. Pasteurization was developed to destroy

Table 3.5.4 Resistance to pasteurization of pathogens present in milk

Microorganism	Pasteurization survival[a]
Bacillus cereus	Yes (spores)
Campylobacter jejuni	No
Clostridium spp.	Yes (spores)
Escherichia coli	No
Listeria monocytogenes	No
Mycobacterium paratuberculosis	Yes (limited)
Salmonella spp.	No
Staphylococcus aureus	No
Yersinia enterecolitica	No

[a] Heat treatment at 72°C for 15 s.
Source: Schuck, 2014.

the microorganisms *Mycobacterium tuberculosis* and *Coxiella burnettii*, the two non-spore-forming pathogens that are most resistant to heat treatment (Jong, 2008).

However, it should be noted that a residual microbial population is able to withstand pasteurization and remain in the final product. These thermoduric microorganisms include *Microbacterium, Micrococcus, Enterococcus, Streptococcus, Lactobacillus* and *Corynebacterium* species (Touch and Deeth, 2009).

Although vegetative cells are inactivated by pasteurization, bacterial spores are highly resistant to heat treatment (Table 3.5.4). *Bacillus* spp. spores such as *B. cereus, B. licheniformis, B. mycoides, B. circulans, B. coagulans, B. stearothermophilus* and *B. thermodurans* are often isolated from pasteurized milk (Vissers and Driehuis, 2009).

During milk powder production, heat treatments that are more intense than pasteurization are often employed depending on industrial preferences and available equipment. In this regard, there is not a unique temperature adopted by all dairy industries; certain companies use 90°C for 15 s while others use 112°C for the same amount of time (De Carvalho et al., 2013).

Microorganisms such as *Streptococcus thermophilus* can resist heat treatments and form biofilms inside heat exchangers. High levels of certain bacteria in the final product are partly explained by their multiplication but they may also be associated with the milk contamination from biofilm release.

Although several bacterial species are resistant to heat treatments commonly applied in industrial settings, bacterial spores are the raw milk microbiota components that cause the most significant problems in dehydrated dairy products. The resistance of these structures to extreme environmental factors such as high temperatures, low pH and low water activity allows spores to survive the milk powder production process. Thermophilic spore-forming bacteria survival in milk powder has already been documented in samples stored for over 90 years (Ronimus et al., 2006).

In addition to their effects on milk microbiota, the heat treatments mentioned above can denature the β-lactoglobulin which, when associated with other proteins and minerals, leads to deposit formation due to the milk flow through the heat exchangers (Jong, 2008). Protein encrustations in heat exchangers or pipes provide support to biofilm development. The adhesion of microorganisms and bacterial spores generally occurs when thermal processes are below 80°C. This adhesion has been evaluated as the main source of contamination for thermally treated milk (Jong, 2008).

3.5.3.3 *Homogenization*

Homogenization is not a mandatory step and may be performed before or after concentration. Homogenization reduces fat globule size by inhibiting separation and preventing the formation of "cream" in concentrated milk. Under unsatisfactory sanitation and hygiene conditions, an increase in milk microbial counts may occur inside the homogenizer due to the detachment of biofilms (Shuck, 2014).

3.5.3.4 *Concentration*

After homogenization, the milk may be concentrated by vacuum evaporation or by membrane filtration to eliminate part of the water present. The decrease in final volume reduces transportation costs and additional drying steps that lead to higher energy costs.

The evaporation process takes place in multi-stage evaporators that produce concentrated milk (Figure 3.5.3). During concentration, the product remains at temperatures ranging between 45 and 55°C for long periods which favors high multiplication of thermophilic bacteria.

Heat pretreatment (pasteurization) is therefore necessary because the temperature range used in the evaporators is not high enough to guarantee the microbiological safety of the final product.

During the membrane separation processes, milk and its constituents are fractionated, concentrated and purified in order to obtain different composition solutions (Carvalho and Maubois, 2010). The milk is then filtered through a membrane with pores of defined size (Cunha et al., 2002; Alves et al., 2014). The process is based on the selective permeability of certain components across the membrane, where only molecules smaller in size than the membrane pore can pass through, while those of larger size are retained (Lira et al., 2009). During the process, the feed tube separates the milk into two streams: filtered or permeate which is the material passing through the membrane, and the concentrate or retentate, which is retained by the membrane (Baldasso, 2008; Goulas and Grandison, 2008).

Membrane separation procedures are classified according to pore size (microfiltration, ultrafiltration, nanofiltration and reverse osmosis) (Habert et al., 2006). The filter pores have diameters smaller than 10 µm; this allows milk components, such as somatic cells, bacteria and spores to be retained and separated (Figure 3.5.4).

Milk can be heat-treated (pasteurized) before filtration or it can be serially filtered. Both processes eliminate spoilage microbiota and possible pathogens. Serial filtration has been shown to eliminate more microorganisms from the retentate than pasteurization (Alves, 2013).

During filtration, milk is first microfiltered through membranes with a mean pore diameter of 0.8 to 1.4 µm (Alves, 2013) in order to remove the microbial load (Rektor and Vatai, 2004). The milk is then filtered through ultrafiltration and nanofiltration membranes to obtain a concentrated and sterile retentate without the presence of "bacterial debris" which contains many still-active enzymes (Carvalho and Maubois, 2010).

Although pasteurization and microfiltration produce retentates with low spoilage and pathogenic microorganism counts, final milk products can still be contaminated during processing stages. The presence of aerobic mesophiles, *E. coli* and psychrotrophic bacteria in the retentate may be directly related to the air quality in the processing

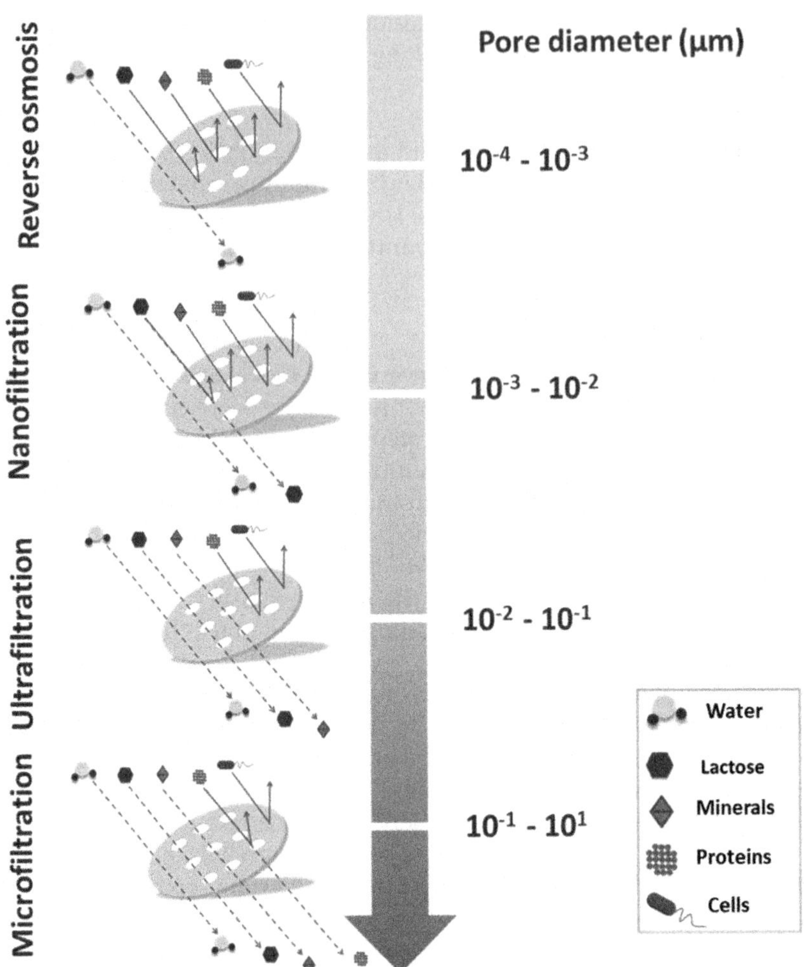

Figure 3.5.4 Membrane separation characteristics. Dotted line: Constituents that flow through the membrane. Continuous line: Constituents that are partially or totally retained by the membrane. Adapted from Bylund, 1995.

environment or to hygienic failure in utensils that come in direct contact with the product (Alves, 2013).

The filtration membranes require special cleaning and microbicidal treatment. The use of detergents containing proteolytic enzymes, which remove substances encrusted on the surface of the membrane, iodophores and hydrogen peroxide microbiocides is recommended (Robinson and Itsaranuwa, 2002). The membrane is cleaned in order to reestablish normal membrane flow and retention characteristics and to prevent the development of microorganisms in the system. Chemical cleaning generally includes rinse steps with good-quality water, alkaline cleaning, alkaline chlorine cleaning and acid cleaning (Baldasso, 2008).

3.5.3.5 *Spray drying*

Several different procedures can be employed to dehydrate milk for milk powder production. Spray drying is the most widely used procedure in the dairy industry (Figure 3.5.2).

To spray-dry milk, a fine dispersion of concentrated milk droplets is atomized inside a drying chamber. These droplets encounter a flow of hot air that fosters the quick evaporation of water to transform the milk into a powder.

The fast evaporation of the water due to the contact with hot air is at the origin of relatively low temperatures at the interface of the particles (~45°C). Consequently, spray drying will not eliminate microorganisms in concentrated milk.

Milk powder's water activity (a_w) around 0.2 has a bacteriostatic effect on the eventual microorganisms, which ensures product safety. However, certain microorganisms may survive the previously mentioned procedures used in milk powder production. These can also produce metabolites that remain in the milk powder. This results in product deterioration and consequential economic loss.

3.5.4 *Contaminating microorganisms in milk powder*

Milk powder contains about ten times more bacteria and thermoresistant toxins per gram than milk post-heat treatment due to the concentration of the liquid that occurs before drying.

Studies have shown that bacteria, including *Enterobacteriaceae* and *L. monocytogenes*, have been able to survive the drying process (Daemen and van der Stege, 1982; Doyle et al., 1985). *L. monocytogenes* have been isolated from milk powder after 12 weeks of storage (Doyle et al., 1985). This result reinforces the need for thermal treatments on raw milk to produce milk powder.

Among these heat-resistant microorganisms, *Bacillus* bacteria are common in milk powder. In the United States, 62.5% of samples of milk powder tested were contaminated with *B. cereus*, while in Brazil, *B. cereus* species were found to be present in 80% of the samples tested (Becker et al., 1994). *B. licheniformis* and *B. subtilis* were the cause of an outbreak in Croatia involving 12 children who consumed reconstituted milk powder (Pavic et al., 2005). *B. cereus* spores can survive in milk powder for several months, and rapid growth has been shown in reconstituted milk powder at room temperature (Becker et al., 1994). Limiting milk powder contamination requires the use of good-quality milk and improvement in processing and hygiene techniques (Robinson and Itsaranuwa, 2002).

During the 1950s, milk powder contamination from staphylococcal enterotoxins was a significant problem, often caused by *Staphylococcus* multiplication and toxin production in raw milk prior to heat treatment or in concentrated milk prior to drying (Anderson and Stone, 1955; Armijo et al., 1957). Improvements in hygiene and temperature control before drying have eliminated this problem. However, in 1986, several contamination outbreaks were associated with skimmed milk powder consumption. Sample analysis showed no viable pathogens, but staphylococcal enterotoxins type A and B were found in concentrations high enough to cause the disease (El-Dairouty, 1989). In 2000, a large outbreak of food poisoning from reconstituted skimmed milk powder was reported in Japan. The outbreak was caused by heat-stable toxins produced by *S. aureus* (Asao et al., 2003).

Outbreaks involving *Cronobacter* and *Enterobacter* spp. associated with the consumption of milk powder used in infant formulas were reported in different countries and resulted in high mortality rates (30–80%) (Biering et al., 1989; Nazarowec-White and Farber, 1997; Raghav and Aggarwal, 2007).

Cronobacter sakazakii has been implicated in numerous cases of meningitis, in which milk powder has been reported as the carrier of the microorganism (Varnam and Sutherland, 1994).

Yeast and filamentous fungi have also been linked to milk powder contamination. Aboul-Khier et al. (1985) isolated *Aspergillus, Penicillium* and *Mucor* spp. from several dried products, including whole and skimmed milk powder.

The presence of fungal toxins, such as aflatoxin M1, has been occasionally reported in milk powder (Galvano et al., 1996). Studies have shown that the drying process is able to reduce the concentration of toxins, though a significant number resist the processing and storage of the finished product (Marth, 1987).

Outbreaks of foodborne diseases associated with milk powder ingestion are usually the result of post-pasteurization contamination from pathogens. Foodborne pathogenic bacteria are unable to grow but can survive for long periods in the product (Fernandes, 2009).

A post-processing contamination can occur in pasteurized milk from *Enterobacteriaceae* bacteria, including *Serratia, Enterobacter, Citrobacter* and *Hafnia*, Gram-negative psychrotrophic bacteria, such as *Pseudomonas, Alcaligens* and *Flavobacterium*, and Gram-positive bacteria of the genera *Lactobacillus* and *Lactococcus* (Touch and Deeth, 2009).

Several outbreaks of salmonellosis associated with milk powder have been reported worldwide; these are often associated with drying equipment contamination (Collins et al., 1968; Rowe et al., 1987; IDF, 1994; Usera et al., 1996; Anonymous, 1997; FAO, 2007). Licari and Potter (1970) demonstrated that *Salmonella* spp. is not eradicated from milk powder during spray drying. These authors also reported that storage at 45°C and 55°C had a lethal effect on *Salmonella* Typhimurium and *Salmonella* Thompson. However, at temperatures of 25°C and 35°C, the strains were reduced but not eliminated (Licari and Potter, 1970).

The presence of the anaerobic microorganism as *C. perfringens* has also been reported in dehydrated foods and is usually associated with post-processing contamination (Varnam and Sutherland, 1994).

The contamination of raw milk from microorganisms producing heat-resistant hydrolytic enzymes may cause technological problems in formulations that use milk powder as an ingredient. Enzymes with proteolytic and lipolytic activity produced by *Bacillus* spp. and psychrotrophic bacteria are not inactivated by pasteurization and remain active in the final product (Table 3.5.5). Conditions of low water activity limit the catalytic activity of these enzymes; thus they do not participate directly in milk powder deterioration. Yet when milk is used as an ingredient in the manufacture of other products, these enzymes can affect the new formulation by reducing its shelf life.

Moreover, heat-resistant microorganisms (spore and non-spore forming) and fungi are responsible for initiating product deterioration when milk powder is subject to conditions that allow it to absorb moisture during prolonged storage (Clarke, 2001).

3.5.5 *Quality control*

Milk powder has low water activity and is rarely spoiled by microbial growth. However, the United States Department of Agriculture (USDA) and the American Dairy Product Institute have established criteria for standard plate, coliform, fungi and yeast counts to classify powdered milk as extra, standard and unassigned (Richter and Vedamuthu, 2001).

Processing and environmental hygienic-sanitary quality can be studied by analyzing post-processing contamination of indicator microorganisms like fungi, yeasts and coliforms. Standard plate counts and aerobic spore-forming mesophilic microorganisms counts may indicate the quality of raw milk to be used (Richter and Vedamuthu, 2001).

The Commission of the European Communities requires milk powder samples to be analyzed for *Enterobacteriaceae* family bacteria and coagulase-positive staphylococcal counts. A total of five samples must be analyzed, and none should present a count of

Table 3.5.5 Residual activity of extracellular enzymes after pasteurization at 72°C for 15 s

Enzymatic activity	Residual activity (%)
Lipase	59
Protease	66
Phospholipase C	30

Source: Schuck, 2014.

Enterobacteriaceae higher than 10 UFC·g^{-1}. Of the five samples analyzed, only two samples can contain between 10 and 100 UFC·g^{-1} of coagulase-positive staphylococcal, according to commission regulations. If any sample shows coagulase-positive staphylococcal counts above 10^5 UFC·g^{-1}, the presence of staphylococcal toxin in the batch must be investigated (Commission of the European Communities, 2005).

Brazilian legislation requires batches of milk powder to be analyzed for *Bacillus cereus*, thermotolerant coliforms, coagulase-positive staphylococci and *Salmonella* (BRASIL, 2001). Each batch of milk powder must be analyzed according to the legislation in force in the producing and/or importing country.

References

Aboul-Khier, F.; El-Bassiony, T.; Hamid, A. E.; & Moustafa, M.K. Enumeration of molds and yeasts in dried milk and ice-cream products [Egypt]. *Assiut Veterinary Medical Journal* (Egypt)., 1985.

Alves, M.P. Aplicação da tecnologia de separação por membranas no beneficiamento do soro de leite. Dissertação (Mestrado em Engenharia) – Universidade Federal de Viçosa, Departamento de Engenharia de Alimentos, Viçosa, MG, 2013, 117p.

Alves, M.P.; Perrone, I.T.; Souza, A.B.; Stephani, R.; Pinto, C.L.O.; Carvalho, A.F. Estudo da viscosidade de soluções proteicas através do analisador rápido de viscosidade (rva). *Revista do Instituto de Laticínios Cândido Tostes*, 69(2), p. 77–88, 2014.

Anderson, P.H.R.; Stone, D.M. *Staphylococcus* food poisoning associated with spray-dried milk. *Journal of Hygiene*, 53(4), p. 387, 1955.

Anonymous. Salmonella anatum infection in infants linked to dried milk. *Communicable Disease Report Weekly*, 7(5), p. 33–36, 1997.

Armijo, R.; Henderson, D.A.; Timothee, R.; Robinson, H.B. Food poisoning outbreaks associated with spray-dried milk: An epidemiologic study. *American Journal of Public Health and the Nation's Health*, 47(9), p. 1093, 1957.

Asao, T.; Kumeday, Y.; Kawai, T.; Shibata, T.; Oda, H.; Haruki, K.; Nakazawa, H.; Kozaki, S. An extensive outbreak of staphylococcal food poisoning due to low-fat milk in Japan: Estimation of enterotoxin A in the incriminated milk and powdered skim milk. *Epidemiology and Infection*, 130(1), p. 33–40, 2003.

Baldasso, C. Concentração, purificação e fracionamento das proteínas do soro lácteo através da tecnologia de separação por membranas. Dissertação (Mestrado em Engenharia) – Universidade Federal do Rio Grande do Sul, Escola de Engenharia, Porto Alegre, RS, 2008, 163p.

Barclay, R.; Threlfall, D.R.; Leighton, I. Haemolysins and extracellular enzymes of *Listeria monocytogenes and L. ivanovii*. *Journal of Medical Microbiology*, p. 111–118, 1989.

Becker, H.; Schaller, G.; Von Wiese, W.; Terplan, G. *Bacillus cereus* in infant foods and dried milk products. *International Journal of Food Microbiology*, 23(1), p. 1–15, 1994.

Biering, G.; Karlsson, S.; Clark, N.C.; Jonsdottir, K.E.; Ludvigsson, P.; Steingrimsson, O. Three cases of neonatal meningitis caused by *Enterobacter sakazakii* in powdered milk. *Journal of Clinical Microbiology*, 27(9), p. 2054–2056, 1989.

BRASIL. Agência Nacional de Vigilância Sanitária. Resolução RDC nº 12, de 02 de janeiro de 2001. Regulamento técnico sobre padrões microbiológicos para alimentos. *Diário Oficial da República Federativa do Brasil*, Brasília, DF, Seção, 1, 10 jan. 2001.

Brito, M.A.V.P.; Brito, J.R.F. Qualidade do leite. *Produção de leite e sociedade: Uma análise crítica da cadeia do leite no Brasil*, Capítulo, p. 61–74, 2001.

Bylund, G. *Dairy processing handbook*. Tetra Pak Processing Systems AB, 1995, S-221, v. 86, 16.

Carvalho, A.F.; Maubois, J.L. Applications of membrane technologies in the dairy industry. In *Engineering aspects of milk and dairy products*. Eds. Coimbra J.S.R., Teixeira J.A. CRC Press, Boca Raton, FL, 2010, 256p.

Clarke, W. Concentrated and dry milk and wheys. In *Applied dairy microbiology*. Eds. Marth E., Steele J. Marcel Dekker, Inc., New York, 2001, p. 77–92.

Collins, R.N.; Trager, M.D.; Goldsby, J.B.; Boring, J.R.; Cohoon, D.B.; Barr, R.N. Interstate outbreak of *Salmonella* new Brunswick infection traced to powdered milk. *JAMA*, 203(10), p. 838–844, 1968.

Comissão das Comunidades Europeias - CE. Regulamento (CE) nº 2073/2005 da Comissão, de 15 de Novembro de 2005, relativo a critérios microbiológicos aplicáveis aos gêneros alimentícios. *Jornal Oficial da União Europeia L*, 338/1, 22 dez. 2005.

Côrrea, A.P.F.; Daroit, D.J.; Velho, R.V.; Brandelli, A. Hydrolytic potential of a psychrotrophic *Pseudomonas* isolated from refrigerated raw milk. *Brazilian Journal of Microbiology*, p. 1479–1484, 2011.

Cunha, C.R.; Spadoti, L.M.; Zacarchenco, P.B.; Viotto, W.H. Efeito do fator de concentração do retentado o rendimento de queijo Minas Frescal de baixo teor de gordura fabricado por Ultrafiltração. *Food Science and Technology*, 22(1), p. 76–81, 2002.

Daemen, A.L.M.; Van Der Stege, H.J. The destruction of enzymes and bacteria during the spray drying of milk and whey (II): The effect of the drying conditions. *Netherlands Milk and Dairy Journal*, 36, p. 211–229, 1982.

De Carvalho, A.F.; Neves, B.S.; Perrone, I.T.; Stephani, R.; Vieira, S.D.A. *Sinópse dos trabalhos apresentados no 1° Curso de Leites Concentrados e Desidratados*. Templo, Juiz de Fora, 2013, 176p.

Doyle, M.P.; Meske, L.M.; Marth, E.H. Survival of *Listeria monocytogenes* during the manufacture and storage of nonfat dry milk. *Journal of Food Protection*, 48(9), p. 740–742, 1985.

Ei-Dairouty, K.R. Staphylococcal intoxication traced to non-fat dried milk. *Journal of Food Protection*, 52(12), p. 901–902, 1989.

Fernandes, R.H.E.A. *Microbiology handbook dairy products*. Leatherhead Publishing, Cambridge, 2009.

Food and Agriculture Organisation; World Health Organisation. *Enterobacter sakazakii* and Salmonella in powdered infant formula: Meeting report, Rome, January 2006. In *Microbiological Risk Assessment Series*, 10th ed. Food and Agriculture Organisation, World Health Organisation, Geneva, 2007.

Galvano, F.; Galofaro, V.; Galvano, G. Occurrence and stability of aflatoxin M1 in milk and milk products: A worldwide review. *Journal of Food Protection*, 59(1a), p. 1079–1090, 1996.

Goulas, A.; Grandison, A.S. Applications of membrane separation. In *Advanced dairy science and technology*, 1st ed. Eds. Britz, T.J., Robinson, R.K. Blackwell Publishing, 2008, 300p.

Habert, A.C.; Borges, C.P.; Nobrega, R. *Processos de separação por membranas*. Série Escola Piloto em Engenharia Química, COPPE/UFRJ. e-papers, Rio de Janeiro, 2006. 180p.

International Dairy Federation. Recommendations for the hygienic manufacture of milk and milk based products. In IDF Bulletin No. 292. Ed. International Dairy Federation. IDF, Brussels, 1994.

De Jong, P. Thermal processing of milk. *Advanced dairy science and technology*, 1-34, Britz, T., & Robinson, R. K. (Eds.). John Wiley & Sons. 2008.

Lewis, M.J.; Deeth, H.C. Heat treatment of milk. In *Milk processing and quality management*, Tamime, A. Y. (Ed.). (2009). Milk processing and quality management. John Wiley & Sons. p. 168–204, 2009.

Licari, J.J.; Potter, N.N. "Salmonella survival during spray drying and subsequent handling of skimmilk powder. II. Effects of drying conditions." *Journal of Dairy Science*, 53.7, 871–876, 1970.

Lira, H.L.; Silva, M.C.D.; Vasconcelos, M.R.S.; Lira, H.L.; Lopez, A.M.Q. Microfiltração do soro de leite de búfala utilizando membranas cerâmicas como alternativa ao processo de pasteurização. *Ciência e Tecnologia de Alimentos*, 29(1), p. 33–37, 2009.

Mallet, A.; Guéguen, M.; Kauffmann, F.; Chesneau, C.; Sesboué, A.; Desmasures, N. Quantitative and qualitative microbial analysis of raw milk reveals substantial diversity influenced by herd management practices. *International Dairy Journal*, 27(1–2), p. 13–21, 2012.
Marth, E.H. Dairy products. In *Food and beverage mycology*. Ed. Beuchat L.R. Avi Publishers, New York, 1987, p. 175–209.
Monsallier, F.; Verdier-Metz, I.; Agabriel, C.; Martin, B.; Montel, M.-C. Variability of microbial teat skin flora in relation to farming practices and individual dairy cow characteristics. *Dairy Science and Technology*, 92(3), 2012, p. 265–278.
Nazarowec-White, M.; Farber, J.M. Incidence, survival, and growth of *Enterobacter sakazakii* in infant formula. *Journal of Food Protection*, 60(3), p. 226–230, 1997.
Pavic, S.; Brett, M.; Petric, I.; Lastre, D.; Smoljanovic, M.; Atkinson, M.; Kovacic, A.; Cetinic, E.; Ropac, D. An outbreak of food poisoning in a kindergarten caused by milk powder containing toxigenic *Bacillus subtilis* and *Bacillus licheniformis*. *Archiv fur Lebensmittelhygiene*, 56(1), p. 20–22, 2005.
Raghav, M.; Aggarwal, P.K. Isolation and characterisation of *Enterobacter sakazakii* from milk foods and environment. *Milchwissenschaft*, 62(3), p. 266–269, 2007.
Rektor, A.; Vatai, G. Membrane filtration of Mozzarella whey. *Desalination*, 162, p. 279–286, 2004.
Richter, R.L.; Vedamuthu, E.R. Milk and milk products. In *Compendium of methods for the microbiological examination of foods*, 4th ed. Yvonne Salfinger and Mary Lou Tortorello (Ed.) APHA, Washington, 2001, p. 483–495.
Robinson, R.; Itsaranuwat, P. The microbiology of concentrated and dried milks. In *Dairy microbiology handbook: The microbiology of milk and milk products*. Ed. Robinson R. John Wiley & Sons, Inc., New York, 2002a, p. 175–212.
Ronimus, R.S.; Rueckert, A.; Morgan, H.W. Survival of thermophilic sporeforming bacteria in a 90(+) year old milk powder from Ernest Shackelton's Cape Royds Hut in Antarctica. *Journal of Dairy Research*, 73(2), p. 235–243, 2006.
Rowe, B.; Hutchinson, D.N.; Gilbert, R.J.; Hales, B.H.; Begg, N.T.; Dawkins, H.C.; Jacob, M.; Rae, F.A.; Jepson, M. *Salmonella* Ealing infections associated with consumption of infant dried milk. *Lancet*, 8564, p. 900–903, 1987.
Schuck, P. Microbiology of dried milk products. In *Encyclopedia of food microbiology*, 2nd ed. Robinson, R. K. (Ed.). Academic press. 2014, p. 738–743.
Shumi, W.; Hossain, M.T.; Anwar, M.N. Production of protease from *Listeria monocytogenes*. *International Journal of Agriculture and Biology*, 6(6), 1097–1100, 2004.
Touch, V.; Deeth, H.C. Microbiology of raw and market milks. In *Milk processing and quality management*, Tamime, A. Y. (Ed.). John Wiley & Sons. 2009, p. 48–71.
Usera, M.A.; Echeita, A.; Alduena, A.; Raymundo, R.; Prieto, M.I.; Tello, O.; Cano, R.; Herrera, D.; Martinez-Navarro, F. Interregional salmonellosis outbreak due to powdered infant formula contaminated with lactose fermenting *Salmonella virchow*. *European Journal of Epidemiology*, 12(4), p. 377–381, 1996.
Varnam, A.H.; Sutherland, J.P. Concentrated and dried milk products. In *Milk and milk products: Technology, chemistry and microbiology*. Eds. Varnam A.H., Sutherland J.P. Chapman and Hall, London, 1994, p. 103–158.
Verdier-Metz, I.; Gagne, G.; Bornes, S.; Monsallier, F.; Veisseire, P.; Delbes-Paus, C.; Montel, M.C. Cow teat skin, a potential source of diverse microbial populations for cheese production. *Applied and Environmental Microbiology*, 78(2), p. 326–333, 2012.
Vilar, M.; Rodriguez-Otero, J.; Dieguez, F.; Sanjuan, M.; Yus, E. Application of ATP bioluminescence for evaluation of surface cleanliness of milking equipment. *International Journal of Food Microbiology*, 125(3), p. 357–361, 2008.
Vissers, M.M.M. Driehuis, F. On-Farm hygienic milk production. In *Milk processing and quality management*, Tamime, A. Y. (Ed.). John Wiley & Sons 2009, p. 1–22.
Walstra, P.; Wouters, J.T.M.; Geurts, T.J. Milk powder. In *Dairy science and technology*, 2nd ed. Walstra, P., Walstra, P., Wouters, J. T., & Geurts, T. J. CRC press. USA, 2006, p. 517–522.

chapter 4

Innovations and prospects

Mark A. Fenelon, Eoin G. Murphy, Evandro Martins, Tatiana Lopes Fialho, Pierre Schuck, Antônio Fernandes de Carvalho, Rodrigo Stephani, Ítalo Tuler Perrone, Thao Minh Ho, Zhengzheng Zou, Bhesh Bhandari, Nidhi Bansal, Gaëlle Tanguy, Serge Méjean, Romain Jeantet, Anne Dolivet, Song Huang and Gwénaël Jan

Contents

4.1 *Infant and follow-on formulae*

Mark A. Fenelon and Eoin G. Murphy

4.1.1 *Introduction to infant milk and follow-on formula*

Infant formulae (IF) are nutritional products which are intended for use by infants from 0 to 6 months, whereas infant follow-on (FO) foods are defined as weaning and/or complementary foods generally introduced from 6 months postpartum. Infant formulations can be manufactured in either liquid, ready-to-feed (RTF) or powder format. There are a number of key steps involved, starting with nutritional calculations and formulation, followed by manufacturing and packaging. Typical manufacturing steps include batch make-up, heating, homogenization and either: (1) filling using UHT and/or retort for RTF formats or (2) evaporation, drying and filling (cans/pouches/recyclable containers) for powder formats. Powdered IF can be further broken down into three typical manufacturing processes: (1) wet-mixing, (2) dry-blending or (3) a combination of the wet-mixing and dry-blending processes (Montagne et al., 2009), each giving different functionality (Schuck et al., 2016). Wet-mixing processes ensure complete dissolution of ingredients into a homogeneous mix, allowing the manufacturer control over the reconstitution properties of the finished powder, compared to dry-blending where individual ingredients are mixed together in powder form. The focus of the discussion in this chapter is on IF and FO powder manufactured using a spray-drying process, incorporating an evaporation step to pre-concentrate the liquid feed prior to drying.

4.1.2 *Overview of wet processing*

Manufacturing parameters used during drying are dependent on the composition of macro (protein, fat and carbohydrate) and micro (minerals, vitamins, nucleotides, probiotics and other fortifications) nutrients, and their interactions. Formulation dynamics, and particularly interactions during heating, affect the stability of IF and FO powders (Fenelon et al., 2018; Murphy et al., 2014; Schuck et al., 2016), and can manifest as quality defects on reconstitution. Sequential heating steps are used during wet processing, evaporation and drying to ensure the microbial quality of finished powder (Figure 4.1.1). The final heat applied to the liquid feed concentrate is via a heat exchanger prior to atomization in the spray-dryer; often, this step is not considered a critical control point due to difficulties in calculating thermal load at high solids content and the changing flowrate needed to control the dryer outlet temperature. Where formulations are dry-blended, ingredients are mixed in powder form under low shear conditions to minimize the breakdown of agglomerates, which can cause increases in bulk density and impaired reconstitution characteristics. A combined process may involve manufacturing a base formulation (protein, fat and carbohydrate–based emulsion) using a wet-mix process, followed by dry-blending of lactose, minerals and vitamins in powder format to complete the required nutrient specification. Pro-oxidants such as iron, zinc and manganese can be added using a dry-blending step to separate them from long-chain fatty acids, e.g., DHA and ARA, which are susceptible to oxidation. The concept of using ingredients to generate a base IF mix allows manufacturers the option of generating a variety of different recipes by altering the final dry-blend composition with added specialized nutrients.

During wet processing, batches are prepared using large jacketed mixing vessels where ingredients are entrained using continuous recirculation under high shear. Once the carbohydrates, fat, protein and minerals have been dispersed in water, the mix is held under agitation to promote hydration. Whey ingredients are often added after or during lactose addition at a temperature of typically between 55 and 65°C and aid casein in providing pH buffering during subsequent mineral addition. Blends of vegetable oils (e.g., sunflower, palm, coconut and soybean) are added to the formulation to simulate the nutritional qualities of human milk fat (Shuck et al., 2016). In general, wet-mix processes can be categorized as low solids where liquid and powder ingredients are reconstituted to between 20 and 40% w/w, or high solids where reconstituted

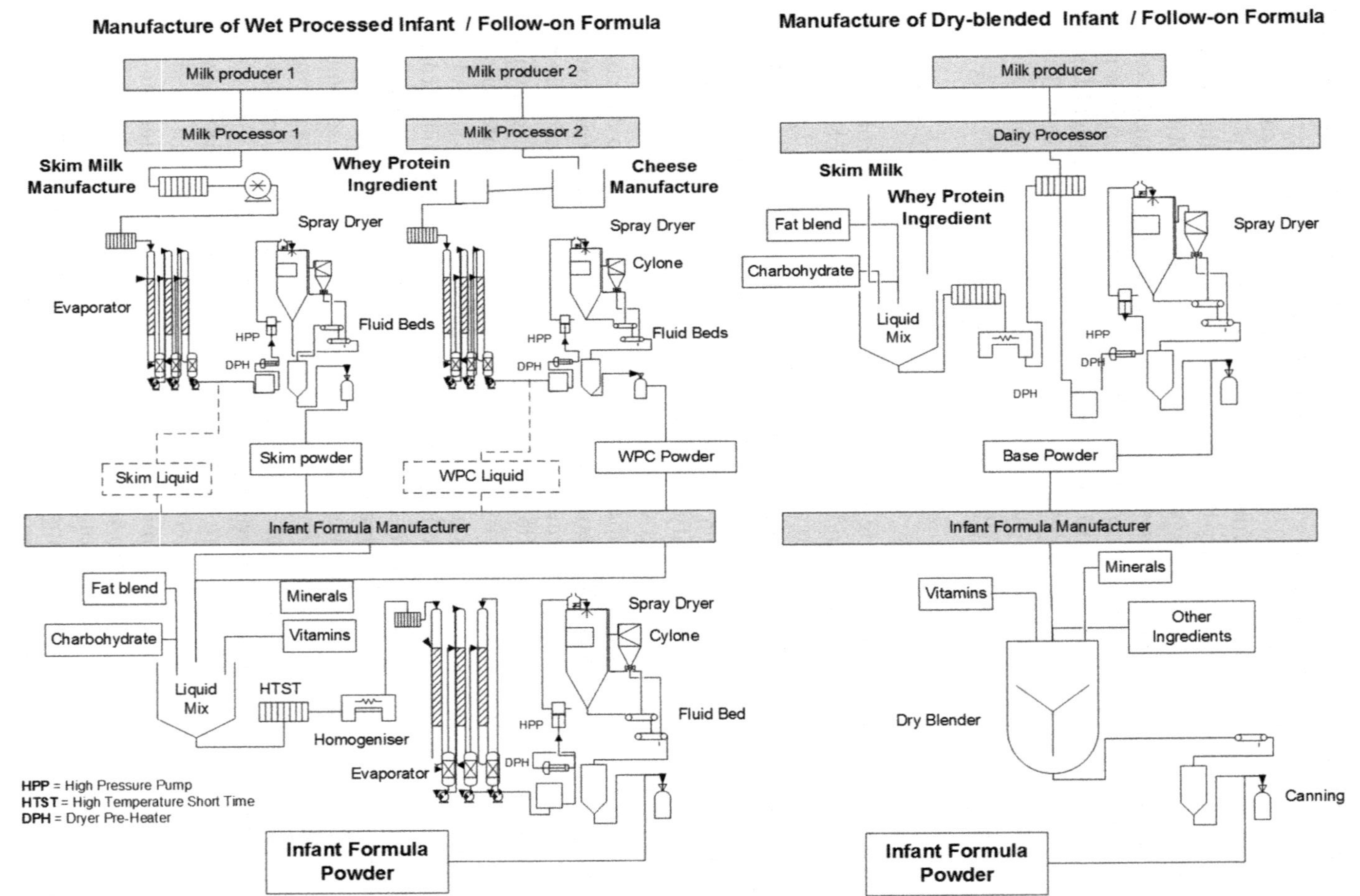

Figure 4.1.1 Generalized wet-processed and dry-blended infant and follow-on formula manufacturing process.

wet mixes are typically greater than 50 total solids (w/w). In the former case, evaporation is required prior to spray-drying (discussed below). The quantity and composition of mineral salts added to wet mixes are dependent on the innate levels provided by the protein ingredients and the finished product label claim. The order of addition of minerals is carefully carried out to achieve a consistent pH throughout the batch make-up – the mix can be adjusted using an alkali or acid (Montagne et al., 2009) if required. The pH throughout batch make-up is a key quality parameter, where low pH can result in protein aggregation and fouling during heat treatment (McSweeney et al., 2004; Simmons et al., 2007), and high pH can lead to a darker color in powders, possibly due to accelerated Maillard reaction. Calcium, in particular, affects the physical stability of formulations during heating; free ionic calcium can promote protein aggregation through electrostatic shielding, ion-specific hydrophobic bonding and covalent crosslinking mechanisms (Hebishy et al., 2019). Therefore, processors employ carefully calculated and verified ratios of soluble:insoluble calcium as well as using calcium-chelating agents to reduce the extent of protein aggregation, and hence viscosity increase, during heating. Fat-soluble vitamins are added to the fat phase and the water-soluble vitamins added directly into the protein, carbohydrate, fat wet mix. Heat-sensitive vitamins are often added in excess, whereby manufacturers calculate the inactivation of these minerals during their processing and add overages accordingly.

The heating time/temperature combinations used for IF and FO formulae can vary between manufacturers and often between individual formulae, depending on composition. High-temperature short-time (HTST) treatment is often used prior to spray-drying for powdered formulae (Schuck et al., 2016). Selecting the required heat treatment is dependent on the balance between achieving a microbial specification and the degradation of nutrients/bioavailability and product quality.

Typical homogenization pressures used during the manufacture of infant formula are ~13.8 and 3.5 MPa for first- and second-stage homogenization, respectively (McCarthy et al., 2012), at 45–66°C. The stability of resultant IF emulsion is strongly influenced by the type of protein system used and the position of the homogenization and heat treatment steps (Buggy et al., 2017). In typical processes, evaporation is used to remove water, thus concentrating the solids content of the liquid IF mix. Evaporation is a relatively economical and low-energy process compared to spray-drying, where water removal is energy-intensive and expensive. The solids content of liquid infant formula concentrates may be increased from ~20–40% total solids (TS) after heat treatment to greater than 50% TS after evaporation. The process of evaporation of IF concentrates can cause increases in viscosity, fat globule size and changes in other functional properties (McCarthy et al., 2012). The evaporated IF concentrate is spray-dried to yield a powder with low residual moisture content (i.e., approximately 2–3% w/w).

4.1.3 Drying infant milk and follow-on formula

The drying of infant and follow-on formula is usually achieved by spray-drying. Much of the innovation in the IF and FO industry tends to be product-related, i.e., the generation of new or improved recipes incorporating specific ingredients that impart a health or nutritional benefit to the consumer. Process-related innovations tend to focus on the wet processing side, for example, high solids heat treatment (Murphy et al., 2011, 2013) or membrane separation followed by the recombination of purified streams (McCarthy et al., 2017; Tobin & Verdurmen, 2016). In contrast, the spray-drying technology used for IF manufacture has remained relatively stationary. However, it is important to understand the effects of drying on the physical and nutritional characteristics of IF products.

There is a high level of variation in dryer configurations due to the large number of processing sites worldwide. Indeed, there can even be significant variation in dryer configuration within a single company or manufacturing site resulting from different equipment manufacturers, old vs. new dryers, modifications and combinations thereof. However, in general, spray-dryers for IF manufacture are similar to dryers for other dairy applications and consist of the following common elements:

- Feed atomization and agglomeration
- Heated and filtered air supply
- Dehydration, generally split between a primary drying chamber and secondary/tertiary drying in fluidized beds
- Air-powder separation and fines return
- Powder conveying

Many of the above elements are reviewed in other parts of this book. However, the following sections will discuss their application in the IF sector.

4.1.3.1 Atomization and agglomeration

Most IF installations utilize pressure nozzle atomization technologies because of the ability to manipulate parameters such as pressure, swirl/core inserts, orifice size and nozzle orientation. This in turn facilitates the optimization of spray angles, forced primary agglomeration between multiple nozzles and secondary agglomeration with returned fine particles. The enhanced control over agglomerate structure and, hence, wettability is a key advantage of pressure nozzles over rotary wheels when drying IF. Nozzles may also provide a homogenization effect, reducing fat globule size and potentially improving the reconstitution properties of the powder. However, the degree of flexibility allowed by nozzle atomization, discussed above, can also bring a potentially problematic level of complexity to drying, especially when taking into consideration other dryer variables such as air temperature, flow patterns and velocities. Therefore, caution should be exercised when changing atomization properties; it is recommended that changes are carried out incrementally and subjected to testing to assess the impact on product reconstitution and quality.

Fines return systems are designed to maximize contact between fine particles and droplets in the wet atomizing zone of the dryer. Various configurations are employed depending on the atomization system used; however, in general, it is easier to achieve the required agglomerate structure in nozzle systems, due to the flexibility mentioned above. Modern fines return systems allow adjustment of the lance positions relative to each other and to the fines line, facilitating a variety of feed spray – feed spray and feed spray – fines contact within the hot zone of the dryer. The latter provides multiple possibilities to adjust agglomerate morphology, strength and subsequent dissolution properties of the powder. It is common practice to have lances in a chasing configuration to increase the overlap of the feed spray with the fines. Lances are often arranged in bundles (typically two or three) at the top of the dryer to reduce the size of the spray circle, increasing the collision between recirculated fine particles and the feed spray. Over-agglomeration is an occasional problem during manufacture, owing to the relatively high carbohydrate content of formulations; this may be further exacerbated by the presence of sticky components (e.g., fructose). In such cases, only a portion of the fines may be returned to the wet atomization zone, with the remainder being redirected to the fluid bed (Písecký, 2012). While not commonly done, manufacturers can remove a portion of carbohydrate, most commonly lactose, from the wet mix in order to introduce this fraction in dry crystalline form to the wet atomization zone. The mesh size of the crystalline lactose is closely controlled in order to enhance agglomerate structure along with wettability and flowability.

4.1.3.2 Air supply

The primary drying air is filtered and heated by an indirect heat exchanger prior to use. In some cases, air dehumidification systems may be used to reduce high air moisture contents which limit the drying capacity of the air. This is most often employed in humid climates (e.g., South-East Asia), where IF production is not feasible without air dehumidification. The treated air is introduced to the spray-dryer through an air-distribution system, which controls the airflow pattern within the chamber. Uniform contact between the air and atomized feed is important when spray-drying all types of material; however, in IF, where high demands are placed on both the nutritional and physical properties of finished products, good spray-air contact is paramount. When contact between air and the atomized feed is poor, insufficient drying can result in wet or partially dried powder adhering to drier surfaces, causing fouling and, in extreme cases, blockages. Conversely, air that has not been cooled sufficiently by the evaporation that takes place during spray-air contact can cause problems during subsequent contact with partially dried powder, resulting in particle overheating and the reduction of solubility. In addition, poorly directed hot air can also come into contact with deposits, which can also result in overheating; however, due to the static nature of the deposits, the effect is more pronounced, resulting in extreme discoloration and insolubility. These deposits can make their way into finished IF powders where they are termed "scorched particles" – leading to defective product (Skanderby et al., 2009).

4.1.3.3 Dehydration

Drying within the primary chamber proceeds at a rapid rate; air residence times are typically in the region of 30 s which means that, in order for sufficient drying to take place in this timeframe, driving forces must be high, i.e., high dryer outlet temperature and low outlet air moisture content. When the entire drying operation takes place within the chamber, energy efficiency is low and conditions can be quite harsh on powder particles as final powder temperatures approach the high dryer outlet air temperature. With this in mind, multi-stage drying, i.e., the splitting of drying into several steps, has a number of benefits for manufacturers of IF. Secondary/tertiary drying steps typically take the form of integrated or external fluid beds which increase the total residence time of drying, allowing for lower driving forces and gentler conditions within the drying chamber, i.e., lower outlet temperatures and higher relative humidity. However, the extent to which drying can be split between primary and secondary stages is very much dependent on the properties of the material to be dried. In the case of IF and FO formulae, the relatively high amorphous carbohydrate and fat contents promote stickiness and fluidized bed blocking under conditions of high relative humidity and powder moisture content. As a result, multi-stage spray-dryers used for IF production are generally operated very close to single-stage drying conditions, with up to 99% of the drying taking place in the drying chamber. Air supply temperatures to fluidized beds are low, typically between 50 and 70°C, which is reflective of the limited dehydration taking place at this stage.

Despite the fact that integrated bed systems typically remove very little water during IF manufacture, they can be beneficial in keeping the overall humidity levels down in spray-dryers. Consider a single-stage dryer with a primary airflow rate of 100,000 kg/h drying 8000 kg/h of 50% w/w IF, operating at 180°C inlet air temperature, an inlet absolute air humidity of 4 g/kg and an outlet air temperature of 91.3°C. Based on evaporative cooling of air by water, these conditions correspond to an outlet air humidity of 42.0 g/kg. Now, if we consider a dryer with an integrated fluidized bed operating with the same evaporative capacity and with fluid bed air supplied at a flow rate of 25,000 kg/h, temperature of 60°C and an absolute humidity of 4 g/kg. The effect of mixing this dry air with the moist air in the dryer is significant; absolute humidity and temperature are reduced to 34.4 g/kg and 85°C, respectively (Bloore & O'Callaghan, 2009). While the above example is simplified – for example, it considers that no drying takes place in the fluidized bed – it illustrates that integrated beds when operated at low temperatures can serve to make the air in spray-drying chambers both cooler and dryer. This can be particularly beneficial for the drying of sticky IF products.

After secondary drying, powders are generally cooled in an external fluid bed. As with all cooling beds used in dairy powder production, cooling air should be dehumidified to reduce the chance of sticking/caking during the operation. Typical air temperatures for drying IF are shown in Table 4.1.1.

4.1.3.4 Air-powder separation

Fine particles are generally removed from exhaust air using cyclones and bag-house filters. These separation technologies are generally operated in series, with the cyclone acting as the primary separator. Cyclones operate with no moving parts and do not utilize filter materials to achieve separation; therefore the fine powder obtained is of high quality and can be re-utilized within the drying process as fines. Bag house filters are used as a subsequent separation step for the purpose of reducing the particulate concentration of exhaust air to within environmental emission limits. This operation involves the physical separation of fine particles using filters and, while CIP-able systems are available, recovered powders are generally not re-introduced to the process due to the potential for contamination with foreign materials.

4.1.3.5 Powder conveying

After drying is complete, powder must be transferred to another location for storage, blending and/or packing. Transport can have a large effect on the microstructure of IF powders; breakage of powder structures can cause rehydration issues with negative consequences for consumer perception of products (Hanley, 2011). This is of particular

Table 4.1.1 Typical air temperatures used during spray drying of infant and follow-on formula

	Temperature
Main chamber drying air	170–200°C
Integrated fluid bed air	50–70°C
External fluid bed cooling air	20–30°C
Exhaust air	80–100°C

Source: Adapted from D. Montagne, H., P. Van DaeL M. Skanderby, & W. Hugelshofer, 2009.

concern to manufacturers exporting to, for example, markets where powders are expected to solubilize easily even under extremely gentle rehydration protocols.

Typically, pneumatic conveyance is used to transport IF powders, whereby air is used to transport the dried product through pipelines. Pneumatic conveyance of IF is generally achieved by dense phase conveying, under which the mass ratio of solids to conveying air is typically in the region of 10–50 (Yanko, 2007). In contrast, dilute phase conveying is sometimes used, which is characterized by mass ratios of 0–15 (although often as low as 1) and high velocities. As a consequence, dense phase is the generally preferred regime of conveyance for IF as it has a much lower impact on important powder properties such as wettability, compared to dilute phase (Hanley et al., 2011). Where dilute-phase conveying is utilized, the effect of the conveyance system on the particle size, microstructure and physical behavior should be investigated. If possible, appropriate changes to conveying velocity, bend radii and piping dimensions should be made to minimizes the effect of dilute-phase conveying. A study by Hanley et al., 2011, showed that bulk densities before conveying and the strength of individual agglomerates were related to the protein content, i.e., agglomerate strength increased with an increasing protein to fat ratio, an important consideration when conveying powder.

4.1.4 Compositional considerations during product development

As mentioned previously, much of the innovation in the IF/FO sector tends to be concerned with the generation of new or improved recipes that impart functional benefits to the finished product. Recipe changes can have significant effects on drying properties, i.e., protein/carbohydrate/fat level and type, mineral profile, fat emulsification, addition of novel ingredients, e.g., human milk oligosaccharides (HMOs). Laboratory and pilot plant experiments can be used to evaluate the effect of changing composition on feed viscosity and/or stickiness, and the ensuing impact on drying and product quality. It is essential that manufacturers understand the functional properties of ingredients when incorporating them into new product formats.

4.1.4.1 Effect of recipe on emulsification/wet processing

The creation of a stable emulsion is central to achieving efficient drying and good powder functionality. Particle size distribution (usually determined using laser light scattering) is an important quality measurement taken throughout processing unit operations and subsequent reconstitution. In a system where an IF powder is fully dissolved, the largest detected particles are fat globules, and for this reason particle size distribution is a good determinant of emulsion stability. If fat coalescence occurs due to poor emulsification or damage during processing, the particle size will increase. Likewise, the flocculation of fat globules, by heat and/or Ca^{2+}-induced interactions of proteins, can increase particle size. The particle size distribution of an IF should be mono-modal after homogenization, and also when the finished powder is reconstituted. Additional peaks (bi- or tri-modal) can be due to coalescence or flocculation as mentioned, and/or protein aggregates or un-dissolved powder particles. McCarthy et al., 2012, measured changes in particle size distribution in a model infant formula emulsion during different stages of processing (homogenization, evaporation and drying). The study was conducted using pilot-scale equipment, with a single effect recirculating evaporator and a low-pressure (3 bar) two-fluid nozzle atomization system. After homogenization, the fat globule size distribution (D(v,0.9)) was 2.4 and 2.2 μm for formulations with 0.21 and 0.37 protein to fat ratios respectively (equating to 0.8 and 1.4 g protein per 100 ml reconstituted formula), and with mono-modal distribution. Evaporation caused the particle size to increase to 4.1 and 2.7 μm, whereas drying reduced the particle size back to 3.5 and 2.1 μm. The effect of evaporation increased with a decreasing protein to fat ratio. A similar increase in particle size upon evaporation was shown by Masum et al., 2020, and by the current authors at commercial scale on a full-size dryer with a high-pressure nozzle system. McCarthy et al. (2012) showed that at a lower protein to fat ratio, the ability of a nozzle system (albeit at low pressure) to decrease the particle size (fat globule size in this instance) is reduced. The same study showed that in a formula with a higher protein to fat ratio, the particle size after reconstitution is similar to that after homogenization, indicating a stable emulsion during processing. It is suggested that the increase in particle size during evaporation is dependent on the number of effects in the evaporator, the shear effect during pumping and the evaporative process itself during concentration. Many other factors affect particle size (fat globule size) distribution including the temperature of the feed at homogenization, the position of the homogenizer (upstream or downstream of the heat exchanger; Buggy et al., 2017) and the surfactant used (in addition to protein), i.e., lecithin, mono- and diglycerides, etc. Any process or recipe change that can reduce emulsion stability will intensify when the formulation is concentrated. This is an important consideration as it is compounded by changes in pH, flow behavior and heat stability, that further add to physical changes in the feed to the dryer.

4.1.4.2 Bulk density/volume and scoop volume

The bulk density of an infant formula powder is a key quality parameter, which ensures the correct delivery of nutrients to the infant. It is directly related to scoop volume, rehydration characteristics, agglomerate strength and friability of the powder particles during transport. The volume that a powder particle occupies is due to its density, which in turn is dependent on the intrinsic physicochemical properties of the concentrate feed and dryer processing parameters. The particle density determines the final bulk volume and ultimately the scoop volume, which translates into the specified weight of nutrients in the reconstituted formula. A further consideration is the headspace in the can, which should remain constant during processing to avoid quality issues related to oxidation during storage. Typically, the residual oxygen level in the headspace of a can, i.e., the volume above the powder, should not exceed 3%. This is relevant during storage, where oxygen levels equilibrate between the powder and the headspace. The amount of interstitial and occluded air, in powder particles/agglomerates, varies depending on the ingredients used to manufacture the formula. In fact, one of the key factors controlling formation of particle structure is feed viscosity which is driven by complex nutrient (protein, fat, mineral interactions) interactions. Understanding the interaction and behavior of infant formulations during processing in concentrated form (>50% solids) is paramount. It is a pre-requisite to the formation of powder agglomerates with good reconstitution properties, that manufacturers measure and control feed viscosity. Processing conditions (e.g., inlet/outlet and fluid bed temperatures, air speed, fines volume, transport air) and formulation characteristics (solids content, viscosity, flow behavior, emulsion quality, etc.) all contribute to powder density, scoop weight and, hence, nutrient delivery. Changes in the secondary structure in whey proteins (particularly β-lactoglobulin) during thermal processing are a major contributor to feed concentrate viscosity (Murphy et al., 2013, 2015). These heat-induced changes can be further influenced by the concentration of free calcium/magnesium ions, citrate, phosphate and other contributors to the ionic environment within an IF. The thermal history of protein ingredients (skim and whey), including the level of denaturation and aggregation, has an effect on formula viscosity during sequential heating steps. Joyce et al., 2017, showed that pre-heating a model IF (5.2% protein at 82°C × 2 min) reduced viscosity, whereas increases in calcium ion activity increased protein aggregation and viscosity. Carbohydrates such as maltodextrin, often used in FO formula, can further increase viscosity, particularly at higher solids content, to levels dependent on their dextrose equivalence value (DE).

4.1.4.3 *Effects of recipe on glass transition and stickiness during drying*

Glass transition is a physical phenomenon whereby amorphous materials undergo a gradual change from a rigid "glassy" state to a flexible "rubbery" state as a result of increased temperature and/or humidity. It is generally reported as the glass transition temperature (T_g) above which the material exists in a rubbery state; however T_g is strongly dependent on atmospheric moisture and will decrease with increasing humidity.

T_g is of importance when drying infant formulations as it can directly affect the rheological properties of powder surfaces which subsequently determine the adhesive, cohesive and overall stickiness properties of IF powders. Many studies have reported the relationship between T_g and sticking point temperature (T_s), and while the exact relationship depends on the analytical methods utilized, the general consensus is that $T_s > T_g$, indicating that the transition of materials from glass to rubbery states is the primary step that enables powder stickiness to occur at higher temperatures (Hogan & O'Callaghan, 2010; Mounsey et al., 2012). Table 4.1.2 shows the typical values of T_g and the associated change in heat capacity ΔC_p for common IF and FO components. It can be seen that, in general, larger molecular weight molecules have higher T_g values

Table 4.1.2 Glass transition temperatures (T_g) of common infant formula components

Component	T_g (°C)	ACp (J/g °C)	Present in	Reference
Casein	132	0.26	Standard formulae	Schuck et al. (2005)
Whey protein	127	0.09	Standard formulae	Schuck et al. (2005)
Lactose	98	0.38	Standard formulae	Schuck et al. (2005)
Glucose	31	0.24	GOS-enriched formulae	Schuck et al. (2005)
Galactose	30	0.24	GOS-enriched formulae	Schuck et al. (2005)
Galactooligosaccharides (GOS), 97% purity	130–140	Not reported	GOS-enriched formulae	Torres et al. (2011)
DE 5 maltodextrin	188	0.3	Specialized formulae	Roos & Karel (1991)
DE 10 maltodextrin	160	0.4	Specialized formulae	Roos & Karel (1991)
DE 20 maltodextrin	141	0.45	Standard and specialized formulae	Roos & Karel (1991)
DE 35 maltodextrin	100	0.3	Standard and specialized formulae	Roos & Karel (1991)
Water	−135	1.94	Residual water present in all formulae	Roos & Karel (1991)

compared to smaller sized molecules. The T_g of multicomponent IF systems can be estimated using relationships such as the Couchman-Karasz (CK) expression (Equation 4.1.1) which calculates overall T_g by taking a weighted average of theoretical T_g and ΔC_p values:

$$T_g = \sum_{i=1}^{n} \frac{w_i \Delta Cp_i T_{gi}}{w_i \Delta Cp_i} \tag{4.1.1}$$

where n is the total number of components in the powder, w_i is the weight fraction of component i, ΔCp_i is the change in heat capacity associated with the glass transition of component I, and T_{gi} is the glass transition temperature of component i.

Due to the high amorphous content of powders and the variable temperature and humidity conditions that can exist in spray-drying systems, CK can be a useful tool during product development or formulation, giving an estimation of the effect of the recipe on glass transition. Figure 4.1.2 shows the effect of moisture content and recipe on model first age IF; the partial substitution of lactose for DE20 maltodextrin can be expected to result in significantly higher T_g which can in turn be expected to reduce stickiness-related difficulties during drying. It should be noted that T_g predicted by Couchman–Karasz should be treated as a "ball-park" estimate, or used to give an indicative reading on the effect of recipe change. To illustrate this, Figure 4.1.2 shows that

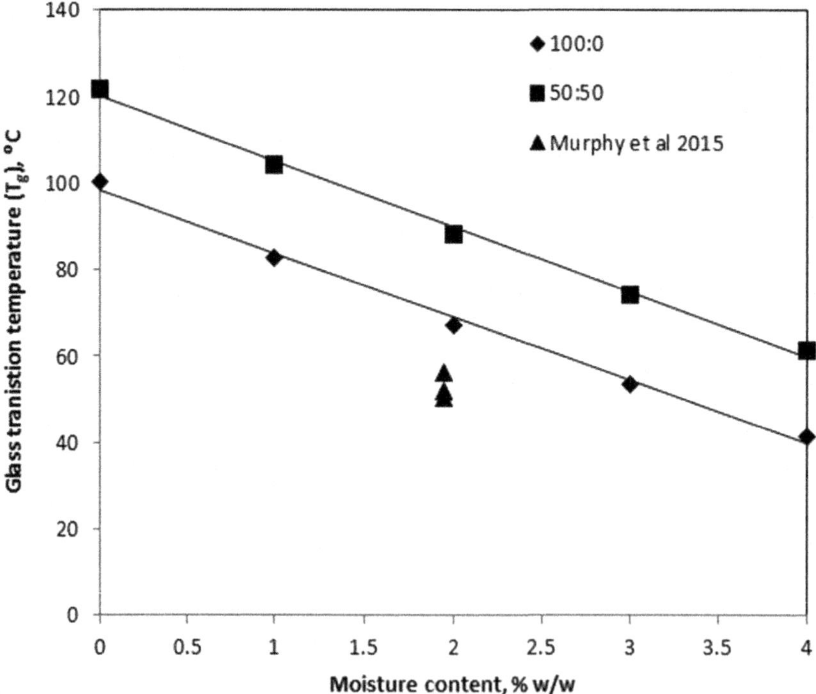

Figure 4.1.2 Effect of moisture content and recipe on glass transition temperature (T_g) as predicted by Couchman–Karasz for infant milk formulae manufactured with (i) lactose as the sole source of carbohydrate (100:0) and (ii) an equal mixture of lactose and DE20 maltodextrin (50:50) compared to measured T_g values for infant formula (100:0) from Murphy et al. (2015).

for a model first age formula identical to one of the formulations used in the prediction, actual T_g values measured by Murphy et al. (2015) are lower than predicted values. It should also be noted that substitution of lactose with low DE maltodextrin can lead to increased dryer feed viscosity, which can, in turn, affect atomisation, dehydration and powder structural properties. Overall, the glass transition and stickiness behavior have a direct effect on IF and FO dryer capacity; recipes rich in high-T_g components (e.g., low DE maltodextrins) allow for greater humidity within the chamber and thus increased capacity.

Hydrolyzed proteins are commonly used as aids to digestive comfort or allergenicity in specialty IF and FO formulae. While hydrolysis has been shown to lower the T_g of various proteins including casein (Netto et al., 1998), the T_g of caseinate–lactose and whey–lactose mixtures were found by Mounsey et al. (2012) to be unaffected by protein hydrolysis. In the case of hydrolyzed protein–lactose mixtures, it was determined that material relaxation was dominated by the amorphous lactose present and the effect of protein molecular weight was negligible. In contrast, the stickiness of these powders significantly increased in the presence of the hydrolyzed proteins. T_g, typically measured by differential scanning calorimetry, represents a thermal transition of the powder as a whole which can be related to the physical relaxation of powder systems; however, stickiness is a surface phenomenon and is therefore dependent on conditions at the surface rather than in the bulk. The hydrolysis of protein alters both the structure formation of powders and the behavior of proteins at the surface, resulting in significant changes in stickiness behavior (Hogan & O'Callaghan, 2010). This highlights the need for caution when inferring stickiness behavior from glass transition data, particularly for IF and FOF containing hydrolyzed proteins.

4.1.4.4 *Fat-induced stickiness*

While the majority of published literature in the area of powder stickiness focuses on the T_g-related stickiness of amorphous carbohydrates, the effect of surface free fat should not be neglected. Various authors have reported high fat coverage on the surface of dried dairy powders, e.g., McCarthy et al. (2013) reported surface fat coverage of up to 78% in an infant milk formula with a bulk fat composition of 32% w/w. Therefore, it is not surprising that excessive quantities of surface fat can contribute to the interaction of powder particles at various stages during manufacture. The effect of surface fat on powder stickiness is related to the fat melting temperature, above which cohesion and stickiness increase due to particle softening and the formation of inter-particulate liquid bridges (Adhikari et al., 2001). The mechanism which causes over representation of fat at powder surfaces is still not fully understood; however, recent work by Foerster et al. (2016) purported that disintegration along the oil–water interface during atomization is the primary reason. This, in combination with the findings of Vignolles et al. (2009) where it was shown that homogenization significantly reduced surface free fat in spray-dried fat-filled milk powders, suggests that the optimization of emulsion quality prior to spray-drying plays an important role in reducing stickiness caused by surface fat coverage.

4.1.5 *Conclusion*

Nutritional powders such as IF and FO are consumer foods, and must reconstitute into a homogenous liquid with no visible flecking, free oil or sediment. Innovation within the field is ingredient-orientated, focusing on achieving the specified nutrition for the target age category; however, equally important is the understanding of interactions between components in recipes and the unit operations used during the manufacture of the finished IF/FO product. These complex interrelationships are driven by physicochemical properties; for example, the concentrate viscosity of the formulation determines spray characteristics, influencing particle density/agglomeration, which in turn affects scoop volume and ultimately the delivery of the nutrients to the infant. A stable emulsion is a pre-requisite for the manufacture of an IF/FO powder with good reconstitution properties. Therefore, when drying IF and FO, it is important not to consider the drying step in isolation but rather as the last step in a sequence, which begins at formulation. That is, however, not detracting from the key influence of correct dryer set-up on powder structure, density, reconstitution properties and overall quality of IF and FO products.

References

Adhikari, B., Howes, T., Bhandari, B., & Truong, V. (2001). Stickiness in foods: A review of mechanisms and test methods. *International Journal of Food Properties, 4*(1), 1–33.

Bloore, C., & O'Callaghan, D. (2009). Process control in evaporation and drying. *Dairy Powders and Concentrated Products*. Chichester, West Sussex: Wiley & Sons. pp 332–350.

Buggy, A. K., McManus, J. J., Brodkorb, A. et al. (2017). Stabilising effect of α-lactalbumin on concentrated infant milk formula emulsions heat treated pre- or post-homogenisation. *Dairy Science and Technology, 96*(6), 845–859.

Fenelon, M. A., Hickey, R. M., Buggy, A. et al. (2018). Whey proteins in infant formula. In *Whey proteins: From milk to medicine*, H.C. Deeth and N. Bansal, eds. Amsterdam: Elsevier. pp 439–478. DOI: 10.1016/B978-0-12-812124-5.00013-8.

Foerster, M., Gengenbach, T., Woo, M. W., & Selomulya, C. (2016). The impact of atomization on the surface composition of spray-dried milk droplets. *Colloids and Surfaces, Part B: Biointerfaces, 140*, 460–471.

Hanley, K. J. (2011). Experimental quantification and modelling of attrition of infant formulae during pneumatic conveying. PhD Thesis, University College Cork, Cork.

Hanley, K. J., Cronin, K., O'Sullivan, C. et al. (2011). Effect of composition on the mechanical response of agglomerates of infant formulae. *Journal of Food Engineering, 107*(1), 71–79.

Hebishy, E., Joubran, Y., Murphy, E., & O'Mahony, J. A. (2019). Influence of calcium-binding salts on heat stability and fouling of whey protein isolate dispersions. *International Dairy Journal, 91*, 71–81.

Hogan, S. A., & O'Callaghan, D. J. (2010). Influence of milk proteins on the development of lactose-induced stickiness in dairy powders. *International Dairy Journal, 20*(3), 212–221.

Joyce, A. M., Brodkorb, A., Kelly, A., & O'Mahony, J. A. (2017). Separation of the effects of denaturation and aggregation on whey-casein protein interactions during the manufacture of a model infant formula. *Dairy Science and Technology, 96*(6), 787–806.

Masum, A. K. M., Huppertz, T., Chandrapala, J. et al. (2020). Physicochemical properties of spray-dried model infant milk formula powders: Influence of whey protein-to-casein ratio. *International Dairy Journal*, January 2020, 104565.

McCarthy, N. A., Gee, V. L., Hickey, D. K. et al. (2013). Effect of protein content on the physical stability and microstructure of a model infant formula. *International Dairy Journal, 29*(1), 53–59.

McCarthy, N. A., Kelly, A. L., O'Mahony, J. A. et al. (2012). Effect of protein content on emulsion stability of a model infant formula. *International Dairy Journal, 25*(2), 80–86.

McCarthy, N. A., Wijayanti, H. B., Crowley, S. V. et al. (2017). Pilot-scale ceramic membrane filtration of skim milk for the production of a protein base ingredient for use in infant milk formula. *International Dairy Journal, 73*, 57–62.

McSweeney, S. L., Mulvihill, D. M., & O'Callaghan, D. M. (2004). The influence of pH on the heat-induced aggregation of model milk protein ingredient systems and model infant formula emulsions stabilized by milk protein ingredients. *Food Hydrocolloids, 18*(1), 109–125.

Montagne, D. H., Van Dael, P., Skanderby, M., et al. (2009). Infant formulae – Powders and liquids. In *Dairy powders and concentrated products*, A.Y. Tamime, ed. Chichester, UK: Wiley-Blackwell. pp 294–331.

Mounsey, J. S., Hogan, S. A., Murray, B. A., & O'Callaghan, D. J. (2012). Effects of hydrolysis on solid-state relaxation and stickiness behavior of sodium caseinate-lactose powders. *Journal of Dairy Science, 95*(5), 2270–2281.

Murphy, E. G., Fenelon, M. A., Roos, Y. H., & Hogan, S. A. (2014). Decoupling macronutrient interactions during heating of model infant milk formulas. *Journal of Agricultural and Food Chemistry, 62*(43), 10585–10593.

Murphy, E. G., Roos, Y. H., Hogan, S. A., et al. (2015). Physical stability of infant milk formula made with selectively hydrolysed whey proteins. *International Dairy Journal, 40*, 39–46.

Murphy, E. G., Tobin, J. T., Roos, Y. H., & Fenelon, M. A. (2011). The effect of high velocity steam injection on the colloidal stability of concentrated emulsions for the manufacture of infant formulations. *Procedia Food Science, 1*, 1309–1315.

Murphy, E. G., Tobin, J. T., Roos, Y. H., & Fenelon, M. A. (2013). A high-solids steam injection process for the manufacture of powdered infant milk formula. *Dairy Science & Technology*, 93(4–5), 463–475.

Netto, F. M., Desobry, S. A., & Labuza, T. P. (1998). Effect of water content on the glass transition, caking and stickiness of protein hydrolysates. *International Journal of Food Properties, 1*(2), 141–161.

Písecký, J. (2012). *Handbook of milk powder manufacture*. Copenhagen, Denmark: GEA Process Engineering A/S.

Roos, Y., & Karel, M. (1991). Water and molecular weight effects on glass transitions in amorphous carbohydrates and carbohydrate solutions. *Journal of Food Science, 56*(6), 1676–1681.

Schuck, P., Blanchard, E., Dolivet, A. et al. (2005). Water activity and glass transition in dairy ingredients. *Le Lait, 85*(4–5), 295–304.

Schuck, P., Jeantet, R., Bhandari, B. et al. (2016). Recent advances in spray drying relevant to the dairy industry: A comprehensive critical review. *Drying Technology, 34*(15), 1773–1790. DOI: 10.1080/07373937.2016.1233114.

Simmons, M. J. H., Jayaraman, P., & Fryer, P. J. (2007). The effect of temperature and shear rate upon the aggregation of whey protein and its implications for milk fouling. *Journal of Food Engineering, 79*(2), 517–528.

Skanderby, M., Westergaard, V., Partridge, A. et al. (2009). Dried milk products. *Dairy Powders and Concentrated Products*. Chichester, West Sussex: Wiley & Sons. pp 180–234.

Sorensen, H., Jorgenen, E., & Westergaatd, V. (1992). Production of powdered baby food. *Scandanavian Dairy Information, 6*, 44–47.

Tobin, J., & Verdurmen, R. E. M. (2016). Improved process for the humanization of animal skim milk. Google Patents, WO2015041529A3.

Torres, D. P., Bastos, M., Maria do Pilar, F. G., et al. (2011). Water sorption and plasticization of an amorphous galacto-oligosaccharide mixture. *Carbohydrate Polymers, 83*(2), 831–835.

Vignolles, M. L., Lopez, C., Madec, M. N., et al. (2009). Fat properties during homogenization, spray-drying, and storage affect the physical properties of dairy powders. *Journal of Dairy Science, 92*(1), 58–70. DOI: 10.3168/jds.2008-1387.

Yanko, E. G. (2007). Hygienic engineering of transfer systems for dry particulate materials. *Trends in Food Science and Technology, 18*(12), 626–631.

4.2 *Lactose hydrolyzed milk powder*

Evandro Martins, Tatiana Lopes Fialho, Pierre Schuck,
Antônio Fernandes de Carvalho, Rodrigo Stephani and Ítalo Tuler Perrone

4.2.1 *Introduction*

Recent surveys predict that 75% of the world population shows some degree of lactose intolerance (Silva et al., 2019). Although this food disorder is widespread around the world, countries from South America, Africa and Asia concentrate a higher number of lactose-intolerant individuals which can be a limiting factor to the consumption of dairy products in these regions (Lule et al., 2016) (Figure 4.2.1).

This food disorder in humans is characterized by the absence or low production of the β-D-galactosidase enzyme, popularly known as lactase, which acts by hydrolyzing the lactose and releasing galactose and glucose monosaccharides for bloodstream absorption (Gerbault et al., 2011).

The inefficient enzymatic hydrolysis of lactose allows this sugar to arrive intact in the intestinal lumen, increasing the local osmolarity besides serving as a substrate to gut microbiota producing acids and gases such as CH_4, CO_2 and H_2 (Lule et al., 2016). The more

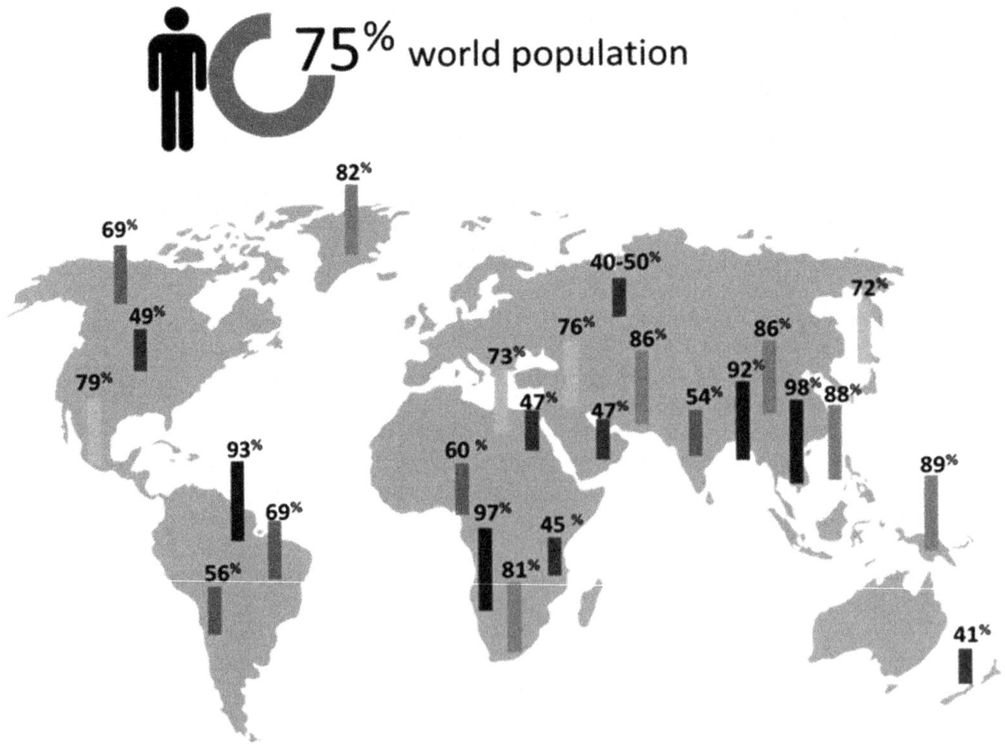

Figure 4.2.1 Distribution of world population with some degree of lactose intolerance.

common clinical consequences of low lactose digestion in intolerant persons are the development of diarrhea, flatulence and abdominal distension (Gerbault et al., 2011).

About 5.7 billion people around the world are potential consumers of dairy products with low lactose content, representing a significant market segment. According to the last report published by Future Market Insights, the global lactose-free dairy products market is estimated at US$17.8 million by 2027 end, and this estimate has driven the development of new lines of products by the dairy industry (Report, 2018).

In this sense, a wide variety of dairy products with lactose-free appeals are available in the market such as pasteurized and UHT milk, yogurt, cheeses, ice-cream, *dulce de leche* and other dairy products (Antunes et al., 2014; Mota et al., 2009; Ruiz-Matute et al., 2012; Vénica et al., 2013).

The technology of milk powder production with low lactose content is quite new, and most of the current industrial processes are based on the enzymatic hydrolysis of this sugar. Although this product is already marketed in some countries, the industry faces several technological drawbacks such as low production yield, equipment operational difficulties and powder adhesion to the drying tower besides problems of the loss of techno-functional properties during the storage of powders (Fialho et al., 2018a, 2018b; Torres et al., 2017).

In this chapter, the main techniques for the production of lactose hydrolyzed milk powder will be approached, emphasizing the challenges and the strategies that can be adopted to overcome the drawbacks associated with production.

4.2.2 *The enzymatic lactose hydrolysis in milk*

The first study involving lactose hydrolysis in raw and pasteurized milks by the application of lactase from *Saccharomyces lactis* was published in 1973 (Kosikowski and Wierbicki, 1973). Eighteen years later, a patent was deposited demonstrating the use of sonicated culture medium containing bacterial cells as a source of lactase to produce milk and dairy products with hydrolyzed lactose (Jackson and Jelen, 1989).

The β-D-galactosidase is isolated from different sources such as plants (almonds, peaches, apricots, apples), animal organs, yeasts, bacteria and filamentous fungi (Richmond et al., 1981). Current enzymes produced from bacteria, yeasts or filamentous fungi are frequently used in foods and can be commercially acquired in a powder or liquid form (Table 4.2.1).

The performance of β-D-galactosidase in industrial applications depends on factors such as substrate and enzyme concentrations, process temperature, media pH, enzyme activity and the structure of the food matrix (Bosso et al., 2016; Zolnere and Ciprovica, 2017). In order to better meet the needs of the industries, companies have commercialized several types of β-D-galactosidase enzymes with action in a broader range of temperatures and pH, as well as reducing the operational cost and making sensorial alterations to the product.

In general lines, the amount of enzyme used to hydrolyze the lactose in milk is defined by the producer, and the ideal process temperature to maintain the maximal enzymatic activity is dependent on the type of enzyme (Zolnere and Ciprovica, 2017). For example, the optimal process temperature for *Kluyveromyces lactis, Kluyveromyces fragilis* and *Kluyveromyces marxianus* is around 40°C (Bosso et al., 2016; Brady et al., 1995; Jurado et al., 2002).

However, this temperature can favor the multiplication of microorganisms present in the milk, causing the premature deterioration of raw material. For this reason, it is advisable that the milk be pasteurized in advance and that the process time does not exceed 4 hours (Ladero et al., 2000).

Another possibility is to pasteurize the raw milk and concentrate in a vacuum evaporator before the enzyme application (Fialho et al., 2018b). The vacuum evaporation promotes

Table 4.2.1 Microbial source of β-D-galactosidase enzymes

Source	Species	Reference
Yeast	*Kluyveromyces lactis*	Fialho et al., 2018a
	Kluyveromyces fragilis	Jurado et al., 2002
	Kluyveromyces marxianus	Brady et al., 1995
Filamentous fungi	*Aspergillus oryzae*	Bosso et al., 2016
	Aspergillus niger	Jones et al., 2017
	Aspergillus sphaericus	Jones et al., 2017
	Claveromycis fragiles	Abbasi & Saeedabadian, 2015
Bacteria	*Bacillus subtilis*	El-Kader et al., 2012
	Bacillus stearothermophilus	Chen et al., 2008
	Bacillus circulans	Yin et al., 2017

the water removal from food matrix by heating under reduced pressure conditions. By this technique, water boils at temperatures lower than 100°C and the evaporation occurs between 55°C and 75°C, which minimizes the thermal damage to the constituents of milk. Milk with 40 to 55% (w/w) of dry matter can be obtained by water evaporation with an energetic cost up to 20 times lower than spray drying (Caric et al., 2009). In this case, the concentrated milk leaves the evaporator, and it is cooled to the optimal temperature for the enzyme and then the enzyme is directly applied (Figure 4.2.2).

To evaluate lactose hydrolysis efficiency in fluid or concentrated milk during or after the process, kits based on the spectrophotometric analysis of the sample can be acquired for quality control by industries (Fialho et al., 2018b). However, these analytical tests are generally expensive and can demand apparatus and equipment not frequently used in daily analysis by the dairy industry. In this sense, some works have proposed lactose quantification in milk by cryoscopy analysis due to the quickness and simplicity of this method for industry (Rodrigues Junior et al., 2016).

Other alternatives for evaluating the degree of lactose hydrolysis in dairy products include chromatographic techniques (Morlock et al., 2014) and more recent analysis by Raman spectrophotometry (Torres et al., 2017).

4.2.3 *Spray drying of concentrated milk with hydrolyzed lactose*

Spray drying allows the conversion of concentrated milk in powder with minimal nutritional losses (Caric et al., 2009). Furthermore, this process allows the production of products with low humidity which guarantees a higher storage time and the reduction of logistic costs in packaging and transportation (Schuck, 2002).

In the production of lactose hydrolyzed milk powder, concentrated milk at approximately 50°C is injected into an atomizer nozzle where it is sprinkled in small droplets. Keeping the concentrated milk temperature at 50°C is important since it reduces the fluid viscosity, contributing to the formation of smaller droplets during atomization and, at the same time, favoring the water evaporation in the drying chamber (Schuck, 2013; Schuck et al., 2009).

The concentrated milk droplets enter in contact with an inlet airflow with low humidity and high temperature (150°C to 300°C) that promotes instantaneous water evaporation (Schuck, 2009). Due to differences of temperature and partial vapor pressure between the hot air and the droplets, energy in heat form is transferred from the air to the droplets while water is removed from the droplets and carried by the airflow.

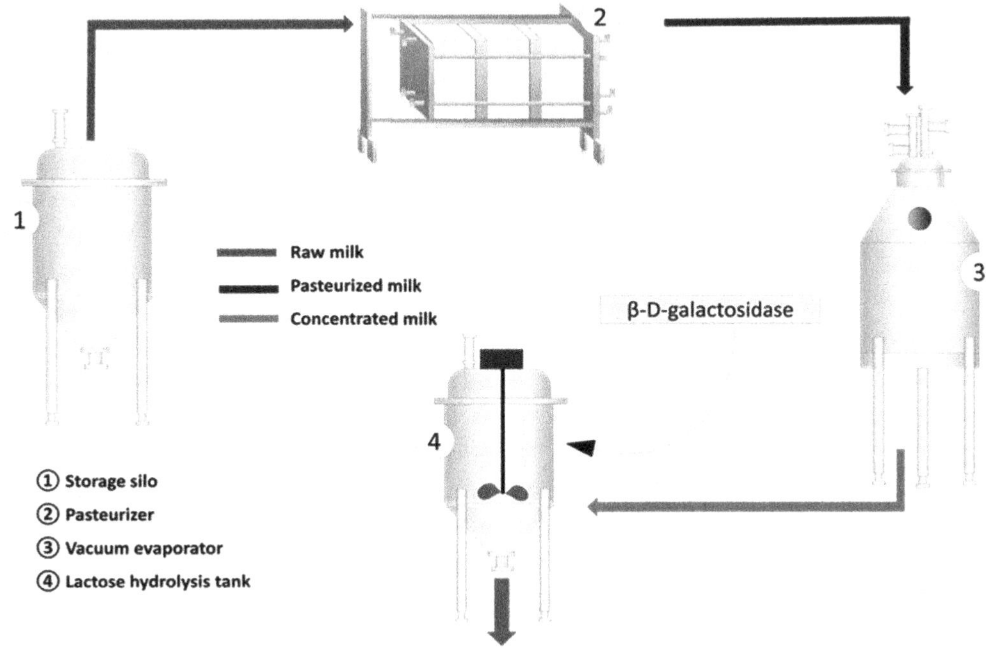

Figure 4.2.2 Schematic representation of lactose hydrolysis in concentrated milk.

The evaporation speed is related to three main factors: (i) superficial evaporation; (ii) difference in water vapor pressure between the food matrix and air; (iii) the migration rate of water from the center of the droplet towards the surface. According to Fourier's law, the larger the exchange area, the faster the heat transfer and the higher the drying speed. Similarly, the higher the partial vapor pressure difference between the droplets and hot air, the faster the drying.

Once the powder particles are formed inside the drying chamber, they can remain attached to the walls of the equipment or be carried by the outlet airflow and collected by a cyclone. The residence time of powder particles inside the drying chamber is short (20 to 60 seconds) and, therefore, there is not a real equilibrium between the humidity of hot air and the product. By consequence, when the outlet air temperature increases, the energetic efficiency of the equipment reduces (Schuck, 2002).

Some drying plants may also contain a fluidizer bed at the end of process with the goal of reducing the temperature of the product and of improving the rehydration of the powder by the agglomeration of particles (Knipschildt, 1896).

The Figure 4.2.3 represents the production of lactose hydrolyzed milk powders by spray drying.

4.2.4 *Problems associated with the production of lactose hydrolyzed milk powder*

Heat-treated dairy products, such as the milk powder, are subjected to sequencing non-enzymatic reactions known as Maillard reactions. These reactions are characterized by chemical interactions between a carbonyl group of the reducing sugar and a free amino group of the protein or amino acid. During prolonged heating or storage of dairy powders,

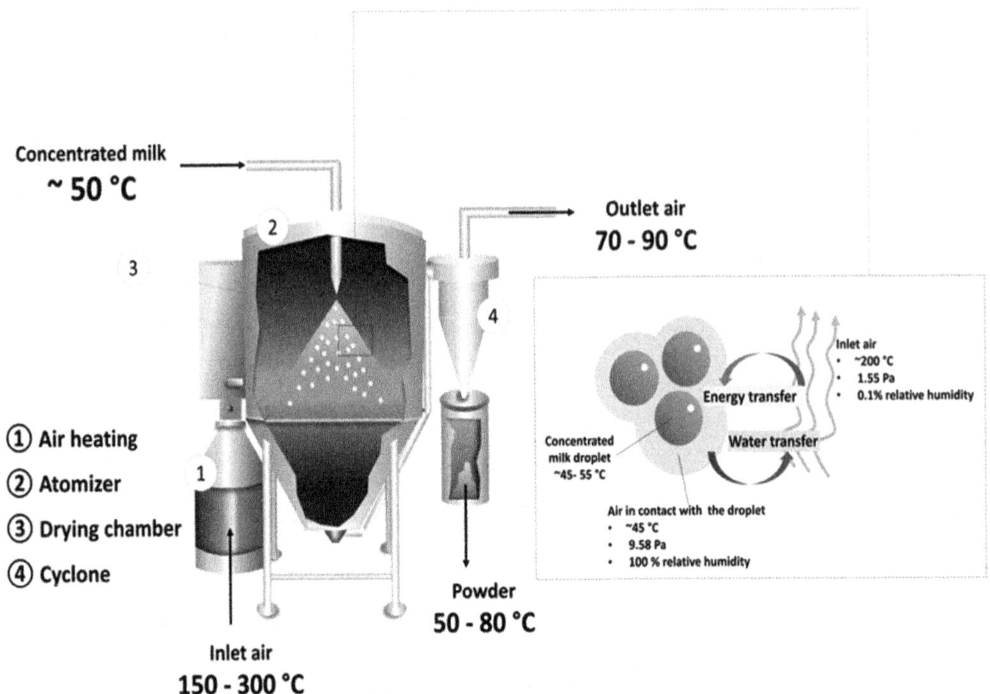

Figure 4.2.3 Schematic representation of production of lactose hydrolyzed milk powder by spray drying.

reactive compounds are formed and polymerize with protein residues, forming dark pigments or melanoidins (Fox et al., 1998).

Lactose is naturally a reducing sugar that participates in the Maillard reaction; however, when hydrolyzed, this disaccharide gives rise to two new reducing sugars molecules (glucose and galactose) that intensify the reaction (Figure 4.2.4a). As a consequence, lactose hydrolyzed milk powders tend to be darker than the traditional product and, in pronounced cases, they can show a lower degree of rehydration (Fernández et al., 2003; Fialho et al., 2018a).

Another effect of lactose hydrolysis is associated with the presence of glucose and galactose molecules that show a more significant sweetening power compared to lactose (Nijpels, 1981). Therefore, dairy products treated with β-D-galactose are sweeter and they may de-characterize the food from the sensorial point of view, resulting in rejection by the consumer.

In addition to the above-mentioned problems (changes in color and taste), the industrial production of lactose hydrolyzed milk powder is subject to some technological drawbacks including low yield, adherence to the drying chamber (Figure 4.2.4b) and caking (Figure 4.2.4d) (Fialho et al., 2018a, 2018b; Torres et al., 2017).

Currently, few reports are dedicated to explaining the causes of inconveniences related to the production of lactose hydrolyzed milk powder; nevertheless, the existing works appoint the glass transition as the key to interpreting the problem (Fernández et al., 2003; Shrestha et al., 2007; Torres et al., 2017).

The glass transition consists of a change of the system from the vitreous state (high-viscosity fluid) to the rubber state. This last physical state is characterized by a low-viscosity solution leading to structural changes such as the agglomeration of particles (Figure 4.2.4c), and consequently adhesion to the chamber and caking (Roos, 2002; Schuck et al., 2005).

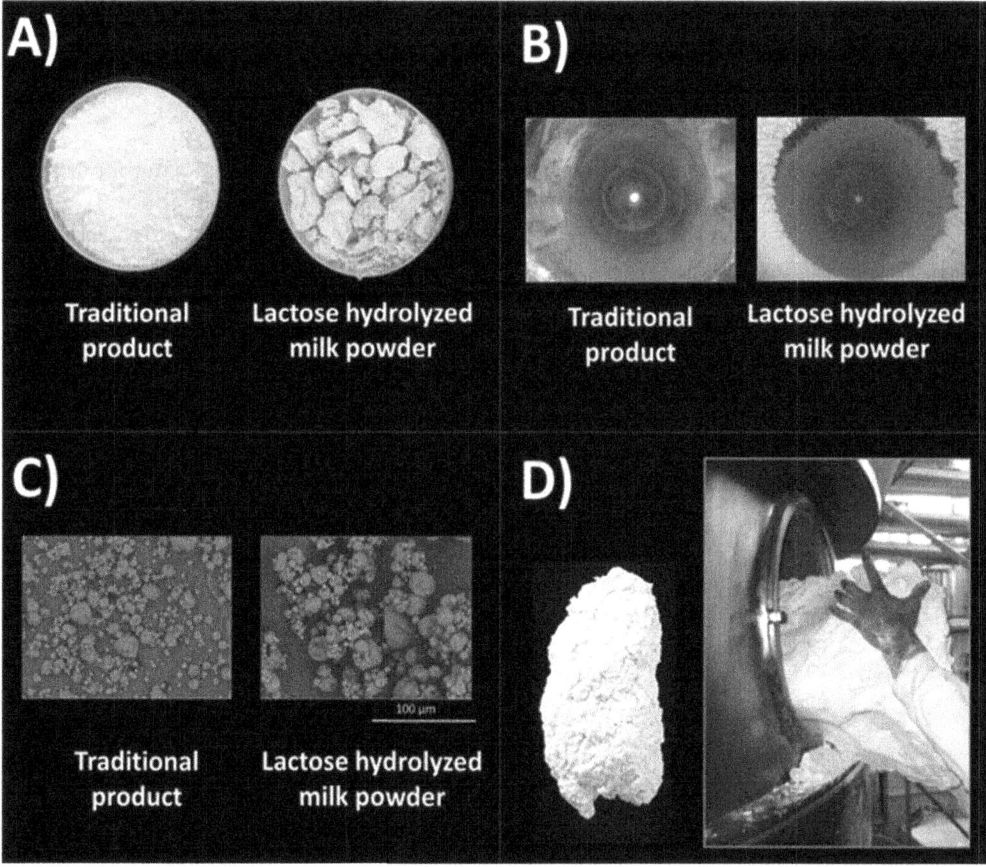

Figure 4.2.4 Drawbacks associated with the production of lactose hydrolyzed milk powder. (A) Darkening of powder; (B) adhesion to drying chamber; (C) agglomeration of powder particles; (D) caking of powder.

The glass transition of milk powders (TTg) occurs at a certain temperature, which is variable with the specific composition of product, as predicted by the equation (Couchman and Karasz, 1978):

$$\mathrm{TTg} = \frac{W_1 \cdot \mathrm{TTg}_1 \cdot \Delta Cp_1 + W_2 \cdot \mathrm{TTg}_2 \cdot \Delta Cp_2 + \cdots + W_n \cdot \mathrm{TTg}_n \cdot \Delta Cp_n}{W_1 \cdot \Delta Cp_1 + W_2 \cdot \Delta Cp_2 + \cdots + W_n \cdot \Delta Cp_n}$$

where W is the percentage of the component in the powdered milk, TTg is the glass transition temperature of the anhydrous component and ΔCp is the specific heat exchange of the component. The subscribed numbers (1, 2, n) correspond to the number of compounds present in milk, including water.

As observed in Figure 4.2.5, the TTg of lactose is approximately 3 times the TTg of glucose and galactose. For example, considering milk powders with 11% w/w of humidity, the traditional product has a TTg = 61°C while the product with hydrolyzed lactose exhibits a TTg = 36°C (Figure 4.2.5) (Fernández et al., 2003). In addition to galactose and glucose, the water content in lactose hydrolyzed milk powder is another factor responsible for TTg reduction (Jouppila et al., 1997). Water shows a very low TTg value and, for this reason, the water content should be kept as low as possible, mainly in lactose hydrolyzed milk powders that are more susceptible to glass transition.

During the spray drying, the particles of milk powder are heated inside the drying chamber. In the case of lactose hydrolyzed milk powders, if this temperature is lower than the TTg of the powder the particles will stay in the vitreous state. By contrast, if this temperature is higher than the TTg of the powder they will readily suffer glass transition (Roos, 2002).

In this situation, the powder undergoes a series of structural transformations such as free volume increase, viscosity decrease, specific heat variation and increase in thermal expansion, culminating in the agglomeration of the powder and adhesion to the surface of the equipment (Jouppila et al., 1997; Jouppila and Roos, 1994; Torres et al., 2017).

In a study carried out by Shrestha et al. (2007), it was demonstrated that the production yield of lactose hydrolyzed milk powder is low mainly because the powder can stick to the equipment. The same work also demonstrated that the TTg of milk powders (moisture close to 0) was 49°C, and only 25% of product was recovered in the cyclone.

The production of this kind of food is still a challenge for the industry, which is still looking for solutions to improve its quality and to reduce the cost of production.

4.2.5 *Perspectives in the production of lactose hydrolyzed milk powder*

Some dairy companies have focused on the incorporation of high-molecular-weight compounds into lactose hydrolyzed milk powder in order to reduce the drawbacks associated

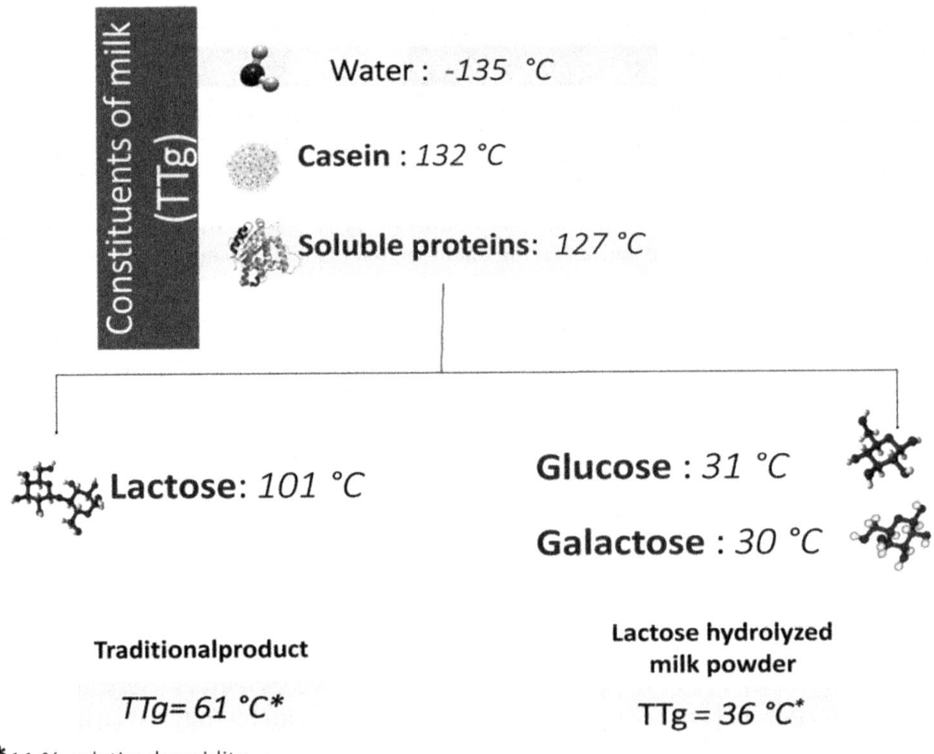

Figure 4.2.5 Glass transition temperature (TTg) of constituents of milk, traditional milk powder and lactose hydrolyzed milk powder. (Source: Kalichevsky et al., 1993; Schuck et al., 2005; Senoussi and Berk, 1995.)

with its production. Substances with high glass transition temperatures, such as maltodextrins, inulin and other polysaccharides, are responsible for the increase of the TTg of powder, thus avoiding the technological inconveniences (Rodrigues Junior et al., 2016).

In a pioneer approach, Fialho et al. (2018a,b) have pointed out that the control of drying parameters is a promising strategy to produce lactose hydrolyzed milk powder with quality and productivity closer to traditional products. According to the authors, when the lactose hydrolyzed milk is dried using the conventional operational parameters applied in the industry, the powder particles are overheated, causing the glass transition.

By simultaneously adjusting the injection rate of the concentrated milk and the inlet air temperature, it is possible obtain a powder with characteristics of color, rehydration, water activity, particle size and morphology very similar to traditional products (Fialho et al., 2018a).

Despite the successful results in terms of sensorial and techno-functional properties found, the spray-dryer settings demand higher thermal energy when compared to traditional milk powder considering the high amount of bound water (Fialho et al., 2018b). In other words, for the industries to obtain a product of good quality and minimize the problems of production, it is necessary to spend more energy during the process.

Another promising strategy consists of physically removing the lactose in milk by using membrane techniques (Figure 4.2.6). The patent n° WO 03/094623 A1 demonstrates the lactose removal by using successive ultrafiltration and nanofiltration of milk (Tossavainen and Sahlstein, 2003). Thus, the exclusion of this sugar increases the TTg powder, instead of reducing it, as discussed, to lactose hydrolysis.

Although membrane filtration is an emerging technique, some companies have not yet incorporated this technology due to implementation costs, and the need for skilled

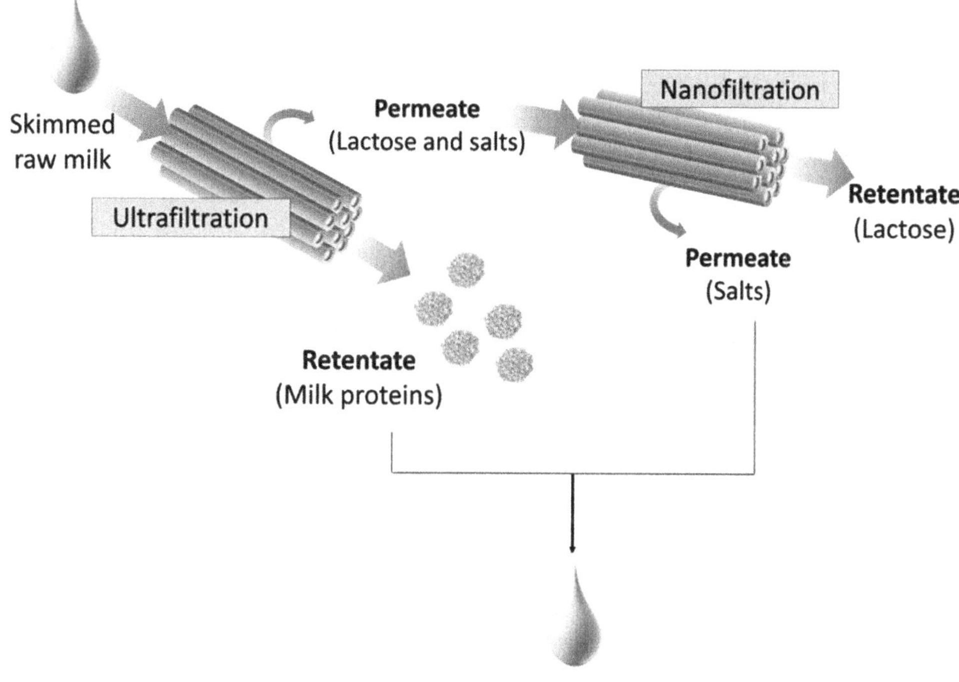

Figure 4.2.6 Schematic representation of physical removal of lactose of milk by membrane filtration. (Source: Tossavainen and Sahlstein, 2003).

laborers due to their country's legislation. The expectation is that in a few years this technology will be more widely disseminated, promoting significant advances in the production of lactose hydrolyzed milk powder.

References

Abbasi, Soleiman; Saeedabadian, Arman. "Influences of lactose hydrolysis of milk and sugar reduction on some physical properties of ice cream". *Journal of Food Science and Technology*, 52.1 (2015): 367–374

Antunes, A. E. C. et al. Development and shelf-life determination of pasteurized, microfiltered, lactose hydrolyzed skim milk. *Journal of Dairy Science*, 97(9), p. 5337–5344, 2014.

Bosso, A.; Morioka, L. R. I.; Santos, L. Fd; Suguimoto, H. H. Lactose hydrolysis potential and thermal stability of commercial β-galactosidase in UHT and skimmed milk. *Food Science and Technology*, 36(1), p. 159–165, 2016.

Brady, D.; Marchant, R.; McHale, L.; McHale, A. P. Isolation and partial characterization of P-galactosidase activity produced by a thermotolerant strain of Kluyveromyces marxianus during growth on lactose-containing media. *Enzyme and Microbial Technology*, 17(8), p. 696–699, 1995.

Caric, M. Milanovic, S., Akkerman, C., Kentish, S. E., Tamime, A. Y., & Tamime, A. Technology of evaporators, membrane processing and dryers. In *Dairy Powders and Concentrated Products*, Wiley VCH, Chichester, UK, p. 99–148, 2009.

Chen, W.; Chen, H.; Xia, Y.; Zhao, J.; Tian, F.; Zhang, H. Production, purification, and characterization of a potential thermostable galactosidase for milk lactose hydrolysis from Bacillus stearothermophilus. *Journal of Dairy Science*, 91(5), p. 1751–1758, 2008.

Couchman, P. R.; Karasz, F. E. A classical thermodynamic discussion of the effect of composition on glass-transition temperatures. *Macromolecules*, 11(1), p. 117–119, 1978.

El-Kader, A. S. S. A.; El-Dosouky, M. A.; Abouwarda, A.; Ali, S. M. A.; Osman, M. I. Isolation, screening, identification and optimization of cultural conditions for selected local bacterial β-galactosidase producer. *Journal of Applied Sciences Research*, (April), p. 2010–2017, 2012.

Fernández, E.; Schebor, C.; Chirife, J. Glass transition temperature of regular and lactose hydrolyzed milk powders. *LWT - Food Science and Technology*, 36(5), p. 547–551, 2003.

Fialho, T. L. et al. Lactose-hydrolyzed milk powder: Physicochemical and technofunctional characterization. *Drying Technology*, 14(14), p. 1688–1695, 2018a.

Fialho, T. L. et al. Lactose hydrolyzed milk powder : Thermodynamic characterization of the drying process. *Drying Technology*, 36(8), p. 922–931, 2018b.

Fox, P. F. , McSweeney, P. L., & Paul, L. H. *Dairy Chemistry and Biochemistry* (No. 637 F6.). London: Blackie Academic & Professional, 1998.

Gerbault, P., Liebert, A., Itan, Y., Powell, A., Currat, M., Burger, J., ... & Thomas, M. G. Evolution of lactase persistence: an example of human niche construction. *Philosophical Transactions of the Royal Society B: Biological Sciences*, 366(1566), p. 863–877, 2011.

Jackson, E.; Jelen, P. Comparison of acid and neutral lactases for batch hydrolysis of lactose in whey. *Milchwissenschajt*, 44, p. 544–546, 1989.

Jones, Grace K.; Hoo, Yuiry; Lee, Kim. Production technology of lactase and its application in food industry application. *Journal of the Science of Food and Agriculture*, 1, p. 46, 2017.

Jouppila, K.; Kansikas, J.; Roos, Y. H. Glass transition, water plasticization, and lactose crystallization in skim milk powder. *Journal of Dairy Science*, 80(12), p. 3152–3160, 1997.

Jouppila, K.; Roos, Y. H. Glass transitions and crystallization in milk powders. *Journal of Dairy Science*, 77(10), p. 2907–2915, 1994.

Jurado, E.; Camacho, F.; Luz6n, G.; Vicaria, J. M. A new kinetic model proposed for enzymatic hydrolysis of lactose by a β-galactosidase from Kluyveromyces fragilis. *Enzyme and Microbial Technology*, 31(3), p. 300–309, 2002.

Kalichevsky, M. T.; Blanshard, J. M. V.; Tokarczuk, P. F. Effect of water content and sugars on the glass transition of casein and sodium caseinate, *Infernational Journal of Food Science and Technology*, 28 p. 139–151, 1993.

Knipschildt, M. E. Drying of milk and milk products, in: Robinson R.K., (Ed.), Modern Dairy Technology. Advances in Milk Processing, Elsevier, London, UK, p. 131–233, 1896.

Kosikowski, F. V.; Wierbicki, L. E. Lactose hydrolysis of row and pasteurized milks by *Succharomyces lactis*. *Journal of Dairy Science*, 56(1), p. 146–148, 1973.

Ladero, M.; Santos, A.; Garcia-Ochoa, F. Kinetic modeling of lactose hydrolysis with an immobilized β- galactosidase from Kluyveromyces fragilis. *Enzyme and Microbial Technology*, 27(8), p. 583–592, 2000.

Lule, V. K. et al. *Food Intolerance: Lactose Intolerance*. First edition. [s.l.]: Elsevier Ltd., 2016.

Market Report. Lactose Free Dairy Products Market: Global Industry Analysis (2012–2016) and Opportunity Assessment (2017–2027). Market Report, January 2018, 253 pages, Future Market Insight Global & Consulting Pvt Ltd 2018.

Morlock, G. E.; Morlock, L. P.; Lemo, C. Streamlined analysis of lactose-free dairy products. *Journal of Chromatography. Part A*, 1324, p. 215–223, 2014.

Mota, K. et al. Produção de doce de leite com teor reduzido de lactose por beta-galactosidase. *Rev. Acad., Ciênc. Agrár. Ambient.*, 7(4), p. 375–382, 2009.

Nijpels, H. H. Lactase and their applications. In G. A. Tucker et al. (eds.) *Enzymes and Food Processing*. London: Applied Science Publishers Ltd., 1981. p. 89.

Richmond, M. L.; Gray, J. I.; Stine, C. M. Beta-galactosidase: Review of recent research related to technological application, nutritional concerns, and immobilization. *Journal of Dairy Science*, 64(9), p. 1759–1771, 1981.

Rodrigues Junior, P. H. et al. FT-Raman and chemometric tools for rapid determination of quality parameters in milk powder: Classification of samples for the presence of lactose and fraud detection by addition of maltodextrin. *Food Chemistry*, 196, p. 584–588, 2016.

Roos, Y. H. Importance of glass transition and water activity to spray drying and stability of dairy powders. *Le Lait*, 82(4), p. 475–484, 2002.

Ruiz-Matute, A. I. et al. Presence of mono-, di- and galactooligosaccharides in commercial lactose-free UHT dairy products. *Journal of Food Composition and Analysis*, 28(2), p. 164–169, 2012.

Schuck, P. Spray Drying of dairy products: State of the art. *Le Lait*, 82(4), p. 375–382, 2002.

Schuck, P. et al. Water activity and glass transition in dairy ingredients. *Le Lait*, 85(4–5), p. 295–304, 2005.

Schuck, P. Understanding the factors affecting spray-dried dairy powder properties and behavior. In *Dairy-Derived Ingredients: Food and Nutraceutical Uses*. Woodhead Publishing Ltd, Cambridge Editors : Corredig, M. Woodhead Publishing Limited, 2009, p. 24–50.

Schuck, P. et al. Drying by desorption: A tool to determine spray drying parameters. *Journal of Food Engineering*, 94(2), p. 199–204, 2009.

Schuck, P. Dairy powders. In *Handbook of Food Powders: Processes and Properties*. Woodhead Publishing Cambridge Editors : Bhesh Bhandari, Nidhi Bansal, Min Zhang, and Pierre Schuck 2013, p. 437–464.

Senoussi, A.; Berk, Z. Retention of diacetyl in milk during spray-drying. *Jounal of Food Science*, 60(5), p. 894–897, 1995.

Shrestha, A. K.; Howes, T.; Adhikari, B. P.; Bhandari, B. R. Water sorption and glass transition properties of spray dried lactose hydrolysed skim milk powder. *LWT - Food Science and Technology*, 40(9), p. 1593–1600, 2007.

Silva, P. H. F.; Oliveira, V. C. D.; Perin, L. M. Cow' s Milk Protein Allergy and lactose Intolerance. In *Raw Milk*. London Editors: Luis Nero Antonio Fernandes De Carvalho Elsevier Inc., 2019. p. 295–309.

Torres, J. K. F. et al. Technological aspects of lactose-hydrolyzed milk powder. *Food Research International*, 101(August), p. 45–53, 2017.

Tossavainen, O.; Sahlstein, J. *Process for Producing a Lactose-Free Milk Product*. Switzerland, 2003.

Vénica, C. L.; Bergamini, C. V.; Zalazar, C. A.; Perotti, M. C. Effect of lactose hydrolysis during manufacture and storage of drinkable yogurt. *Journal of Food & Nutrotional Disorders*, 2, p. 1–7, 2013.

Yin, H.; Pijning, T.; Meng, X.; Dijkhuizen, L.; van Leeuwen, S. S. Engineering of the *Bacillus circulans* β-galactosidase product specificity. *Biochemistry*, 56(5), p. 704–711, 2017.

Zolnere, K.; Ciprovica, I. β-galactosidases for dairy industry : Review. *Research for Rural Development*, 1, p. 215–222, 2017.

4.3 *Properties of functional camel milk powder*

Thao Minh Ho, Zhengzheng Zou, Bhesh Bhandari and Nidhi Bansal

4.3.1 *Introduction*

Unlike other non-bovine mammals, camels have a remarkable ability to adapt to the harsh living conditions of hot climate, scarce water and lack of pastures. According to FAO statistics in 2016, there were about 28 million camels domesticated around the world. Of these the African countries (e.g. Somalia, Sudan, Nigeria, Kenya, Chad, Mauritania, Ethiopia and Mali) had the largest proportion of domesticated camels (about 84% of the world's camel population), followed by Asian countries (India, Yemen, Saudi Arabia, United Arab Emirates, China and Afghanistan) which accounted for about 15% (FAO, 2016). Unlike in other countries where camels can be used for milk, meat, wool, transport, recreation, tourism and agricultural work, in Australia, camels have caused many environmental problems as there are about 1 million wild camels, which is the largest herd of feral camels in the world (Faye, 2015; Saalfeld and Edwards, 2010). Due to the health benefits of and growing demand for camel milk, in many countries (especially in the United States, Australia and Asia) milk production is the main purpose of camel raising. Under the same living conditions, camels provide a much higher quantity of milk per year than any other animals (El-Agamy, 2008). Camel milk has high nutritional value and has many components and properties homologous to human milk; thus it can be a good replacement for human milk in cases in which the source of human milk is limited. In terms of proteins, camel milk has a satisfactory balance of essential amino acids for the human diet, containing a high percentage of β-casein (65% of total casein) and a lack of β-lactoglobulin (El-Agamy, 2009; Hinz et al., 2012; Kappeler et al., 2003). β-casein is less resistant to peptide hydrolysis than other casein components (α-casein and κ-casein); thus its abundance in camel milk is considered to be one of the major reasons for the easy digestibility of camel milk (El-Agamy et al., 2009). Moreover, β-lactoglobulin is one of the major allergens in bovine milk. The absence of β-lactoglobulin in camel milk makes it a promising alternative in infant formula. Furthermore, camel milk also contains many protective proteins (e.g. immunoglobulins, lactoferrin, lysozyme and lactoperoxidase) which have anticancer, anti-diabetic and anti-bacterial properties (Barłowska et al., 2011; Konuspayeva et al., 2009). Another striking characteristic of the nutritional value of camel milk is high vitamin C and iron content, which are five and ten times, respectively, higher than in bovine milk (Farah et al., 1992; Stahl et al., 2006). Therefore, camel milk could serve as a good alternative source of vitamin C for the people living in arid areas, where fruits and vegetables are scarce (Sawaya et al., 1984).

In order to protect the bioactive compounds in camel milk during powder production, low-temperature dehydration approaches such as freeze-drying are the best choice. The low-temperature operation of freeze-drying prevents the deterioration of heat-sensitive components and the occurrence of chemical reactions common at high-temperature dehydration (Bhushani and Anandharamakrishnan, 2017). However, freeze-drying is a time-consuming process and has high operation costs due to batch operation, high resident time and low energy efficiency (Ortega-Rivas et al., 2005). These disadvantages possibly make freeze-drying unsuitable for the large-scale production of camel milk powder, and consequently an alternative dehydration approach is required. In this regard, spray drying, with its specific characteristics (a fast, continuous, one-step and reproducible process; manually controlled or fully automatic operation; and a wide range of dryer designs and configurations with different capacities and production rates), is the most promising method to replace freeze-drying in the production of camel milk powder. In this chapter, the characteristics of the bioactive compounds in camel milk, compared to those in cow and human milk, and the potential application of spray drying in the production of camel milk powder are described.

4.3.2 Composition and properties of camel milk relevant to its drying

4.3.2.1 Important bioactive compounds in camel milk

Camel milk is known for its rich content of bioactive compounds such as lactoferrin, immunoglobulins and lysozyme. These bioactive compounds play a critical role in the development of the immune system of camel neonates as well as enhancing the shelf-life of camel milk. Preservation of the functional properties of these bioactive compounds during drying is the major goal for camel milk powder production.

4.3.2.1.1 Lactoferrin Camel lactoferrin (L_f) is an 80 kDa glycosylated protein consisting of 689 amino acids, with an isoelectric point at pH 8.14 (Kappeler et al., 1999). The protein shares 74.9% amino acid sequence identity with bovine L_f and 74.0% with human L_f. The camel L_f molecule has two independent metal-binding sites and can exist as the ferric-iron (Fe^{3+})-rich hololactoferrin or the iron-free apolactoferrin (Khan et al., 2001). The antimicrobial activity of L_f was the first activity discovered, and it is currently the most widely studied function. The antimicrobial activity of L_f is due to two different mechanisms. Firstly, the iron-binding ability of L_f can locally create iron deficiency so that bacteria are prevented from growing and forming biofilms (Arnold et al., 1977). The second mechanism involves a direct interaction between L_f and the infectious agent. The positively charged N-terminus of L_f interacts with the lipopolysaccharide (LPS) of Gram-negative bacterial walls and prevents the interaction between LPS and bacterial cations (Ca^{2+} and Mg^{2+}). This causes a release of LPS from the cell wall, leading to an increase in the membrane's permeability and subsequently damage to bacteria (Coughlin et al., 1983). For Gram-positive bacteria, L_f can bind to anionic molecules on the bacterial surface (e.g. lipoteichoic acid) and reduce negative charge on the cell wall, thus favoring the action of lysozyme on the underlying peptidoglycan (Leitch and Willcox, 1999). In addition to anti-bacterial activity, L_f is now known to have a long list of other beneficial biological properties such as anticancer activity (Parikh et al., 2011) and immunoregulatory activity (van Der Does et al., 2012).

4.3.2.1.2 Immunoglobulins Immunoglobulins (Igs), also called antibodies, are large and Y-shaped proteins with two heavy chains and two light chains. These proteins play a critical role in the immune system by neutralizing pathogens such as exogenous bacteria and viruses. Based on biological properties and functionalities, immunoglobulins in humans are classified into five isotypes, IgA, IgD, IgE, IgG and IgM (Solomon and Weiss, 1995). Among them, IgA, IgG, IgM and even IgD have been confirmed to be present in camel sera according to western blot results between camel sera and human immunoglobulins (Abu-Lehia, 1997). Unlike bovine IgG, which is the only subclass belonging to this antibody isotype, three subclasses (IgG_1, IgG_2 and IgG_3) have been isolated from camel sera (Hamers-Casterman et al., 1993). Interestingly, camel IgG_2 and IgG_3 have been found to be unique due to the natural absence of light chains. Since these antibodies contain heavy chains only, they have a much lower molecular weight than conventional antibodies

(~95 kDa vs ~160 kDa) (Riechmann and Muyldermans, 1999). The smaller size of camel immunoglobulins enables them to be passed into milk and combat autoimmune diseases. In addition, they can be readily transferred from camel milk to human blood and target specific antigens which are not reachable by larger human immunoglobulins (Yadav et al., 2015). Camel heavy-chain antibodies interact with enzyme active sites more efficiently than conventional antibodies, making them very promising enzyme inhibitors. The relatively simple paratopes in these antibodies also give insight into the design of small enzyme inhibitors (Muyldermans et al., 2001).

4.3.2.1.3 Lysozyme Lysozyme (LZ, N-acetylmuramidase, muramidase) is a protein occurring in various plants, animals, bacteria and viruses. It catalyzes the hydrolysis of the peptidoglycan polymers of bacterial cell walls at the $\beta_{1\text{–}4}$ bond between N-acetylmuramic acid and N-acetylglucosamine residues to exert its anti-bacterial activity (Masschalck and Michiels, 2003; Parisien et al., 2008). Lysozyme can directly lyse sensitive bacteria. It can also work as part of a complex immunological system to enhance the phagocytosis of bacteria by macrophages (Varaldo et al., 1989). Camel milk lysozyme shows anti-microbial activity towards both Gram-positive bacteria (e.g. *Micrococcus lysodeikticus*) and Gram-negative bacteria (e.g. *E. coli*) (Duhaiman, 1988; Elagamy et al., 1996). The peptidoglycan in the cell wall of Gram-positive bacteria is readily accessible to lysozyme. Gram-negative bacteria are generally more resistant to lysozyme due to the presence of a lipopolysaccharide layer in the outer membrane, which shields the peptidoglycan from lysozyme (Masschalck and Michiels, 2003; Vaara, 1992). Some Gram-negative bacteria have been found to possess genes coding for Ivy (inhibitor of vertebrate lysozyme) and its homologs (Monchois et al., 2001), which also contribute to their resistance to lysozyme. In addition to anti-bacterial activity, lysozyme has been proven to have anti-viral, anti-parasitic and anti-fungal activities (Knorr et al., 1979; León-Sicairos et al., 2006; Sylvia et al., 1999).

4.3.2.2 Heat stability of camel milk

The heat stability of milk is an important parameter to be considered during its thermal processing. The heat coagulation time (HCT) of camel milk at 100, 120 and 130°C was investigated at pH 6.3–7.1 (Farah and Atkins, 1992). At 120 and 130°C, milk was very unstable at all pH values with HCT below 2–3 min. At 100°C, the HCT initially increased to 12 min, then remained constant between pH 6.4 and 6.7 and increased progressively with increasing pH, reaching approximately 33 min at pH 7.1. Bovine milk showed much better heat stability at 130°C, with a maximum HCT of ~40 min around pH 6.7 and a minimum near pH 6.8. The heat stability of bovine milk increased above pH 6.9 (Farah, 1993). Kouniba et al. (2005) claimed that heat preservation of camel milk could be done only by pasteurization since it coagulated within 2 min at 100–130°C. In another study, the HCTs at 140°C for bovine, buffalo and camel milks were determined to be 30.1, 26.2 and 2.2 min, respectively (Shyam et al., 2016).

The presence of κ-casein and β-lactoglobulin and their interaction during heating are believed to be critical in maintaining milk stability (Fox and Hoynes, 1976). Therefore, the reduced amount of κ-casein (5% of total casein in camel milk compared with 13.6% in bovine milk) and the absence of β-lactoglobulin might be responsible for the poor stability of camel milk at high temperatures.

4.3.2.3 Thermostability of camel milk proteins

Many studies on the effect of different heating conditions on the properties of bovine and camel whey proteins have been reported (Elagamy, 2000; Farah, 1986; Felfoul et al., 2015;

Laleye et al., 2008). Inconsistent results were observed in these studies possibly due to the differences in milk origin and assay methods. The thermostability of milk proteins during heating can be determined by measuring the changes of nitrogen content (Farah, 1986), size and molecular weight (Elagamy, 2000), free thiol group concentration (Felfoul et al., 2015) and solubility (Laleye et al., 2008) of whey proteins before and after heating. It was reported that whey proteins of camel milk had a higher heat-resistant ability than those of bovine and buffalo milk when the milk samples were heated at 63, 80 and 90°C for 30 min (Farah, 1986); and at 65, 75, 85 and 100°C for 10, 20 and 30 min (Elagamy, 2000). Laleye et al. (2008) reported that the effect of heating temperature on the solubility of bovine and camel whey proteins depended on the pH. At pH 4.5, which is the isoelectric point of many whey proteins, there was a major change in the solubility as the milk was heated at 60–100°C for 1 h. However, at pH 7.0, both bovine and camel whey proteins were the most stable because the aggregation process is inhibited by electrostatic repulsion between the unfolding globular structures at this pH level. At pH 4.5 and 100°C, the solubility of camel and bovine whey proteins decreased by 55 and 52%, respectively. Camel whey proteins were found to be more susceptible to acid denaturation than their bovine counterparts as their solubility decreased by 16 and 9%, respectively, when the pH dropped from 7.0 to 4.0.

The thermostability of some individual bioactive proteins in camel milk has also been investigated. The heat resistance among camel whey proteins was ranked as lysozyme > lactoferrin > immunoglobulin G (Elagamy, 2000). Camel α-lactalbumin was found to be more stable than its bovine counterpart. Also, the secondary structure of camel α-lactalbumin was better preserved than that of bovine α-lactalbumin during heat denaturation (Atri et al., 2010). Lactoperoxidase in camel milk exhibited lower heat stability compared with that in bovine milk when heated at 67–73°C (Tayefi-Nasrabadi et al., 2011).

Differential scanning calorimetry (DSC) has sometimes been used to measure the denaturation temperature of camel milk proteins. Concentrated camel and bovine milk showed denaturation peaks at 77.8 and 81.7°C, respectively (Felfoul et al., 2015). However, it is difficult to draw a conclusion on the whey protein stability by comparing the denaturation peaks, as a mixture of proteins existed in these samples.

Recently, three proteomics studies on the effect of heat treatment on camel milk proteins have been reported (Benabdelkamel et al., 2017; Felfoul et al., 2017; Zhang et al., 2016). Compared with traditional methods, a proteomics method using mass spectrometry provides a much more sensitive and accurate alternative for analyzing the heat denaturation of camel milk proteins. Instead of identifying the proteins of interest one by one, a method which usually requires tedious purification steps, a proteomics method is able to efficiently quantify a large set of proteins simultaneously. In the first study by Zhang et al. (2016), the changes in camel milk proteins after freezing, pasteurization and spray drying were investigated along with those in bovine and caprine milk. A total of 129, 125 and 74 proteins was quantified in bovine, camel and caprine milk sera, respectively. The concentration of these proteins changed with varied processing steps as well as among different species. Some of the immune-related proteins were heat-sensitive such as lactoferrin, glycosylation-dependent cell adhesion molecule 1 and lactadherin, with losses of approximately 25 to 85% after pasteurization and 85 to 95% after spray drying. α-Lactalbumin, osteopontin and whey acidic protein were relatively heat-stable with losses of 10 to 50% after pasteurization and 25 to 85% after spray drying.

In the second study by Felfoul et al. (2017), liquid chromatography coupled with tandem mass spectrometry (LC–MS/MS) was used to identify bovine and camel milk proteins before and after heat treatment at 80°C for 60 min. It was reported that the heat stability of α-lactalbumin (α-la), peptidoglycan recognition protein (PGRP) and serum albumin (CSA) increased in the following order: α-la < PGRP < CSA, and, after heating, a decrease of 100, 68 and 42% in protein abundance was observed for α-la, PGRP and CSA, respectively. For α-la and β-lactoglobulin (β-lg) in bovine milk, 0 and 26% remained, respectively, after heat treatment. A total of 19 protein bands, which were separated using SDS-PAGE and identified by LC–MS/MS, confirmed the vulnerability of camel α-la and PGRP, as well as bovine α-la and β-lg, to heat treatment at 80°C. Meanwhile, casein fractions in both camel and bovine milk remained intact after being heated at 80°C for 60 min.

The third study was a comprehensive and systematic investigation of the denaturation of whey proteins in camel milk after heating at 63 and 98°C for 1 h (Benabdelkamel et al., 2017). Quantitative 2D-difference in gel electrophoresis (DIGE)-mass spectrometry was used for whey protein determination. The abundance of 105 proteins decreased after the heat treatment according qualitative gel image analysis results. The major fraction of proteins affected by heat treatment was found to include 61% enzymes, 20% binding proteins, 10% cell-adhesion proteins, 5% proteins involved in the immune response, 2% transport proteins and 2% others.

4.3.3 Spray-dried camel milk powder

4.3.3.1 Potential application of spray drying in the production of camel milk powder

Despite the high nutritional and medicinal value of camel milk, its global supply is limited because camels are typically raised in arid places like deserts. Therefore, the production of dried powder from camel milk without impairing its bioactive components is highly desirable, not only to make it available worldwide but also to extend its shelf-life, reduce the transportation cost and expand its applications. Many camel milk powders are available on the market (Table 4.3.1). Most of them are produced by freeze-drying techniques due to the fact that the low drying temperature in the freeze-drying process helps to protect bioactive compounds in camel milk, especially the functional properties of its proteins. However, freeze-drying is well-known as a time-consuming and expensive dehydration technique and is not suitable for the large-scale production of dried milk powders (Ortega-Rivas et al., 2005). The high price of camel milk (which is almost ten times more expensive than cow milk), together with the high cost of freeze-drying operation lead to a high cost of camel milk powder. Moreover, after freeze-drying, milk powders need to be ground and sieved to obtain desirable homogeneity in powder particle size.

For several decades, with advances and innovation in spray-drying technology, it has been considered as the most suitable unit operation to produce milk powders with strict food safety and quality control at a large scale. Moreover, among the common approaches to changing the molecular structure of milk powder, spray drying is the most effective approach to induce the complete phase transformation of milk powders from the crystalline to the amorphous state. The evaporation of water from droplet surfaces and subsequent droplet solidification during spray drying occur too rapidly, preventing the molecules of solutes from rearranging into the ordered structure of crystals. Thus, spray-dried milk powder typically exists in amorphous structure (Ho et al., 2017). In the food and pharmaceutical industries, milk powders with an amorphous structure are preferred over crystalline ones due to their very useful properties. Amorphous powders have higher solubility, a higher dissolution rate and better compression ability than their crystalline counterparts due to the random distribution of molecules in amorphous powders resulting in a loosely packed and porous structure, and consequently possessing more sites for external interactions (Yu, 2001).

However, as compared to bovine milk powder, the production of camel milk powder using spray drying is still in an early stage of research, although spray drying at low temperatures (< 60°C) could be applicable for the production of camel milk powder, and there are a few spray-dried camel milk powders available on the market (Table 4.3.1). Due to the lack of information about the nutritional facts and differences in camel milk sources, a comparison of nutritional values of camel milk powder produced by spray drying and freeze-drying is impossible. There are only a few studies about camel milk powder produced by spray drying reported in literature. So far, there is only one study dedicated to the effect of feed direction and spray-drying temperature on the physiochemical properties of whole camel milk powder (Sulieman et al., 2014). It was reported that co-current spray drying gave camel milk powders lower water activity, more whiteness, less browning, a lower insolubility index and better flowability than the powder produced by counter-current spray drying. Similar to spray-dried whole cow milk powder, spray-dried whole camel milk powder has a very poor flowability. Possible reasons could be very fine powder particles, along with high fat coverage on the power particle surface, which leads to an increase in cohesion and inhibition of the powder flow (Kim et al., 2005). In addition, the angle of repose (a characteristic related to the inter-particulate friction or resistance to movement between particles) of spray-dried whole camel milk powder decreased from 50 to 47° as the inlet air temperature increased from 210 to 220°C. The increase of the spray-drying temperature as well as the pre-concentration temperature resulted in a significant increase in the insolubility index. The denaturation of caseins at high temperatures facilitates the formation of insoluble casein–whey–lactose complexes.

There are many possible reasons for limited application of spray drying in the production of camel milk powders both at the laboratory and commercial scale. Firstly, there is a shortage of raw camel milk for research and production as it is only available in a few countries in Asia, Africa and Australia. Importantly, proteins in camel milk, which are the main nutritional and functional components in camel milk, are much less heat-stable than cow milk proteins. Thus the high temperatures in pre-concentration to increase the solids concentration of camel milk and in spray-drying operations possibly denature these proteins (Laleye et al., 2008). Due to the low solids concentration (~10%, w/w), it will not be economical to spray dry camel milk in "as-is" form. In the production of milk powder, the pre-concentration of milk to increase its solids concentration up to 40–50% (w/w) is an integral stage not only to reduce the energy consumption of the drying process, but also to help impart desirable characteristics to the dried powders (Roy et al., 2017). Recent advances in concentration and spray-drying techniques will possibly allow the pre-concentration and spray drying of camel milk powder at low temperatures, by which the functional components in camel milk can be preserved.

4.3.3.2 Physiochemical properties of spray-dried camel milk powder and their changes during storage: a case study

This section describes the physiochemical properties of camel milk powder produced using a pilot spray dryer (an Anhydro co-current spray dryer, water evaporation capacity of 3–4 L/h) equipped with a twin fluid nozzle, and alterations of these properties under accelerated storage conditions (11–33% RH, 37°C) over 18 weeks. Spray drying was carried out at 160 and 70°C inlet air and outlet air temperature, respectively.

4.3.3.2.1 Physiochemical properties of spray-dried camel milk powder The physiochemical properties of spray-dried camel milk powder are illustrated in Table 4.3.2 and Figure 4.3.1. After spray-drying, camel milk powder could be dissolved almost completely into

Table 4.3.1 Current commercial camel milk powder products available on the market

Product image	Origin
	• The Camel Milk Co. Australia Pty Ltd., Australia • www.camelmilkco.com.au/index.php/our-products/20-dairy
	• Auslink Investment, Australia • www.auslinkinv.com.au/
	• Camilk Dairies, Australia • www.camilkdairy.com.au/
	• Desert Farms Inc., United States • https://desertfarms.com/products/camel-milk-powder-200g
	• Desert Farms Inc, United States • https://desertfarms.com/
	• DromeDairy Naturals™, United States • https://dromedairy.com/collections/all
	• Emirates Industry for Camel Milk & Products, United Arab Emirates • https://camelicious.ae/
	• Urban Platter, India • https://urbanplatter.in/
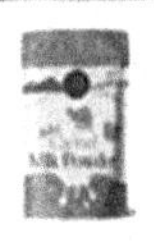	• Aadvik Foods and Products Pvt. Ltd., India • www.aadvikfoods.com/camel-milk-powder
	• Nutra Vita, India • http://nutra-vita.com/product/freeze-dried-camel-milk-powder/
	• Xinjiang Wangyuan Camel Milk Industrial Co., Ltd, China
	• Xinjiang Luo Gan Lin biological Co., Ltd, China

water at normal temperatures (e.g. 30°C) with very high solubility of 98.6%. High solubility of spray-dried camel milk powder indicates the integrity of the protein structure during spray drying (Thomas et al., 2004) and is essential for its applications in food production, especially for reconstitution into water.

X-ray analytical results indicated that spray-dried camel milk powder had an amorphous structure with a very small degree of crystallinity, as a very large hump and a few tiny sharp peaks were observed on its X-ray diffractogram (Figure 4.3.1a). Despite the amorphous structure of spray-dried camel milk powder, it was impossible to detect its glass transition temperature (T_g) by differential scanning calorimetry (DSC). On its DSC curve (Figure 4.3.1b), there were four endothermic events in which the thermal events at the A, C and D peaks were caused by the melting of fat, the melting of non-fat solids and the formation of a more ordered structure after solid-melting, respectively (Rahman et al., 2012). The temperature corresponding to a shifting B peak (~126±0.73°C) was too high to be assigned to T_g of lactose. It has been reported that T_g of lactose in spray-dried whole cow milk powder equilibrated at 11% RH was 62.0°C (Jouppila and Roos, 1994) and that of pure lactose component separated from freeze-dried camel milk powder was about 62.0±7.0 and 116±0.9°C (Rahman et al., 2012). Spray-dried whole camel milk powder contains many components such as water, fat, protein and lactose, and therefore their thermal events under DSC scan could be overlapping. T_g of lactose in milk powder is highly dependent on moisture content; a decrease in moisture content induces an increase in T_g (Jouppila and Roos, 1994). At a moisture content of ~5% (w/w), T_g of spray-dried camel milk powder possibly fell within the melting temperature range of fat.

The particle size of spray-dried camel milk powers had a bimodal distribution as two distinct peaks were observed on the particle size distribution curve (Figure 4.3.1c), with an average particle diameter based on surface area (D[3,2]) being 3.34±0.08 μm. The small particle size of spray-dried camel milk powders contributes to an increase in the cohesion and agglomeration of the powder (Kim et al., 2005). For morphology (Figure 4.3.1d), the powder exhibited mostly spherical particles with a wrinkled and folded surface on which there were some dents and large vacuoles containing minute dried milk particles. A typical Fourier-transform infrared spectroscopy (FTIR) spectrum of spray-dried camel milk powder (Figure 4.3.1e) was similar to that of whole cow milk powder (Ye et al., 2017). All functional groups representing the main components in spray-dried camel milk powder were clearly observed on its FTIR spectrum.

Table 4.3.2 Physicochemical properties of spray-dried camel milk powder

Properties	Values
Moisture content, % (w/w)	4.97±0.14
Water activity (-)	0.34±0.00
True density, g/cm^3	1.261±0.043
Solubility, %	98.62±1.47
Particle size – D[3,2]*, μm	3.34±0.08
Colour (-)	
L*	89.16±1.18
a*	−1.02±0.02
b*	4.43±0.06
Whiteness	88.24±1.01

* *D*[3,2] – particle size based on surface area.

4.3.3.2.2 Changes in the physiochemical properties of spray-dried camel milk powder during storage During storage in a low-RH environment (11.15, 22.26 and 32.27% RH, 37°C) over 18 weeks, many properties of spray-dried camel milk powder remained almost unchanged (e.g. true density, lightness and whiteness) or exhibited a slight alteration such as crystallinity (X-ray analytical results) and morphology (SEM images). At the end of the storage period (week 18), a small development of crystallinity was observed for all spray-dried camel milk powders as the number of small sharp peaks on the XRD hump slightly increased, and the agglomeration of camel milk powders was more obvious. The shape of the powder particles was not affected by storage. Examples of an XRD curve and an SEM image of spray-dried camel milk powder kept at 32.27% RH and 37°C for 18 weeks are shown in Figure 4.3.2a and b, respectively. The changes in crystallinity and morphology primarily happened in the initial period of storage as the RH level in the surrounding environment of the storage container was high. The slight crystallization of lactose, and the agglomeration and caking of camel milk powders during storage led to a decrease in solubility of the powder during storage (Figure 4.3.2c) and an increase in the particle size of the powder (Table 4.3.3). These changes in solubility and particle size of the powder were enhanced at higher storage RH levels. After week 18 of storage at 32.27% RH and 37°C, spray-dried camel milk powder still exhibited a very high solubility (~92%, w/w). Moreover, during storage, especially at 32.27% RH, b* value increased steadily, reflecting the color change of powder towards yellowness or browning (Figure 4.3.2d). The browning of milk powder is caused by non-enzymatic browning reactions, especially Maillard reactions, lactose crystallization and the migration of free fat to the powder surface (Chudy et al., 2015).

As expected, along with structural relaxation of the powder during storage, a reduction in moisture content resulted in an increase of T_g beyond the melting temperature range of milk fat. After 3 weeks of storage, the moisture content of spray-dried camel milk powders stored at 11.15, 22.26 and 32.27% RH reduced from 4.97±0.14% (w/w) at week 0 to 2.97±0.29, 3.96±0.47 and 4.96±0.17%, respectively. From week 3, the signal of glass transition could be witnessed on the DSC curve of all camel milk powders (Figure 4.3.2e). The temperature for this thermal event was similar to the reported T_g of lactose, and it varied with changes in the moisture content of spray-dried camel milk powder (Table 4.3.3). Depending on the moisture content of spray-dried camel milk powder, its T_g of lactose ranged from 57.49±3.43 to 71.68±2.52°C. For spray-dried milk powder, the glass transition temperature is an important factor controlling many functional properties such as its processability, handling properties and stability (Roos, 2002). In order to prevent unwanted changes in the dried milk powder during storage, the powder should be kept at a temperature lower than its T_g.

The amount of unsaturated fatty acids in camel milk accounts for more than 30.5% (w/w) of total triacylglycerols, much higher than that in cow milk (28.3%) (Gorban and Izzeldin, 2001); therefore, during the storage of spray-dried camel milk powder at 11.15, 22.26 and 32.27% RH (37°C), the oxidation of milk fat can occur to produce off-flavor compounds. The GC–MS results showed that the fresh spray-dried camel milk powders almost did not contain any peaks representing the volatile aldehyde compounds, and these compounds were only formed during storage. The prolonging of storage time accelerated the rate of fat oxidation. Fat oxidation of camel milk powders kept at 11.15% RH occurred much faster than that of samples kept at 22.26 and 32.27% RH. At a high RH level, a monolayer of water may have formed to cover the surface of fat, protecting it from direct exposure to air (Roos, 2002).

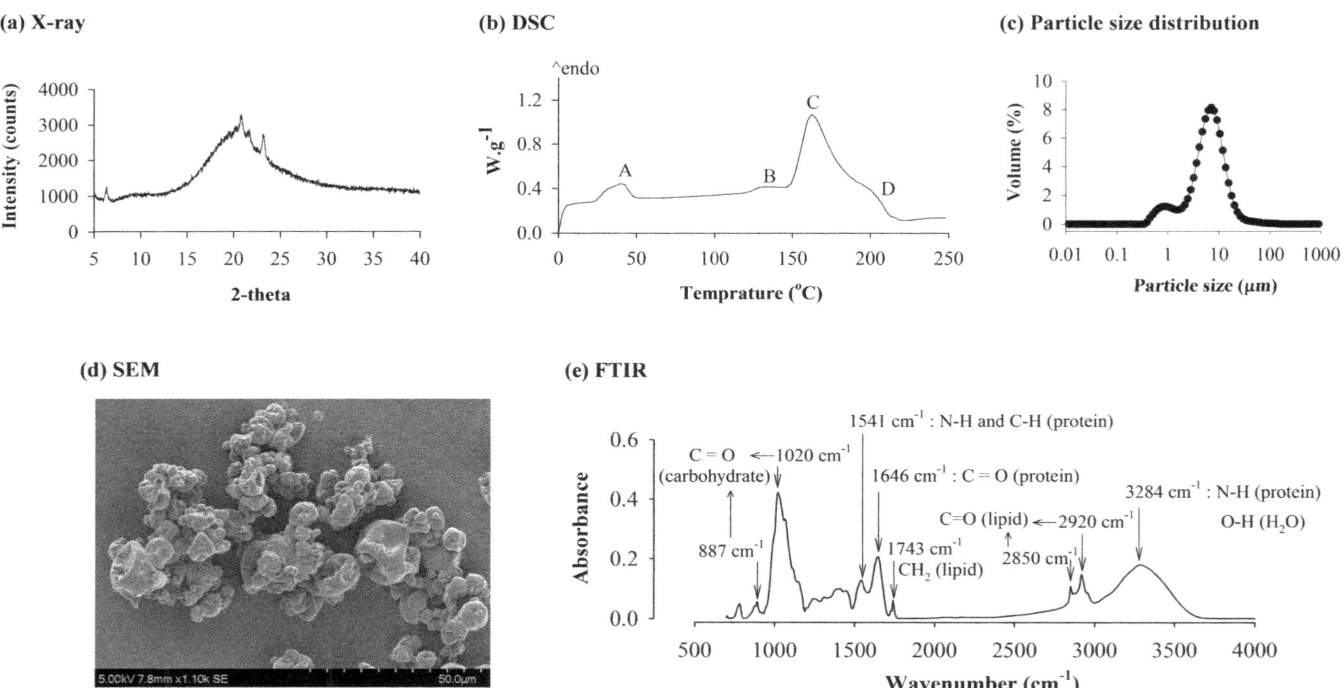

Figure 4.3.1 X-ray diffractogram (a), differential scanning calorimetry (b), particle size distribution (c), scanning electron microscope (d) and Fourier-transform infrared spectroscopy (e) of spray-dried camel milk powder.

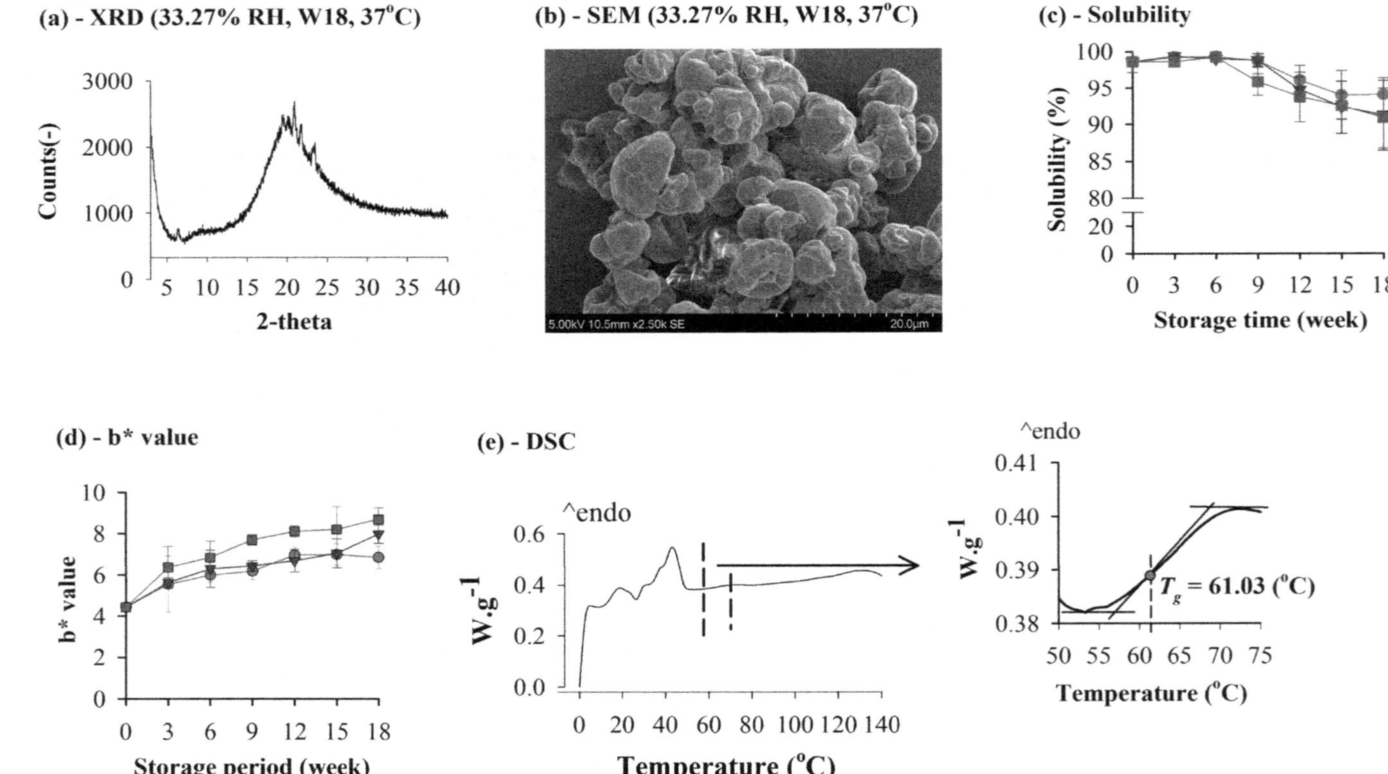

Figure 4.3.2 XRD (a) and SEM (b) of spray-dried camel milk powder kept at 33.27% RH, 37°C for 18 weeks, changes in solubility (c) and b* value (d) of spray-dried camel milk powder during storage at 11.15% RH (—○—), 22.26% RH (—▽—) and 32.27% RH (—■—) and DSC curve (e) of spray-dried camel milk powder kept at 33.27% RH at week 3 in which a signal of glass transition could be observed.

Table 4.3.3 Changes in T_g and D[3,2] of spray-dried camel milk powder during storage

RH (%)	Storage time (week)	T_g (°C)	D[3,2]**, μm
11.15	0	n.d.*	3.34±0.08
	3	68.77±2.73	n.d.
	6	68.96±0.82	4.15±0.03
	9	70.01±1.95	n.d.
	12	69.82±0.91	4.49±0.31
	15	70.06±0.85	n.d.
	18	71.68±2.52	4.32±0.05
22.26	0	n.d.	3.34±0.08
	3	61.03±1.96	n.d.
	6	60.43±1.78	4.13±0.04
	9	62.54±2.33	n.d.
	12	62.17±1.71	4.80±0.06
	15	61.62±1.16	n.d.
	18	64.45±2.63	4.70±0.03
32.27	0	n.d.	3.34±0.08
	3	57.49±3.43	n.d.
	6	59.20±0.54	4.84±0.01
	9	59.03±0.82	n.d.
	12	60.70±0.99	5.45±0.13
	15	62.19±2.80	n.d.
	18	63.28±2.35	5.39±0.31

* n.d. – not determined.

** *D*[3,2] – particle size based on surface area.

4.3.4 *Conclusion and further outlook*

Due to the high content of bioactive compounds in camel milk but the limited geographical distribution of camels, it is highly desirable to produce camel milk powder so that it can be available all over the world. Although spray drying is the most favorable dehydration technique for the industrial-scale production of camel milk powder and research results on the physiochemical properties of spray-dried camel milk powders are promising, practical application still requires further extensive investigation, especially on the retention of bioactive compounds during spray drying as well as the stability of spray-dried camel milk at different storage conditions. One of the important properties of camel milk powder produced by spay drying was its amorphous structure with a very high solubility. With recent innovations in spray-drying technique, it is possible to combine spray drying with other existing novel techniques to minimize the long exposure of camel milk to high temperatures as well as performing spray drying at low temperatures. Moreover, low-temperature dewatering prior to spray drying and the addition of bioactive-protecting materials into camel milk could be possible solutions to help avoid unwanted changes in bioactive compounds during spray-drying operation.

References

Abu-Lehia, I. H. (1997). Composition of camel milk. *Milchwissenschaft* **42**, 368–371.

Arnold, R. R., Cole, M. F., and McGhee, J. R. (1977). A bactericidal effect for human lactoferrin. *Science* **197**(4300), 263–265.

Atri, M. S., Saboury, A. A., Yousefi, R., Dalgalarrondo, M., Chobert, J.-M., Haertlé, T., and Moosavi-Movahedi, A. A. (2010). Comparative study on heat stability of camel and bovine apo and holo α-lactalbumin. *Journal of Dairy Research* **77**(1), 43–49.

Barłowska, J., Szwajkowska, M., Litwińczuk, Z., and Król, J. (2011). Nutritional value and technological suitability of milk from various animal species used for dairy production. *Comprehensive Reviews in Food Science and Food Safety* **10**(6), 291–302.

Benabdelkamel, H., Masood, A., Alanazi, I. O., Alzahrani, D. A., Alrabiah, D. K., Alyahya, S. A., and Alfadda, A. A. (2017). Proteomic profiling comparing the effects of different heat treatments on camel (*Camelus dromedarius*) milk whey proteins. *International Journal of Molecular Sciences* **18**(4), 721–735.

Bhushani, A., and Anandharamakrishnan, C. (2017). Freeze drying. *In "Handbook of drying for dairy products"* (C. Anandharamakrishnan, ed.). John Wiley & Sons, Chichester.

Chudy, S., Pikul, J., and Rudzińska, M. (2015). Effects of storage on lipid oxidation in milk and egg mixed powder. *Journal of Food and Nutrition Research* **54**, 31–40.

Coughlin, R., Tonsager, S., and McGroarty, E. (1983). Quantitation of metal cation bound to membranes and extracted lipopolysaccharide of *Escherichia coli. Biochemistry (Washington)* **22**(8), 2002–2007.

Duhaiman, A. S. (1988). Purification of camel milk lysozyme and its lytic effect on Escherichia coli and Micrococcus lysodeikticus. *Comparative Biochemistry and Physiology. Part B: Comparative Biochemistry* **91**(4), 793–796.

El-Agamy, E. I. (2008). Camel milk. *In "Handbook of milk of non-bovine mammals"* (Y. W. Park and G. F. Haenlein, eds.). John Wiley & Sons, Blackwell Publishing, Iowa, USA.

El-Agamy, E. I. (2009). Bioactive components in camel milk. *In "Bioactive components in milk and dairy products"*, Vol. 107 (W. P. Young, ed.). John Wiley & Sons, Blackwell Publishing, Iowa, USA, pp. 159–192.

El-Agamy, E. I., Nawar, M., Shamsia, S. M., Awad, S., and Haenlein, G. F. W. (2009). Are camel milk proteins convenient to the nutrition of cow milk allergic children? *Small Ruminant Research* **82**(1), 1–6.

Elagamy, E. I. (2000). Effect of heat treatment on camel milk proteins with respect to antimicrobial factors: A comparison with cows and buffalo milk proteins. *Food Chemistry* **68**(2), 227–232.

Elagamy, E. I., Ruppanner, R., Ismail, A., Champagne, C. P., and Assaf, R. (1996). Purification and characterization of lactoferrin, lactoperoxidase, lysozyme and immunoglobulins from camel's milk. *International Dairy Journal* **6**(2), 129–145.

FAO, Food and Agriculture Organization of the United Nations. (2016). FAOSTAT database. FAO, Rome, Italy. http://www.fao.org/faostat/en/#home.

Farah, Z. (1986). Effect of heat treatment on whey proteins of camel milk. *Milchwissenschaft* **41**, 763–765.

Farah, Z. (1993). Composition and characteristics of camel milk. *Journal of Dairy Research* **60**(4), 603–626.

Farah, Z., and Atkins, D. (1992). Heat coagulation of camel milk. *Journal of Dairy Research* **59**(2), 229–231.

Farah, Z., Rettenmaier, R., and Atkins, D. (1992). Vitamin content of camel milk. *International Journal for Vitamin and Nutrition Research* **62**(1), 30–33.

Faye, B. (2015). Role, distribution and perspective of camel breeding in the third millennium economies. *Emirates Journal of Food and Agriculture*, 318–327.

Felfoul, I., Jardin, J., Gaucheron, F., Attia, H., and Ayadi, M. A. (2017). Proteomic profiling of camel and cow milk proteins under heat treatment. *Food Chemistry* **216**, 161–169.

Felfoul, I., Lopez, C., Gaucheron, F., Attia, H., and Ayadi, M. (2015). Fouling behavior of camel and cow milks under different heat treatments. *Anais an International Journal* **8**(8), 1771–1778.

Fox, P. F., and Hoynes, M. C. T. (1976). Heat stability characteristics of ovine, caprine and equine milks. *Journal of Dairy Research* **43**(3), 433–442.

Gorban, M. S. A., and Izzeldin, O. M. (2001). Fatty acids and lipids of camel milk and colostrum. *International Journal of Food Sciences and Nutrition* **52**(3), 283–287.

Hamers-Casterman, C., Atarhouch, T., Muyldermans, S., Robinson, G., Hammers, C., Songa, E. B., Bendahman, N., and Hammers, R. (1993). Naturally occurring antibodies devoid of light chains. *Nature* **363**(6428), 446.

Hinz, K., Connor, P. M., Huppertz, T., Ross, R. P., and Kelly, A. L. (2012). Comparison of the principal proteins in bovine, caprine, buffalo, equine and camel milk. *Journal of Dairy Research* **79**(2), 185–191.

Ho, T. M., Truong, T., and Bhandari, B. (2017). Spray-drying and non-equilibrium states/glass transition. *In "Non-equilibrium states and glass transitions in foods : Processing effects and product-specific implications"*. Woodhead Publishing, Sawston, UK, pp. 111–136.

Jouppila, K., and Roos, Y. (1994). Glass transitions and crystallization in milk powders. *Journal of Dairy Science* **77**(10), 2907–2915.

Kappeler, S. R., Ackermann, M., Farah, Z., and Puhan, Z. (1999). Sequence analysis of camel (*Camelus dromedarius*) lactoferrin. *International Dairy Journal* **9**(7), 481–486.

Kappeler, S. R., Farah, Z., and Puhan, Z. (2003). 5′-Flanking regions of camel milk genes are highly similar to homologue regions of other species and can be divided into two distinct groups. *Journal of Dairy Science* **86**(2), 498–508.

Khan, J. A., Kumar, P., Paramasivam, M., Yadav, R. S., Sahani, M. S., Sharma, S., Srinivasan, A., and Singh, T. P. (2001). Camel lactoferrin, a transferrin-cum-lactoferrin: Crystal structure of camel apolactoferrin at 2.6 Å resolution and structural basis of its dual role. *Journal of Molecular Biology* **309**(3), 751–761.

Kim, E. H. J., Chen, X. D., and Pearce, D. (2005). Effect of surface composition on the flowability of industrial spray-dried dairy powders. *Colloids and Surfaces, Part B: Biointerfaces* **46**(3), 182–187.

Knorr, D., Shetty, K. J., and Kinsella, J. E. (1979). Enzymatic lysis of yeast cell walls. *Biotechnology and Bioengineering* **21**(11), 2011–2021.

Konuspayeva, G., Faye, B., and Loiseau, G. (2009). The composition of camel milk: A meta-analysis of the literature data. *Journal of Food Composition and Analysis* **22**(2), 95–101.

Kouniba, A., Berrada, M., Zahar, M., and Bengoumi, M. (2005). Composition and heat stability of Moroccan camel milk. *Journal of Camel Practice and Research* **12**, 105–110.

Laleye, L. C., Jobe, B., and Wasesa, A. A. H. (2008). Comparative study on heat stability and functionality of camel and bovine milk whey proteins. *Journal of Dairy Science* **91**(12), 4527–4534.

Leitch, E. C., and Willcox, M. D. (1999). Elucidation of the antistaphylococcal action of lactoferrin and lysozyme. *Journal of Medical Microbiology* **48**(9), 867.

León-Sicairos, N., López-Soto, F., Reyes-López, M., Godínez-Vargas, D., Ordaz-Pichardo, C., and de La Garza, M. (2006). Amoebicidal activity of milk, apo-lactoferrin, sIgA and lysozyme. *Clinical Medicine &Amp; Research* **4**(2), 106.

Masschalck, B., and Michiels, C. W. (2003). Antimicrobial properties of lysozyme in relation to foodborne vegetative bacteria. *Critical Reviews in Microbiology* **29**(3), 191–214.

Monchois, V., Abergel, C., Sturgis, J., Jeudy, S., and Claverie, J. M. (2001). Escherichia coli ykfE ORFan gene encodes a potent inhibitor of C-type lysozyme. *The Journal of Biological Chemistry* **276**(21), 18437.

Muyldermans, S., Cambillau, C., and Wyns, L. (2001). Recognition of antigens by single-domain antibody fragments: The superfluous luxury of paired domains. *Trends in Biochemical Sciences* **26**(4), 230–235.

Ortega-Rivas, E., Juliano, P., and Yan, H. (2005). *"Food powders: Physical properties, processing, and functionality"*. Springer Food Engineering Book Series, Springer, Boston, MA.

Parikh, P. M., Vaid, A., Advani, S. H., Digumarti, R., Madhavan, J., Nag, S., Bapna, A., Sekhon, J. S., Patil, S., Ismail, P. M., Wang, Y., Varadhachary, A., Zhu, J., and Malik, R. (2011). Randomized, double-blind, placebo-controlled phase ii study of single-agent oral talactoferrin in patients with locally advanced or metastatic non–small-cell lung cancer that progressed after chemotherapy. *Journal of Clinical Oncology* **29**(31), 4129–4136.

Parisien, A., Allain, B., Zhang, J., Mandeville, R., and Lan, C. Q. (2008). Novel alternatives to antibiotics: Bacteriophages, bacterial cell wall hydrolases, and antimicrobial peptides. *Journal of Applied Microbiology* **104**, 1–13, Oxford, UK.

Rahman, M. S., Al-Hakmani, H., Al-Alawi, A., and Al-Marhubi, I. (2012). Thermal characteristics of freeze-dried camel milk and its major components. *Thermochimica Acta* **549**, 116–123.

Riechmann, L., and Muyldermans, S. (1999). Single domain antibodies: Comparison of camel VH and camelised human VH domains. *Journal of Immunological Methods* **231**(1–2), 25–38.

Roos, Y. H. (2002). Importance of glass transition and water activity to spray drying and stability of dairy powders. *Le Lait* **82**(4), 475–484.

Roy, I., Bhushani, A., and Anandharamakrishnan, C. (2017). Techniques for the preconcentration of milk. *In "Handbook of drying for dairy products"* (C. Anandharamakrishnan, ed.). JohnWiley & Sons, Chennai, pp. 23–37.

Saalfeld, W., and Edwards, G. (2010). Distribution and abundance of the feral camel (Camelus dromedarius) in Australia. *The Rangeland Journal* **32**(1), 1–9.

Sawaya, W. N., Khalil, J. K., Al-Shalhat, A., and Al-Mohammad, H. (1984). Chemical composition and nutritional quality of camel milk. *Journal of Food Science* **49**(3), 744–747.

Shyam, P. S., Bhavbhuti, M. M., Wadhwani, K. N., Darji, V. B., and Aparnathi, K. D. (2016). Evaluation of camel milk for selected processing related parameters and comparisons with cow and buffalo milk. *International Journal of Health* **3**, 27–37.

Solomon, A., and Weiss, D. T. (1995). Structural and functional properties of human lambda-light-chain variable-region subgroups. *Clinical and Vaccine Immunology* **2**(4), 387.

Stahl, T., Sallmann, H. P., Duehlmeier, R., and Wernery, U. (2006). Selected vitamins and fatty acid patterns in dromedary milk and colostrums. *Journal of Camel Practice and Research* **13**, 53–57.

Sulieman, A. M. E., Elamin, O. M., Elkhalifa, E. A., and Laleye, L. (2014). Comparison of physicochemical properties of spray-dried camel's milk and cow's milk powder. *International Journal of Food Science and Nutrition Engineering* **4**, 15–19.

Sylvia, L.-H., Paul, L. H., Yongtao, S., Philip, L. H., Hsiang-Fu, K., Diana, L. B., and Hao-Chia, C. (1999). Lysozyme and RNases as anti-HIV components in β-core preparations of human chorionic gonadotropin. *Proceedings of the National Academy of Sciences of the United States of America* **96**(6), 2678.

Tayefi-Nasrabadi, H., Hoseinpour-Fayzi, M. A., and Mohasseli, M. (2011). Effect of heat treatment on lactoperoxidase activity in camel milk: A comparison with bovine lactoperoxidase. *Small Ruminant Research* **99**(2–3), 187–190.

Thomas, M. E. C., Scher, J., Desobry-Banon, S., and Desobry, S. (2004). Milk powders ageing: Effect on physical and functional properties. *Critical Reviews in Food Science and Nutrition* **44**(5), 297–322.

Vaara, M. (1992). Agents that increase the permeability of the outer membrane. *Microbiological Reviews* **56**(3), 395.

van Der Does, A. M., Hensbergen, P. J., Bogaards, S. J., Cansoy, M., Deelder, A. M., van Leeuwen, H. C., Drijfhout, J. W., van Dissel, J. T., and Nibbering, P. H. (2012). The human lactoferrin-derived peptide hLF1-11 exerts immunomodulatory effects by specific inhibition of myeloperoxidase activity. *Journal of Immunology* **188**(10), 5012.

Varaldo, P., Valisena, S., Mingari, M., and Satta, G. (1989). Lysozyme-induced inhibition of the lymphocyte response to mitogenic lectins. *Proceedings of the Society for Experimental Biology and Medicine* **190**(1), 54–62.

Yadav, A., Kumar, R., Priyadarshini, L., and Singh, J. (2015). Composition and medicinal properties of camel milk: A review. *Journal of Dairying, Foods and Home Sciences* **34**(2), 83.

Ye, M. P., Zhou, R., Shi, Y. R., Chen, H. C., and Du, Y. (2017). Effects of heating on the secondary structure of proteins in milk powders using mid-infrared spectroscopy. *Journal of Dairy Science* **100**(1), 89–95.

Yu, L. (2001). Amorphous pharmaceutical solids: Preparation, characterization and stabilization. *Advanced Drug Delivery Reviews* **48**(1), 27–42.

Zhang, L., Boeren, S., Smits, M., van Hooijdonk, T., Vervoort, J., and Hettinga, K. (2016). Proteomic study on the stability of proteins in bovine, camel, and caprine milk sera after processing. *Food Research International* **82**, 104–111.

4.4 *How to produce dairy powders without the use of a drying tower*

Pierre Schuck, Gaëlle Tanguy, Serge Méjean and Romain Jeantet

4.4.1 *Introduction*

Whey is mainly derived from cheese manufacture. Its dry matter (DM) represents 6% w/w, and is mainly composed of lactose (70–72% w/w DM), proteins (8–10% w/w DM) and minerals (12–15% w/w DM). Whey and whey permeate constituents have attracted considerable interest over the past decade due to the growing demand for infant milk powders and functional ingredients in food and pharmaceutical applications, as well as nutrients in dietetic and health foods. The production of these latter includes, after various pretreatments such as membrane filtration and heat treatments, the elimination of water, which is usually realized in two processing steps: vacuum concentration up to 60% w/w DM in multiple-stage falling-film evaporators, then drying in spray dryers up to 96–97% w/w DM. Additionally, the highly hygroscopic amorphous lactose may be crystallized in between these two steps to produce non-hygroscopic powders (Schuck, 2011).

In practice, vacuum concentration is configured to reduce the energy cost of the operation as much as possible (multiple stage, implementation of thermal and mechanical vapor recompression). As mentioned by various authors, the mean specific energy consumption can fall to as low as 75 $kJ{\cdot}kg^{-1}$ of water removed (Westergaard, 2004). On the other hand, spray drying is more energy-consuming than falling-film concentration. For example, Schuck et al. (2015) estimated the specific energy consumption of drying as 13 times higher than the that of evaporation in the case of whey permeate. One way to reduce the energy cost of the whole process is thus to push forward the vacuum concentration of the product up to the higher dry matter content, i.e., to evaporate as much water as possible during the evaporation step. However, one limit exists in the viscosity of the concentrate (whey and permeate concentrates becoming highly viscous at over 60% w/w DM), that affects the ability of current spraying systems to spray such viscous products.

In the last decade, new technologies have been developed to expand the limit of the concentration step, by integrating equipment able to handle highly viscous/shear thinning concentrates. As an example, the Tixotherm® process (Henningfield & Dinesen, 2004; Pisecky, 2005, 2012) consists of concentrating a 60% w/w DM permeate concentrate produced by falling-film evaporation up to 86% w/w DM in a double-jacketed horizontal tube heated by steam, in which the increasing viscosity of the product is controlled by a vigorous mechanical treatment.

Recently, an innovative technological scheme has been proposed and developed jointly by INRAE and VOMM (Rozzano, Italy) in order to obtain whey and permeate powders from highly concentrated products and without the use of a spray-drying step. It consists of using a three-step process combining two thin-film horizontal rotary equipment, also known as turbo concentrator and turbo mixer, with a thin-film horizontal rotary dryer, also known as turbo dryer, to carry out the three main processing steps leading to the final powder: ultra-high concentration, granulation and drying. The turbo concentrator is able to handle highly viscous concentrates (namely, in the range of 60 to 80% w/w for whey concentrates) by implementing high shear forces in the vicinity of the wall: the concentration limit is dependent only on the product properties and not the technology as in the case of falling-film evaporators. The concept of this innovative process has been validated with various tests carried out on a VOMM continuous pilot plant producing 50 kg·h^{-1} of powder for several hours; then it was patented (Schuck et al., 2016).

4.4.2 *Principle*

The innovative process is based on turbo concentrators, and it is used on whey and permeate concentrates, previously prepared by using conventional concentration from 6 to 60% DM in falling-film evaporators and lactose crystallization in tanks.

The first step is the ultra-high concentration of the concentrates from 60 to 80% DM with the turbo concentrator. Then the resulting, paste-like, ultra-high concentrated product is granulated in a second turbo mixer by mixing with a portion of the complete dried product coming out of the process, in a mass ratio of 1 kg of ultra-high concentrated product for 1 kg of powder. Lastly granules at 88% DM are dried up to 96–97% DM using the turbo dryer (Figure 4.4.1).

4.4.3 *Results*

The innovative process was tested for the production of permeate powders. The properties of the resulting powder (experimental powder) were compared to a standard powder obtained using a conventional process and whose properties are given in Schuck et al. (2012). The results of the study are detailed in Tanguy et al. (2017).

Regarding the physical properties (Table 4.4.1), the experimental and standard (control) powders showed comparable bulk, true and packed densities: these latter were equal to 527, 1554 and 616 kg·m^{-3} for the new process compared to 568, 1514 and 686 kg·m^{-3} for the control, respectively. Both powders also presented nearly the same flowability and floodability indices: 78 and 50 for the experimental powder and 74 and 49 for the control powder. According to the classification related to these indices, both powders could be considered as having a fair flowability and being well-inclined to flood. Among the physical properties studied, only the size of the particles could differentiate the experimental powder from the control one. Indeed, the powders produced using

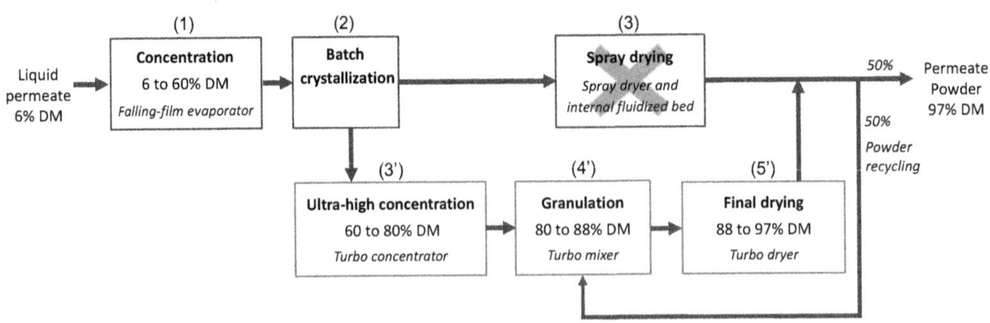

Figure 4.4.1 New process scheme for the production of permeate powders (adapted from Tanguy et al., 2017).

Table 4.4.1 Dry matter, total dry matter, water activity, densities, hygroscopicity, granulometry, span, dispersibility, solubility and wettability of permeate powders (Tanguy et al., 2017)

	Experimental powder	Standard powder (control)
Dry matter ($g \cdot kg^{-1}$)	979.1 ± 0.9	977.5 ± 0.9
Total dry matter ($g \cdot kg^{-1}$)	941.1 ± 0.4	951.5 ± 0.6
Water activity at 25°C (-)	0.20 ± 0.01	0.23 ± 0.01
Bulk density ($kg \cdot m^{-3}$)	527 ± 3	568 ± 3
True density ($kg \cdot m^{-3}$)	1554 ± 1	1514 ± 3
Packed density ($kg \cdot m^{-3}$)	616 ± 1	686 ± 1
Hygroscopicity (%) at 85% RH	16.33	14.5
d(0.5) (μm)	200 ± 4	123 ± 2
Dispersibility (%)	96.5 ± 0.0	93.3 ± 0.1
Solubility (%)	100 ± 0.0	96 ± 0.0
Wettability (s)	2 ± 1	4 ± 1

the new process had larger particles than those obtained using a conventional process. As an example, the d(0.5) values of the experimental powder were 200 μm compared to 123 μm for the control. This result could be inherent to granulation, favored in the new processing scheme. Despite these differences in particle size distribution, the rehydration properties of the experimental and control powders were close to each other, with a solubility of over 95.0%, a dispersibility of over 90.0% and a wettability of less than 5 seconds.

The energy cost was estimated on the sole basis of the energy required for the removal of water. The partial energy cost of spray drying from 60 to 97% DM was evaluated at 3392 $kJ \cdot kg^{-1}$ of permeate powder compared to 2311 $kJ \cdot kg^{-1}$ for the three-step process using the turbo concentrator, the turbo mixer and the turbo dryer. Depending on the type of falling-film evaporator used (including or not vapor recompression systems) and its energy efficiency, the overall energy cost of the innovative processing scheme was estimated in the range of 3513–8986 $kJ \cdot kg^{-1}$ of powder compared to 4594–10,067 $kJ \cdot kg^{-1}$ for the conventional one. Consequently, the new three-step process provides 10.7 to 23.5% energy savings when it is calculated for the entire process (from raw material "whey" to the final product) and up to 32% for the sole post-vacuum concentration step (from 60 to 97% w/w DM).

Additional savings regarding the investment needed for equipment and tailor-made building were estimated around 40%, in relation to the difference in size of a set of horizontal turbo equipment compared to a conventional spray-drying chamber and set of externalities (in the range of 25 to 40 meters high, not including the ventilation and filtration systems). However, this needs to be confirmed by further investigation, and completed with other data regarding CIP solutions and water needs, for which benefits are expected in relation to the smaller dead volume of the turbo equipment compared to conventional spray-drying systems.

4.4.4 Conclusions and perspectives

This breakthrough technology offers the opportunity to produce at low cost and with regular physicochemical characteristics both permeate and whey powders. Indeed, the first experimental results showed that the resulting powders have a comparable quality to that of commercial powders obtained with spray drying, in terms of physical and rehydration properties, except for their density (higher densities for the experimental powders). Moreover, the potential energy savings were estimated between 10 and 32%. It also offers the possibility to create innovative powder products. One prospect would be to granulate the ultra-high concentrated product with a powder having a different composition (maltodextrin, caseins, etc.) and to apply the process to other thixotropic food products such as egg whites.

This new technology has been tested up to now only at pilot scale, although for long runs (4 to 5 h) and showing constant parameters. Further developments are thus needed to optimize operating conditions, and some are now in progress. For example, the ultra-high concentration step was carried out at atmospheric pressure. Today, the manufacturer is operating the ultra-high concentration tests under vacuum to protect heat-sensitive components (proteins, vitamins, etc.) from heating damage. Likewise, the higher the dry matter content, the faster the lactose crystallization, and therefore a part of the lactose crystallization could occur inside the turbo concentrator; it would then reduce the long-term crystallization step. Lastly, the possibility of carrying out the ultra-high concentration at even higher DM content than the current 80% DM is under consideration. It seems feasible and promising for some products, strongly reducing the ultra-high concentrated product/powder ratio or even going so far as to remove the granulation step (step 4′ in Figure 4.4.1).

Acknowledgments

The authors acknowledge VOMM Impianti e Processi SpA for its technical support, valuable advice and discussions.

References

Henningfield, T.D., Dinesen, R.A. (2004). Process and plant for evaporative concentration and crystallization of a viscous lactose containing aqueous liquid. U.S. Patent No 6,790,288. Washington, DC: U.S. Patent and Trademark Office.

Pisecky, J. (2005). Spray-drying in the cheese industry. *International Dairy Journal*, 15(6–9), 531–536. doi:10.1016/j.idairyj.2004.11.010.

Pisecky, J. (2012). *Handbook of Milk Powder Manufacture*. Copenhagen, Denmark: GEA Process Engineering A/S.

Schuck, P. (2011) Lactose and oligosaccharides | Lactose: Crystallization. In Fuquay JW, Fox PF and McSweeney PLH (eds.), *Encyclopedia of Dairy Sciences*, 2nd Edition, vol. 3. San Diego: Academic Press, 182–195.

Schuck, P., Dolivet, A., Jeantet, R. (2012). *Analytical Methods for Food and Dairy Powders*. Oxford, UK: Wiley-Blackwell.

Schuck, P., Jeantet, R., Tanguy, G., Mejean, S., Gac, A., Lefebvre, T., Labussiere, E., Martineau, C. (2015) Energy consumption in the processing of dairy and feed powders by evaporation and drying. *Drying Technology*, 33(2), 176–184. doi:10.1080/07373937.2014.942913.

Schuck, P., Garreau, D., Dolivet, A., Tanguy-Sai, G., Mejean, S., Jeantet, R., Vezzani, M. (2016). Milk powder. French Patent Application (Poudre laitière, published at FR3024331), International Patent Application (published at WO2016016397).

Tanguy, G., Dolivet, A., Méjean, S., Garreau, D., Talamo, F., Postet, P., Jeantet, R., Schuck, P. (2017) Efficient process for the production of permeate powders. *Innovative Food Science and Emerging Technologies*, 41, 144–149.

Westergaard, V. (2004). *Milk Powder Technology*. Copenhagen, Denmark: Niro A/S.

4.5 *Prediction of spray-drying parameters: SD²P® software*

Pierre Schuck, Anne Dolivet, Serge Méjean and Romain Jeantet

4.5.1 *Introduction*

Over the past decade, the dairy market has experienced a considerable development of increasingly targeted and complex dry dairy products. Among these latter and as an example, infant formulae show nowadays highly complex formulations, involving up to 50 different constituents that should be mixed together before either the evaporation or spray-drying steps. Such a variability greatly influences the binding of water in the product, and consequently its drying behavior. Indeed, the outlet gas temperature and humidity need to be adapted depending on the stickiness–solubility pattern, affecting in turn the production costs (Zhu et al., 2009). The outlet gas characteristics are usually adjusted by setting the inlet gas temperature and the concentrated flow rate, and the optimum operating point for a given concentrate can be determined empirically by resorting to pilot-plant or industrial-scale trials. In the case of pilot-plant trials, scale-up remains a difficult task most of the time, and industrial-scale trials are risky and costly. Moreover, such trials need to be performed for each new product developed by a company, resulting in a time-consuming task without making it possible to obtain generic results.

On the other hand, simulation is considered to be a smart way to optimize the process conditions in terms of performance and product specifications while minimizing experiments in validation tests, provided that it is sufficiently reliable (Patel et al., 2009). However, knowledge-based models are still far from completely addressing this issue, given the current difficulty in feeding them with relevant data of the above-mentioned diversity of the concentrates.

In order to fill this gap, two effective spray-drying simulation methods have been developed in the past decades, both based on lab-scale drying experiments to consider the drying behavior of the concentrate: namely, the reaction engineering approach (combining single- droplet drying kinetics and 1D simulation approaches; Lin and Chen, 2002) and a drying-by-desorption approach (combining a desorption behavior and a black-box simulation approach). This chapter will focus on the second one, which is based on global mass and energy balances over the spray-drying system and was originally developed at INRAE in France (Schuck et al., 2009).

This approach considers the drying behavior of the concentrate through an original drying by desorption test: in practice, a small sample of the concentrate is dried by desorption with previously dehydrated zeolite in excess, the whole being placed in a specially designed cell. The water transfer from the concentrate to the atmosphere is tracked by measuring the changes in relative humidity (RH) over time using a thermodynamic sensor.

From these data, the drying behavior of the concentrate (ratio of free or "easily evaporated" water to bound water or "water that is not easily evaporated") can be extrapolated; it is then combined with several characteristics of the drying system considered to determine the expected key drying parameters of this concentrate, first of all the inlet gas temperature for a given outlet gas target. In other words, the prediction relies on the consideration of both the specific drying behavior (water–constituent interactions in the concentrate) and drying equipment characteristics. For that purpose, the SD²P® software (Spray Drying Parameter Simulation and Determination; N° IDDN.FR.001.480 002.003.R.P.2005.000.30100, 2005) has been developed by UMR STLO INRAE-Institut Agro (Rennes, France).

This drying-by-desorption approach has been successfully applied to numerous dairy and non-dairy concentrates, dried in the frame of various drying tower designs (one, two or three stages, from 5 kg to 6 tons of water evaporated/hour). A good match between measured and predicted parameters (in the range 1–5% error) was observed. Note that this program considers only the drying chamber and the internal fluid bed in its calculation, the external fluid bed not being considered as it represents a negligible part of the drying process and is more dedicated to cooling the dried particles below the glass transition temperature.

In the following, the drying-by-desorption method is first described. Then the key input and output parameters of the SD²P® software are discussed.

4.5.2 *Drying by desorption*

As mentioned above, a desorption test is necessary to establish the "dryability" of a milk product before it is possible to characterize its behavior in the drying equipment. This desorption test makes it possible to determine, among other things, the availability of the water to be eliminated.

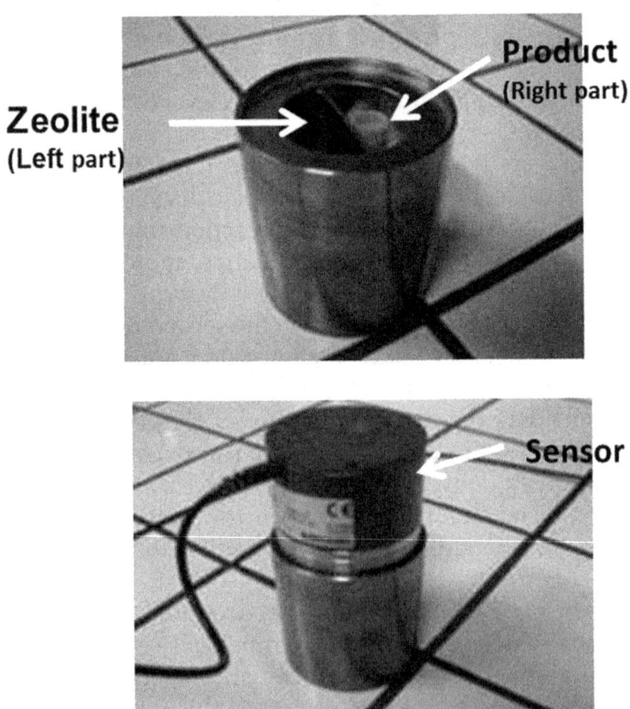

Figure 4.5.1 Device for drying by desorption.

For this purpose, 160 ± 1 mg of prepared concentrate is placed in a two-part device (Figure 4.5.1), the first part (right) containing a capsule in which the product is placed and the second (left), larger, part in which the zeolite (≈100 g) is placed.

A thermohygrometric sensor is positioned on the upper part of the device in order to monitor the water transfer between the product and the zeolite over time, expressed through the relative humidity (RH) of the internal device atmosphere. The entire apparatus is placed inside an oven at 45 ± 1°C. A diagram of the procedure can be seen in Figure 4.5.2.

On the basis of the desorption results, it is possible to establish the drying kinetics of the concentrate considered, which reveals the difficulty (or ease) of eliminating the water during spray drying (Figure 4.5.3). Integrating the kinetics curve of the drying process over time makes it possible to identify two categories of water, differing by their binding behavior:

- The blue area represents the mass of water in the concentrate that is easily evaporated (≈ free water).
- The red area represents the mass of water that is difficult to evaporate from the concentrate (≈ bound water).

Speaking of energy rather than mass, the red area represents the additional energy required by the system in order to free bound water from the constituents and evaporate it, in addition to the amount of energy needed for pure/free water. As shown in Figure 4.5.4, it is therefore necessary to increase the temperature of the drying air (from 1 to 1′) in order to correct the curve and achieve the enthalpy necessary to evaporate both free water

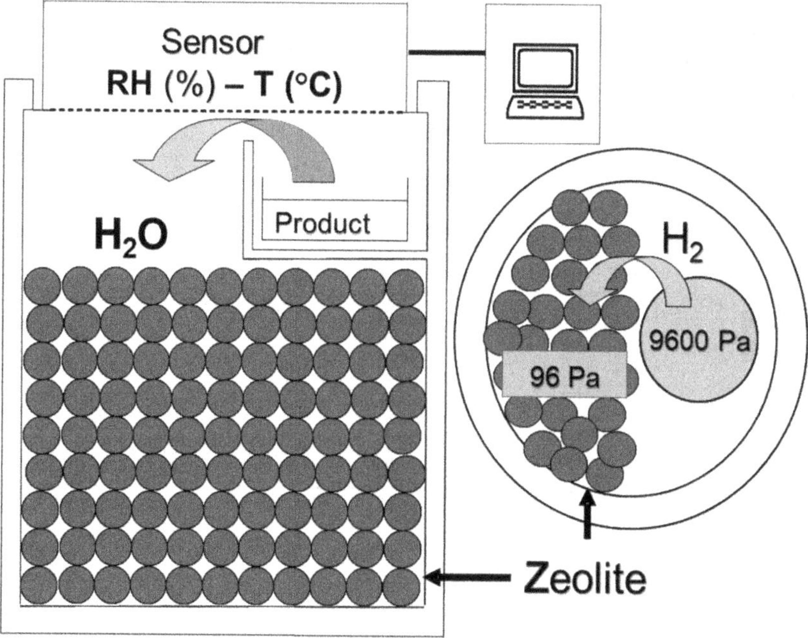

Figure 4.5.2 Diagram of device for drying by desorption.

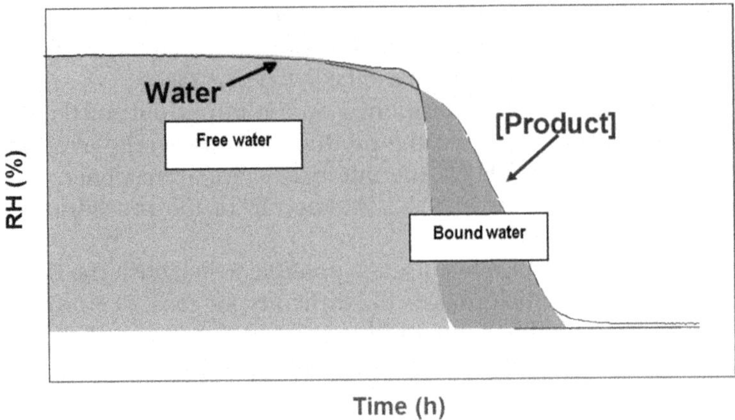

Figure 4.5.3 Theoretical kinetics of drying by desorption.

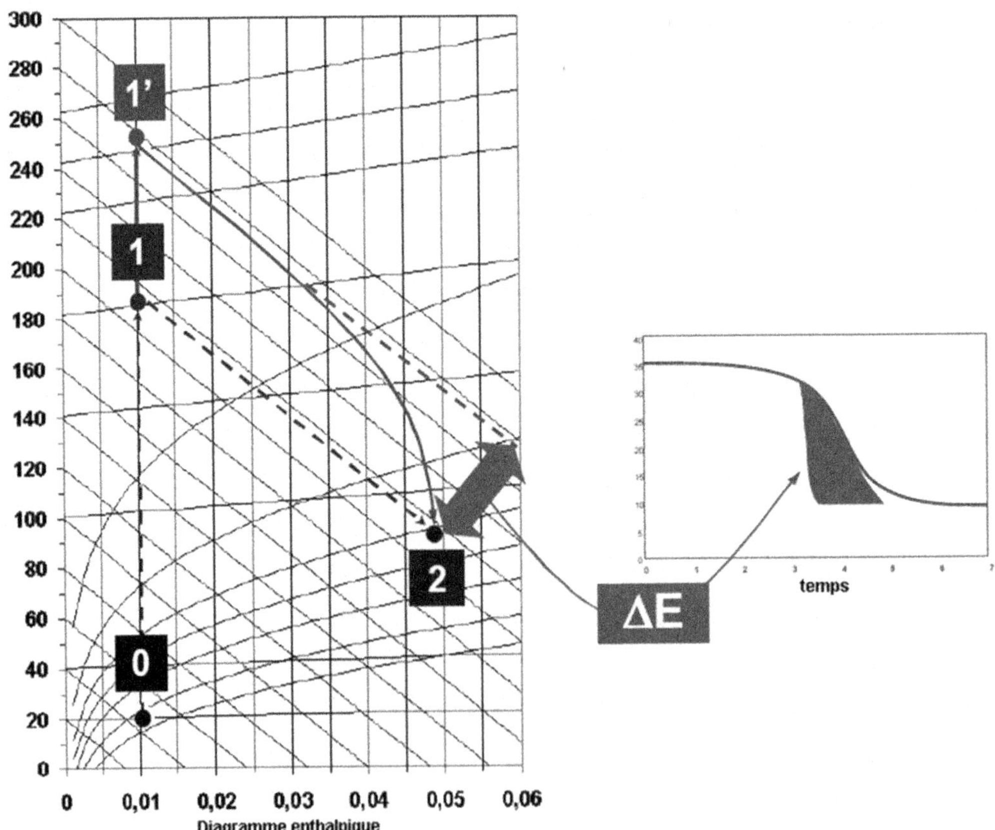

Figure 4.5.4 Enthalpy diagram.

and bound water in order to reach point 2. This reasoning is at the basis of the drying-by-desorption approach, and further integrated by the SD²P® program.

4.5.3 *Input parameters of the SD²P® program*

Once the sample has been analyzed, the application software can be used. The basis of the calculation is the data obtained through the desorption process (RH in relation to time), and this information can be processed using a computer equipped with Excel® and the SD²P® program.

The "Setting of parameters" window contains two tabs: the first tab, "Drying by desorption" enables you to enter the parameters related to inlet gas, expected features of the outlet product and characteristic performances of the spray-drying device (Figure 4.5.5). The second tab (Advanced) makes it possible to access the constant parameters of the program and the mass flow rates of the drying air (Figure 4.5.6). For the sake of clarity, they are defined below.

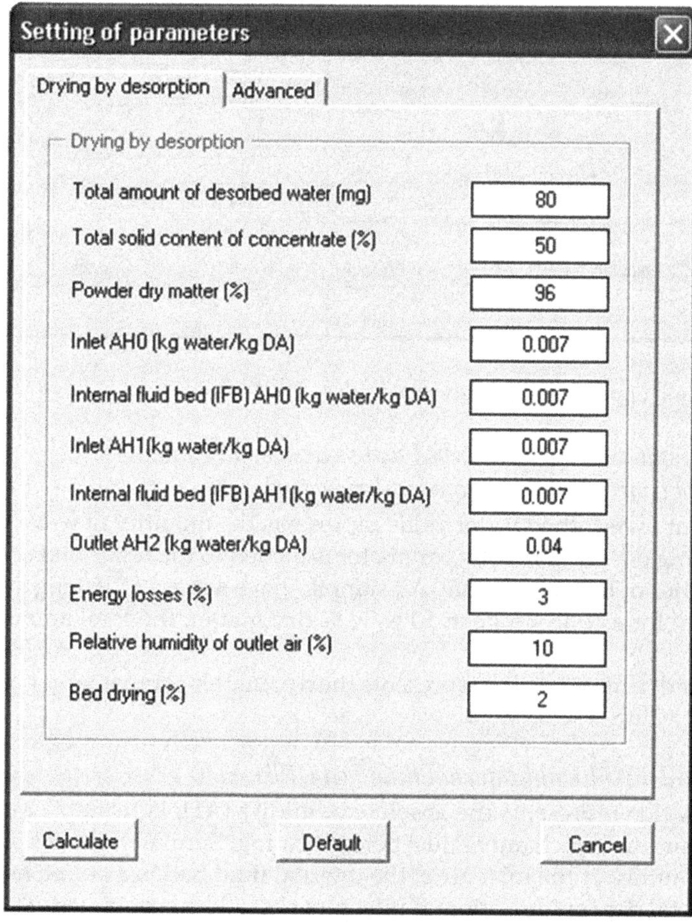

Figure 4.5.5 Setting of parameters ("Drying by desorption").

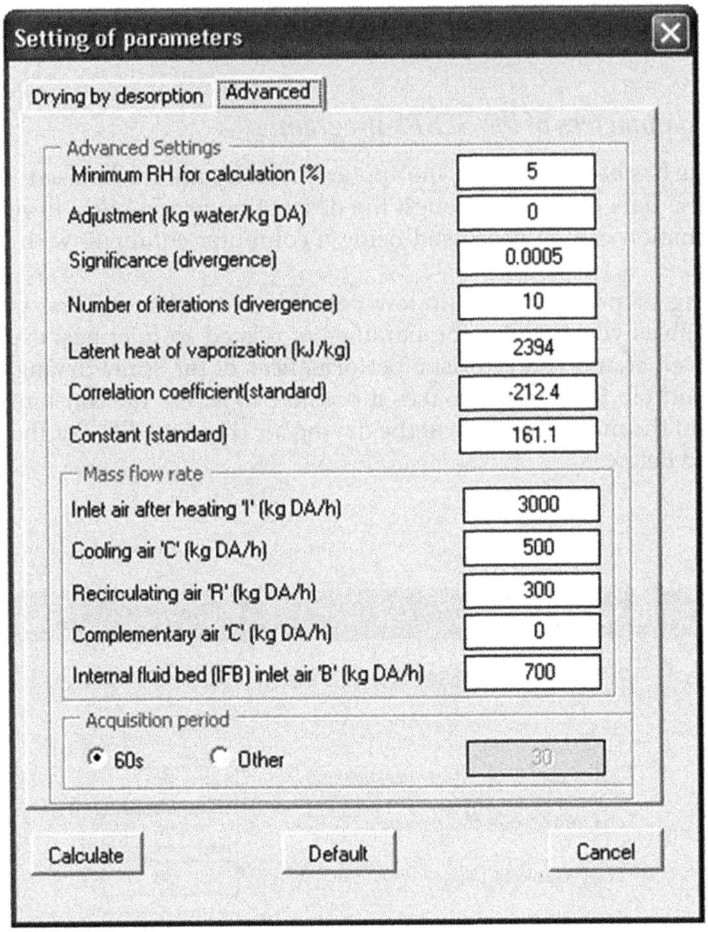

Figure 4.5.6 Setting of parameters ("Advanced").

4.5.3.1 Total amount of desorbed water, total solid content of concentrate and powder dry matter

The total amount of desorbed water value represents the quantity of water desorbed from the sample during desorption. This parameter is related to the total solid content, that is to say, the dry matter of the concentrate. All samples have a mass of 160 mg at the beginning of the test. Thus, for a concentrate at 50 w/w % dry matter, the total amount of desorbed water is 80 mg.

The powder dry matter value represents the dry matter content target of the final powder. The default value is 96 w/w %.

4.5.3.2 Inlet AH0 and internal fluid bed (IFB) AH0

The inlet AH0 value represents the absolute humidity (AH; kg water/kg dry air (DA)) of the inlet main air (drying chamber) just before heating. Similarly, the IFB AH0 represents the absolute humidity of the inlet air of the internal fluid bed just before heating.

Partial or total dehumidification of inlet airs should be considered. The default value for each one is 0.007 kg water/kg DA.

4.5.3.3 *Inlet AH1 and internal fluid bed (IFB) AH1*

Similarly, inlet AH1 and IFB AH1 values (kg water/kg DA) represent the absolute humidity of the inlet main air just after heating and of the inlet air entering the internal fluid bed just after heating, respectively. They differ from their corresponding AH0 values if heating is direct, to an extent of +45 mg water/(°C.kg DA). If the heating is indirect then AH1 = AH0. Their default value is 0.007 kg water/kg DA.

4.5.3.4 *Outlet AH2*

Last, the outlet AH2 value (kg water/kg DA) represents the absolute humidity of the air leaving the drying system. It is related to the evaporation capacity of the equipment. The default value is 0.040 kg water/kg DA.

4.5.3.5 *Energy losses*

This value represents energy losses (%) of the drying system, varying according to the drying tower design. The default value is 3%, but it can reach up to 10%.

4.5.3.6 *Relative humidity of outlet air*

The relative humidity of the outlet air value (RH; %) represents the imbalance with the full thermodynamic equilibrium. In an ideal situation, meaning for an infinite drying time, the RH of the outlet air should directly correspond to the water activity (a_w) of the powder coming out of the dryer: as an example, reaching the thermodynamic equilibrium would give an outlet air RH of around 20% for a powder at a_w of 0.2.

However, and given the residence time of the product, the air and the powder do not have time to reach this equilibrium in industrial towers. The RH of the outlet air is thus lower than 20% for a target a_w of 0.2. A preliminary study, by introducing a thermo-hygrometer into the outlet air (after the cyclone and/or bag filters), can refine this value (Schuck et al., 2005, 2007, 2008). The default value is 10% for a target a_w of 0.2.

4.5.3.7 *Bed drying*

This value represents the percentage of the water that is removed by the internal fluid bed. The default value is 3%, which means that 97% of the water removal comes from the drying chamber.

4.5.3.8 *Airflow rates*

The Advanced tab makes it possible to define the values of the different airflow rates (kg/h) entering the drying system (Figure 4.5.6). It should be mentioned that the accuracy of the final prediction of this drying-by-desorption approach depends to a very large extent on the accuracy with which these different flow rates are measured; therefore, special attention must be paid to the collection of these values. Depending on the drying device, one should consider:

- Inlet air after heating, which represents the main inlet airflow (default value is 3000 kg/h)
- Cooling air "C," which represents the cooling airflow of the nozzle line or of the atomizer (default value is 500 kg/h)
- Recirculating air "R," which represents the airflow of the suppressor used for the reincorporation of the fine particles in the chamber (default value is 300 kg/h)
- Complementary air "C," which represents the complementary airflow that can be introduced into the chamber as back-up (e.g., "wall sweep"; default value is 0 kg/h)
- Internal fluid bed (IFB) inlet air, which represents the inlet airflow of the internal fluid bed (default value is 700 kg/h)

4.5.3.9 *Constant parameters of the program*

The Advanced tab makes it possible to access several parameters related to the software (Figure 4.5.6). The most important are:

- The minimum RH for calculation (%) corresponds to the minimum RH of the desorption curve considered by the SD2P® program (default value is 5% RH).
- The significance determines the threshold RH value (%) above which the drying kinetics of the product diverge from those of free water (default value is 0.0005).
- The number of iterations determines the number of successive steps considered by the algorithm to identify the breakpoint of divergence: at each step, the significance should be reached (default value is 10).
- The latent heat of vaporization represents the latent heat used by the SD2P® program. The default value is 2394 kJ/kg, corresponding to the latent heat of vaporization of water at the temperature at which desorption occurs (45°C).
- The correlation coefficient and constant value correspond to the linear regression coefficient and intercept of the breakpoint value as a function the total solid content of a reference concentrate. It makes it possible to compare the "dryability" of a given product with that of a standard (by default, a skim milk concentrate) having the same total solid content (see 4.5.4). Their default values are –212.4 and 161.1, respectively.

NB: the skim milk standard may be replaced by another standard. In this case it is necessary to perform desorption on a new standard concentrate with a different total solid content (between 10 and 60%), to establish the amount of cumulated water at the breakpoint of divergence, and then perform linear regression and thus establish the new correlation coefficient and its constant, in order to introduce it into the "Setting of parameters" (Advanced tab).

4.5.4 *Output parameters of the SD²P® program*

Once the parameters have been entered, the spray-drying parameters are calculated and gathered in a synthesis tab (Figure 4.5.7). This latter can be broken down into three parts:

- Air: summarizing all the parameters concerning air entering and leaving the drying chamber and IFB (temperature, absolute humidity, relative humidity).
- Product: summarizing all the parameters concerning the concentrate (mass and volume water flow rate) and powder (mass flow rate).
- Energy: summarizing economic (kWh cost, cost per ton of removed water or per ton of produced powder) and energy (droplet temperature during spraying, dew point temperature at the air outlet, energy balance, energy consumption ratio and yield) features for the equipment.

Finally, there are two active yellow cells for inlet air temperature following heating "I," and the corresponding standard breakpoint.

From the time the temperature of the inlet air following heating "I" proposed by the SD²P® program is not achievable for various reasons related to the equipment (e.g., insufficient power of the air batteries), it is possible by clicking on this cell to specify a maximum temperature, lower than that predicted. The SD²P® program will then recalculate all parameters so that the temperature of the inlet air following heating "I" will be as close as possible to this maximum temperature.

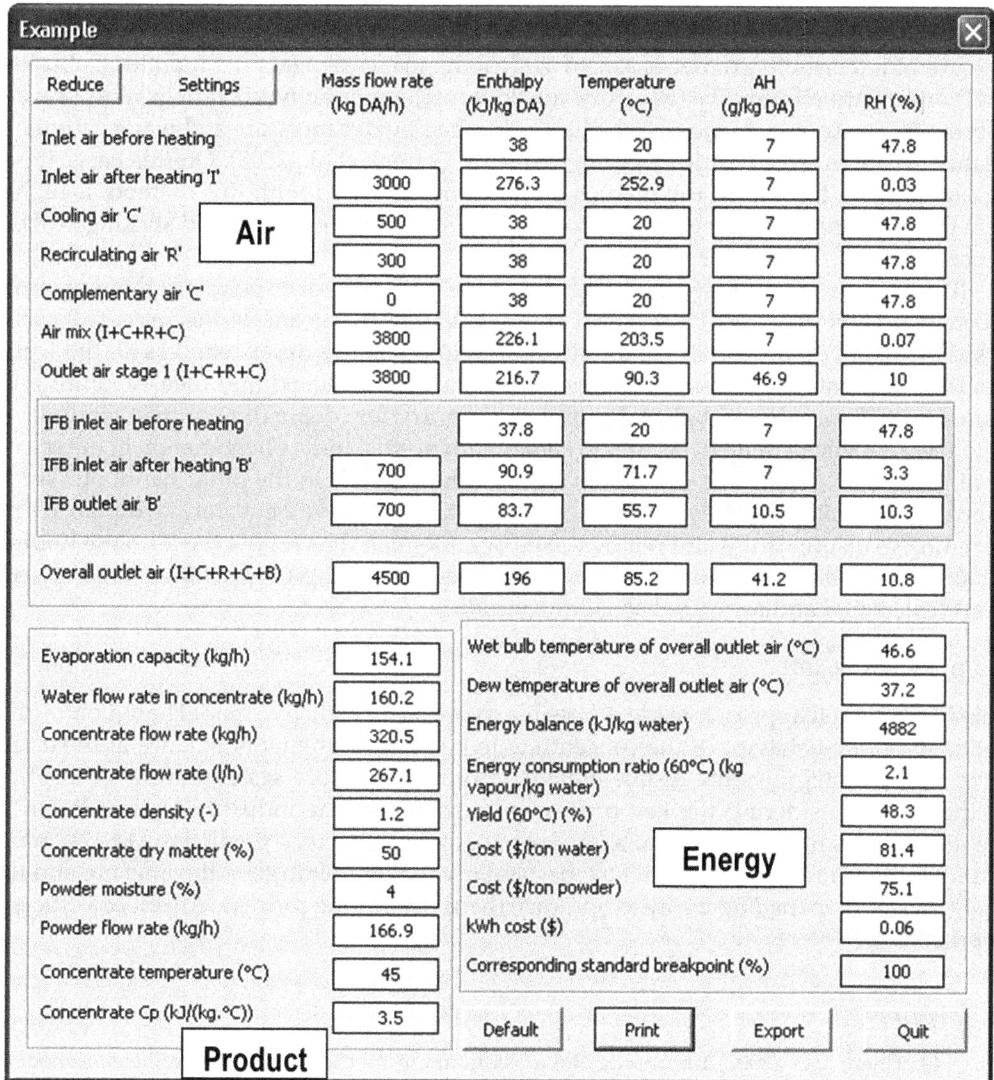

Example

Reduce | Settings

Air

	Mass flow rate (kg DA/h)	Enthalpy (kJ/kg DA)	Temperature (°C)	AH (g/kg DA)	RH (%)
Inlet air before heating		38	20	7	47.8
Inlet air after heating 'I'	3000	276.3	252.9	7	0.03
Cooling air 'C'	500	38	20	7	47.8
Recirculating air 'R'	300	38	20	7	47.8
Complementary air 'C'	0	38	20	7	47.8
Air mix (I+C+R+C)	3800	226.1	203.5	7	0.07
Outlet air stage 1 (I+C+R+C)	3800	216.7	90.3	46.9	10
IFB inlet air before heating		37.8	20	7	47.8
IFB inlet air after heating 'B'	700	90.9	71.7	7	3.3
IFB outlet air 'B'	700	83.7	55.7	10.5	10.3
Overall outlet air (I+C+R+C+B)	4500	196	85.2	41.2	10.8

Product

Evaporation capacity (kg/h)	154.1
Water flow rate in concentrate (kg/h)	160.2
Concentrate flow rate (kg/h)	320.5
Concentrate flow rate (l/h)	267.1
Concentrate density (-)	1.2
Concentrate dry matter (%)	50
Powder moisture (%)	4
Powder flow rate (kg/h)	166.9
Concentrate temperature (°C)	45
Concentrate Cp (kJ/(kg.°C))	3.5

Energy

Wet bulb temperature of overall outlet air (°C)	46.6
Dew temperature of overall outlet air (°C)	37.2
Energy balance (kJ/kg water)	4882
Energy consumption ratio (60°C) (kg vapour/kg water)	2.1
Yield (60°C) (%)	48.3
Cost ($/ton water)	81.4
Cost ($/ton powder)	75.1
kWh cost ($)	0.06
Corresponding standard breakpoint (%)	100

Default | Print | Export | Quit

Figure 4.5.7 Synthesis page.

The "corresponding standard breakpoint" provides an idea of the "dryability" of a product in comparison with a skim milk concentrate with the same dry extract. This is an indirect measurement of the ability to stick to the drying chamber. Thus, the lower this value, the higher the stickiness tendency. Moreover, it makes it possible to correct the input evaporation capacity in order to reach optimized drying conditions. For example, if the equivalent divergence standard is 80% in relation to the desorption of the product, it means that it is recommended to lower the evaporation capacity to 80% of its original value. Clicking on this cell will automatically correct the input AH2 as defined in the first analysis, and re-determine the whole set of parameters.

4.5.5 Process improvement to avoid stickiness

The use of a thermohygrometric sensor with some measurements (temperature, absolute (AH) and relative humidity (RH), dry air flow rate, water activity) is helpful to prevent sticking in the drying chamber and to optimize the powder moisture and water activity in relation to the relative humidity of the outlet air (Schuck et al., 2005). On this basis, it can be extrapolated that when the predicted AH value, obtained with this method, is higher than the measured AH, it means that water is retained in the chamber and sticking is likely to occur.

It should be remembered here that the calculated AH corresponds to the maximum theoretical value that can be reached: this calculation by means of the mass balance is based on the hypothesis that the air circulating in the spray dryer removes all the water from the concentrate. Thus, a difference between the calculated and measured absolute humidity of the outlet air below 2 g of water/kg dry air (depending on the spray dryer with regard to measurement accuracy) comes with no stickiness phenomenon in the spray-dryer chambers, regardless of the dairy concentrate used. On the other hand, our study showed that sticking is likely to occur for differential AH above 2 g water/kg dry air, corresponding to insufficient water removal. The operator can thus follow the absolute humidity and anticipate a variation in drying parameters according to the differences between SD^{2}P® calculated and measured absolute humidity.

4.5.6 Conclusion

This fundamental approach made it possible to obtain a predictive model based on results obtained on the behavior of the concentrate to be dried and energy balance at the entire drier scale; it integrates the results obtained on the laboratory scale with the desorption method and considering the key process parameters for the industrial extrapolation of the results. This model permitted a very precise and satisfactory prediction (±1–5% maximum) of the main drying parameters, particularly the temperature of the inlet air, and the energy costs. It opened up a way to optimize the spray-drying process with an eco-efficient approach.

References

Lin, X.Q., Chen, X.D. (2002) Improving the glass-filament method for accurate measurement of drying kinetics of liquid droplets. *Transactions of the Institution of Chemical Engineers* 80(4), 401–410.

Patel, K.C., Chen, X.D., Jeantet, R., Schuck, P. (2009) One-dimensional simulation of co-current, dairy spray drying systems—Pros and cons. *Dairy Science & Technology* 90(2–3), 181–210.

Schuck, P., Méjean, S., Dolivet, A., Jeantet, R. (2005) Thermohygrometric sensor: A tool for optimizing the spray drying process. *Innovative Food Science & Emerging Technologies* 6(1), 45–50.

Schuck, P., Méjean, S., Dolivet, A., Jeantet, R., Bhandari, B. (2007) Keeping quality of dairy ingredients. *Le Lait* 87(4–5), 481–488.

Schuck, P., Dolivet, A., Méjean, S., Jeantet, R. (2008) Relative humidity of outlet air: Key parameter to optimize moisture content and water activity of dairy powder. *Dairy Science & Technology* 88(1), 45–52.

Schuck, P., Dolivet, A., Méjean, S., Zhu, P., Blanchard, E., Jeantet, R. (2009) Drying by desorption: A tool to determine spray drying parameters. *Journal of Food Engineering* 94(2), 199–204.

Zhu, P., Jeantet, R., Dolivet, A., Méjean, S., Schuck, P. (2009) Characterization of a spray-drying pilot plant in relation to mass and energy bal ances and to quality of the obtained powders. *Industries Alimentaires & Agricoles* 126, 23–29.

4.6 Spray drying of probiotics: towards a controlled and efficient process

Song Huang, Pierre Schuck, Gwénaël Jan and Romain Jeantet

4.6.1 Introduction

Over the past 20 years, the dairy industry has been increasing the share of valuable milk protein fractions by means of milk cracking. Following this trend, the nature of dairy powders has changed to more specific intermediate dried products, and high-protein powders have been developed in order to match functional and nutritional properties target. In the meantime, the market for infant formulae has shown a continuous increase, supported by the growth of the world population and the rise in the standard of living in emerging countries. Given the considerable economic weight of these high added-value products for the dairy sector, it is nowadays of prime importance to develop research routes that will make it possible to better control the above-mentioned nutritional and functional properties, while complying with sustainable production mode requirements. To address this issue, and as infant formulae should incorporate live micro-organisms with probiotic properties in the near future (Martin et al., 2016; Radke et al., 2017), we developed and scaled up an innovative process for achieving the one-step growth and drying of bacterial culture.

The use of spray drying to produce viable cultures, especially when dealing with "sensitive" probiotic strains, is still challenging given the loss of viability due to high temperature, oxidative stress and high shear exposure. However, extensive investigations have been carried out in the last two decades in order to find strategies to protect probiotics during spray drying because of spray drying's key advantages, compared to freeze-drying, in terms of productivity and energy costs (Huang et al., 2017b).

4.6.2 Results: discussion

In this work, a novel process has been proposed in order to produce probiotics simply and sustainably via spray drying. In this novel and now patented process (Jeantet et al., 2015), the highly concentrated sweet whey was first used as a two-in-one medium for the growth and spray drying of probiotics. Two probiotic strains, one fragile (*Lactobacillus casei* BL23) and the other robust (*Propionibacterium freudenreichii* CIRM-BIA129), were used as test strains. Compared to the usual process, the intermediate operation steps from growth to spray drying (e.g. cell washing, harvesting and re-suspending) were avoided in this novel process (Figure 4.6.1). Meanwhile, probiotic metabolites could be retained in the final powders as the food-grade nature of the culture medium made it possible to eliminate these steps.

To achieve such a two-in-one process, the total solids (TS) of the sweet whey medium were increased from 5 wt % to 40 wt %, and the optimal range of TS was explored in relation to its effect first on biomass production after growth, and then on viability after spray drying and during storage (Huang et al., 2016a). It was clearly shown that culturing *L. casei* and *P. freudenreichii* in 20% ~30 wt % TS sweet whey resulted in greater biomass production after spray drying and sustained viability upon long-term storage (Figure 4.6.2A to C, *L. casei* as an example). Besides, it is interesting to note that

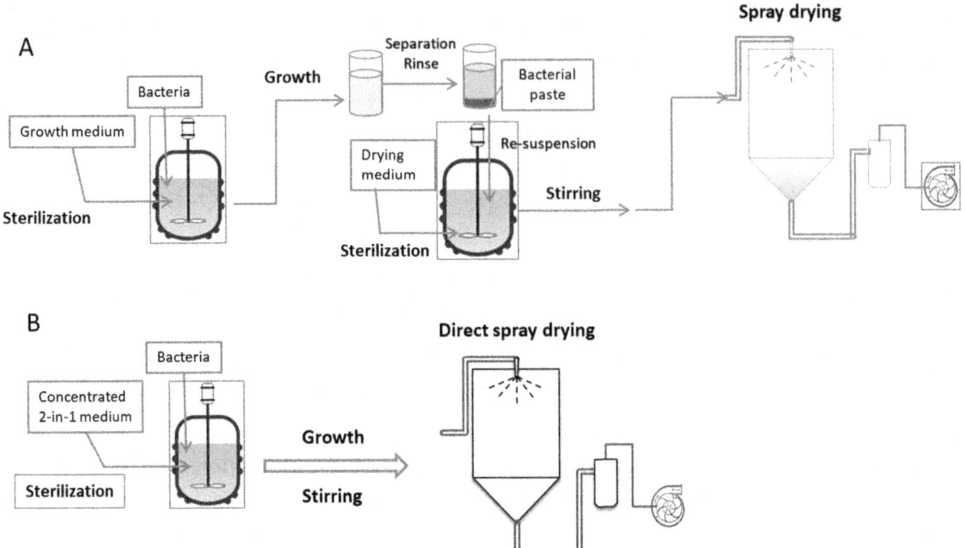

Figure 4.6.1 Diagrams of (A) the conventional process used in spray drying of probiotic bacteria and (B) the innovative two-in-one drying process.

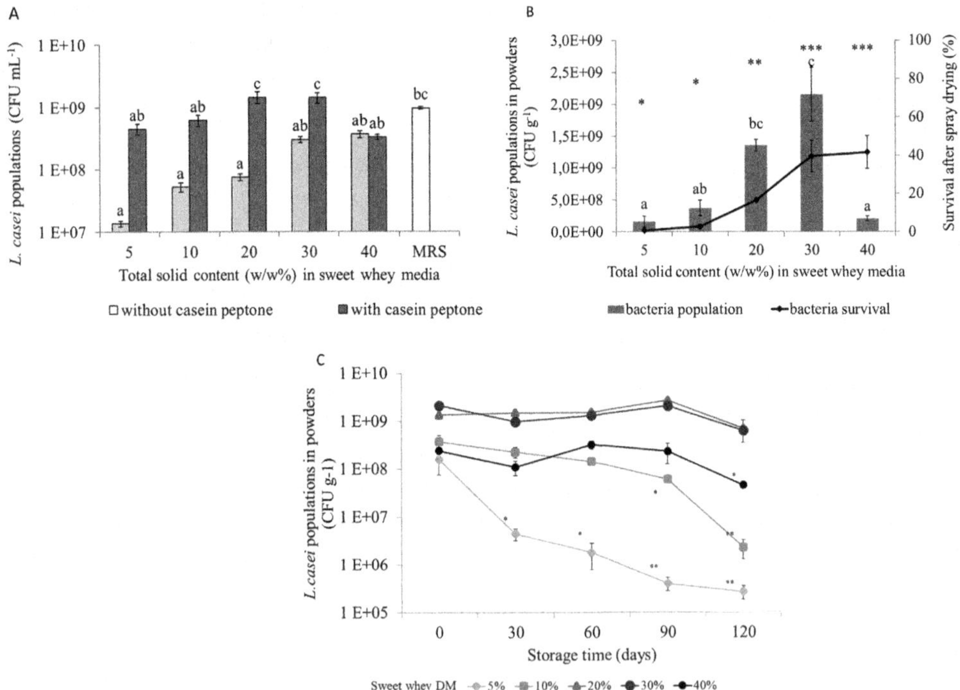

Figure 4.6.2 The behavior of *L. casei* in the two-in-one process: (A) final *L. casei* populations in the sweet whey culture media with different TS, with or without casein peptone supplementation (compared to the standard culture media, i.e. MRS broth). (B) Remaining bacteria populations (left *y*-axis) and survival (right *y*-axis) of *L. casei* in the sweet whey culture media with different TS after spray drying. (C) Change in bacteria populations in spray-dried powders of *L. casei* from sweet whey culture media with different TS during storage at 4°C for 120 days. Different numbers of * mean significant differences in dependency ($p < 0.05$).

the growth of *L. casei* was less dependent on casein peptone supplementation in sweet whey with high TS (Figure 4.6.2A).

The probiotic protection mechanisms enhanced by the two-in-one process were mainly related to bacterial osmoregulation in highly concentrated sweet whey, which triggered stress cross-protection, thus leading to the multi-stress tolerance of bacteria.

More specifically, the ability of bacteria to adapt to sub-lethal stresses may later induce their tolerance towards the same type of stress (stress adaptation) or different type of stresses (multi-stress response, also termed cross-protection) (Desmond et al., 2001). In our two-in-one process, the large final bacterial population yield suggests that bacteria adapt to the osmotic stress during growth in 30 wt % sweet whey, instead of being inhibited by this latter. Osmotic adaptation is believed to induce multi-stress tolerance against different stresses occurring in the spray-drying process (Wood, 2011). We confirmed this hypothesis by evidence acquired from proteomic analysis, and enzymatic and microscopic techniques. Highly concentrated sweet whey triggered different stress tolerance pathways, including the overexpression of chaperones and the accumulation of key compatible solutes, namely trehalose and polyphosphates (Figure 4.6.3C, F compared to 4.6.3A, B and 4.6.3D, E, respectively). It led to live and stable probiotics with induced multitolerance against heat, acid, bile-salt and spray-drying stresses (Huang et al., 2016b).

To validate the scale up feasibility of an innovation process is of prime importance for its industrial application and further development. As scaling factors and laws are generally poorly known, this often represents the bottleneck of innovation. In this study, the feasibility of scaling up the two-in-one process was validated with a semi-industrial pilot-scale spray dryer (Figure 4.6.4).

In order to minimize the heat inactivation of bacteria during spray drying, a multi-stage drying process can be used to lower the spray-drying temperature while increasing the overall

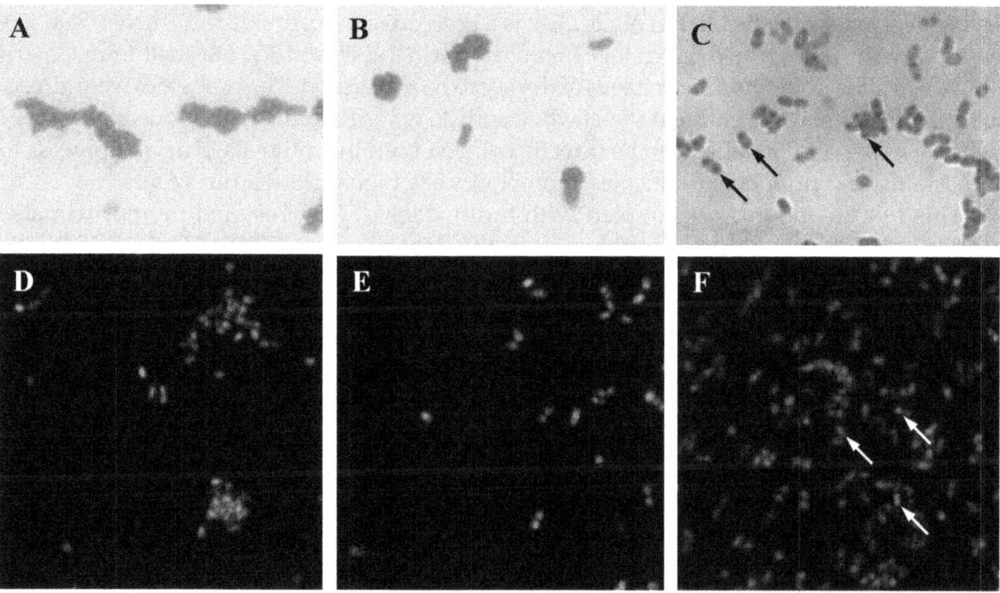

Figure 4.6.3 Concentrated sweet whey culture triggers intracellular accumulation of carbohydrates (Periodic acid-Schiff staining followed by light microscopy observation, A to C; dark granules, black arrow in C, indicate carbohydrates) and polyphosphate (DAPI staining followed by confocal microscopy, D to F; Green fluorescence, white arrow in F, indicates co-localization of cytosolic polyphosphate and DNA). Propionibacteria were cultivated either in YEL medium (A, D), in isotonic 5% sweet whey (B, E) or in hyper-concentrated 30% sweet whey (C, F).

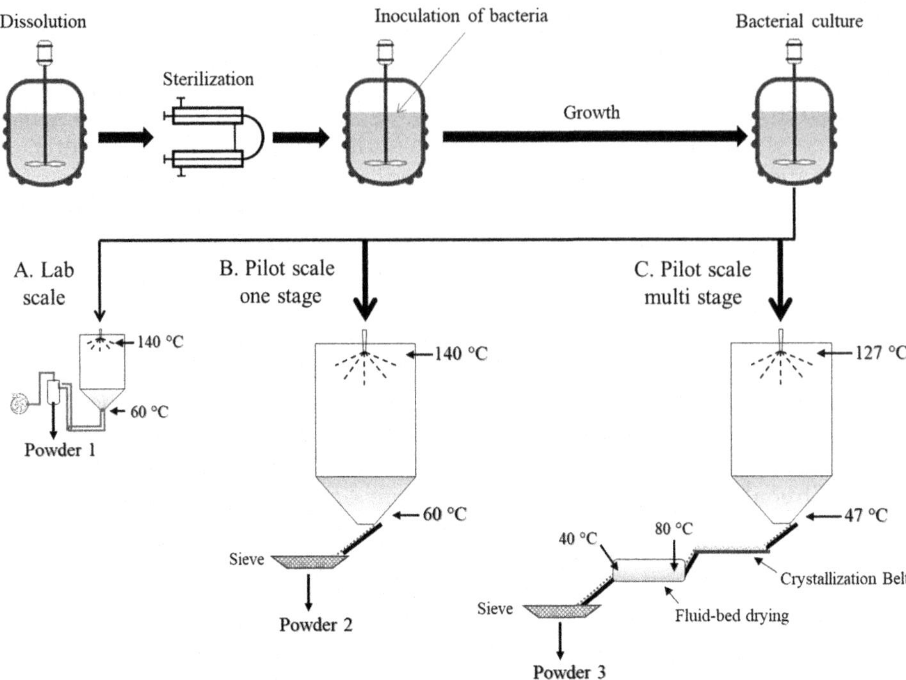

Figure 4.6.4 Schematic diagram of three drying process: (A) lab-scale spray drying (for obtaining Powder 1), (B) semi-industrial pilot-scale one-stage spray drying (for obtaining Powder 2) and (C) semi-industrial pilot-scale multi-stage spray drying (for obtaining Powder 3).

drying/residence time (Schuck et al., 2013). It is a prerequisite to provide the concentrated feed material for the first spray-drying stage in order to form the semi-dried but shell-formed particles, which allow the subsequent stages of drying to be performed. The probiotics were already cultivated in highly concentrated sweet whey with 30 wt % TS in the two-in-one process. This high dry-matter fermentation can be directly coupled with the multi-stage drying process for the sustainable continuous production of probiotics at a high yield (Figure 4.6.4).

This two-in-one process coupled with multi-stage drying resulted in approximately 100% (>10^9 CFU g^{-1}) survival of probiotics in the final powders (Figure 4.6.5). Besides, the powder stability was also monitored during a 6-month storage, which indicated that the storage temperature and powder moisture content both played crucial roles in probiotic stability. Spray drying also afforded a strain-dependent enhancement of bacterial resistance against *in vitro* simulated digestion (Huang et al., 2017a).

4.6.3 *Conclusion*

In this work, we focused on how to obtain live probiotic cells through spray drying. Indeed, addressing this issue is mandatory for providing a new generation of infant formulae, with enhanced nutritional value. A simplified process involving the direct spray drying of a culture performed on a preconcentrated culture medium was proposed, scaled up and further patented. Tested on fragile as well as robust probiotic strains, the process has demonstrated an efficiency comparable to that of freeze-drying, with respective survival rates between 40 and 100%, depending on the micro-organism. The dry matter concentration of the growth and drying medium was optimized in order to obtain a protective effect against bacterial

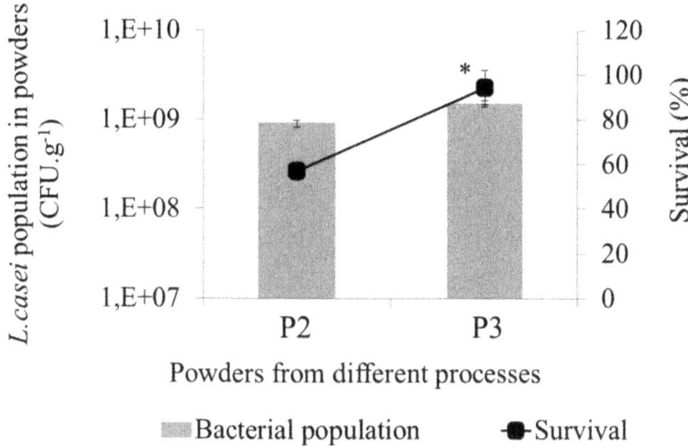

Figure 4.6.5 Population (left *y*-axis) and survival (right *y*-axis) of *L. casei* BL23 after one-stage spray drying or multi-stage drying at semi-industrial scale. P2 refers to one-stage spray drying and P3 multi-stage drying process.

cells during atomization. The stability of micro-organisms over time was verified at 4 and 6 months, with results comparable to those of freeze-drying but at a markedly lower production cost, which notably includes much less consumption of energy. This study opens new avenues for incorporating live and functional probiotics in infant formulae in the near future.

References

Desmond, C., C. Stanton, G.F. Fitzgerald, K. Collins and R.P. Ross (2001) Environmental adaptation of probiotic lactobacilli towards improvement of performance during spray drying, *International Dairy Journal*, 11(10), pp. 801–808.

Huang, S., C. Cauty, A. Dolivet, Y. Le Loir, X.D. Chen, P. Schuck, et al. (2016a) Double use of highly concentrated sweet whey to improve the biomass production and viability of spray-dried probiotic bacteria, *Journal of Functional Foods*, 23, pp. 453–463.

Huang, S., H. Rabah, J. Jardin, V. Briard-Bion, S. Parayre, M.-B. Maillard, et al. (2016b) Hyperconcentrated sweet whey, a new culture medium that enhances *Propionibacterium freudenreichii* stress tolerance, *Applied and Environmental Microbiology*, 82(15), pp. 4641–4651.

Huang, S., S. Méjean, H. Rabah, A. Dolivet, Y. Le Loir, X.D. Chen, et al. (2017a) Double use of concentrated sweet whey for growth and spray drying of probiotics: Towards maximal viability in pilot scale spray dryer, *Journal of Food Engineering*, 196, pp. 11–17.

Huang, S., M.-L. Vignolles, X.D. Chen, Y. Le Loir, G. Jan, P. Schuck and R. Jeantet (2017b) Spray drying of probiotics and other food-grade bacteria: A review, *Trends in Food Science and Technology*, 63, pp. 1–17.

Jeantet, R., S. Huang, G. Jan, P. Schuck, Y. Le Loir and X.D. Chen (2015) Method for preparing a probiotic powder using a two-in-one whey-containing nutrient medium, Patent application filed in Europe on 21 September 2015, EP no. 15 306465.4.

Martin, C.R., P.R. Ling and G.L. Blackburn (2016) Review of infant feeding: Key features of breast milk and infant formula, *Nutrients*, 8(5), pp. 279–289.

Radke, M., J.C. Picaud, A. Loui, G. Cambonie, D. Faas, H.N. Lafeber et al. (2017) Starter formula enriched in prebiotics and probiotics ensures normal growth of infants and promotes gut health: A randomized clinical trial, *Pediatric Research*, 81(4), pp. 622–631.

Schuck, P., A. Dolivet, S. Méjean, C. Hervé and R. Jeantet (2013) Spray drying of dairy bacteria: New opportunities to improve the viability of bacteria powders, *International Dairy Journal*, 31(1), pp. 12–17.

Wood, J.M. (2011) Bacterial osmoregulation: A paradigm for the study of cellular homeostasis, *Annual Review of Microbiology*, 65, pp. 215–238.

chapter 5

The drying of milk at the laboratory scale

From the industrial need to the scientific challenge

Luca Lanotte

Contents

5.1 Limitations of the drying process at the industrial scale

The production of powders is nowadays considered at the forefront in the dairy industry, thanks to the continuous improvement of the process engineering techniques in the last decades, as widely stressed in the different sections of this book. Due to the reliability and the efficiency of such techniques, new products are increasingly conceived, tested and put on the market to answer the high demand of end-users. However, despite the considerable recent advances, the operations characterizing the transformation from raw milk to dairy powders still present some technological limitations that prevent a further optimization of the drying process. For example, as it concerns the spray-drying technique, the high viscosity of some products at the inlet of the drier may affect the pulverization of the feeding solutions in monodisperse microdroplets, thus affecting the stability of the overall process.[1] Moreover, non-complete evaporation of water during the droplet trajectory towards the exit of the drying chamber can cause the onset of unsolved problems, such as the so-called caking and sticking on the walls of the drier.[2–6] Lastly, the onset of these inconveniences during the drying stages may frequently influence the packaging and storage of the powders,[7,8] causing an alteration of the final product characteristics. Such well-known technological problems represent a limitation to the tight control of the functional properties (e.g. rehydration,[9–11] nutritional characteristics[12,13]) of the dairy powders, which is a crucial task due to the versatile and wide range of possible users.

Reviewing the limitations associated with the drying operations leads to a complex and challenging question: how thoroughly do we know the mechanisms governing the drying of dairy products? In fact, from a very simplistic point of view, the towers used to obtain dairy powders can be considered as a black box. The main characteristics of the final product, such as the shape and size of the dry particles, merely depend on the tuning of the feeding parameters (e.g. solute concentration, flow rate) and the process environmental conditions (temperature, pressure, relative humidity). In contrast, a direct observation of the droplet-to-particle transition in a drying chamber and, thus, on-line monitoring of the process efficiency are far from being achieved. Therefore, exploring new experimental approaches to couple process engineering competences with the study of the physico-chemical mechanisms related to droplet evaporation has become increasingly important in optimizing the process at the industrial scale and the quality of the products. In the light of these considerations, in recent years the research on the drying of dairy solutions has broadened its horizons, progressing from the industrial large-scale approach to the laboratory approach. The differences between the two strategies in terms of the quantity of elaborated fluids, experimental conditions and temporal and spatial scales are absolutely glaring. Nevertheless, the investigation of the drying mechanisms at the micro-scale using a physico-chemical point of view represents a useful opportunity for shedding light on the dynamics of particle formation. At the conclusion of this book, we propose an overview of the more recent pioneering works performed on the evaporation of dairy solutions using essentially microscopy and microfluidic approaches. At the same time, we aim to highlight the high potential of the laboratory scale in order to strengthen the connection between the functional research related to the dairy industry and the fundamental research (physics, chemistry). To this end, we introduce an overview of promising techniques used on other biological fluids or model systems that could contribute to the better understanding of the drying-induced droplet-to-particle transition in dairy systems.

5.2 Laboratory investigation of evaporation in dairy systems

5.2.1 State of the art

Milk is a complex biological fluid[14] consisting of three phases: (i) lipids,[15] mainly triglycerides, organized in globular structures (average size d $\approx$ 5 μm), (ii) proteins,[16] i.e. whey proteins and caseins, with different size and physico-chemical properties (d $\leq$ 300 nm) and (iii) an aqueous medium as solvent. Thus, from a physical point of view, milk can be defined as a microemulsion and a colloidal dispersion at the same time. In the last decades, various works have been performed on the evaporation of biological fluids[17,18] (i.e. blood, plasma, urine, saliva) to detect the presence of pathologies and to develop diagnostic devices. The drying dynamics have also been investigated in very popular drinks (i.e. tea,[19] coffee[20]) to understand and optimize their nutritional properties. Such works have been carried out using mainly a so-called single-droplet approach, coupling the direct visualization of the droplet morphological evolution with other techniques suitable for characterizing the sol–gel transition (ultrasounds,[17] mass measurement,[21] electrophoresis,[22] etc.).

As it concerns milk or, more generally, dairy products, the study of the drying process at the lab scale has consisted recently of two main strategies.[23] To find a compromise between the reduced laboratory scale and the larger one of the industrial technology, model spray driers have been designed and manufactured in recent years.[24–26] Despite the significant differences in terms of production rate and powder particle size compared to industrial facilities, such pilot systems are very useful for the development of new formulas and innovative products, due to the rapidity of the tests and the easier control of the experimental parameters. Nevertheless, this kind of experimental setup is not conceived for highlighting the drying process mechanisms, thus stressing the need for finding alternative approaches. In this perspective, in analogy with already mentioned works on other biological fluids, the single-droplet approach has proved to be very valuable. Some pioneering theoretical works have been published to model the evaporation of a milk droplet in a free-falling condition[23] or in the presence of a substrate.[27,28] In the former case, some research groups have proposed the possibility of studying the evaporation of a droplet using the so-called "levitated droplet" approach.[29–31] In fact, they have managed to suspend drops by the application of acoustic waves or using air/gas flow, and they observed their drying-induced morphological evolution. However, these methods imply strong deformations and non-uniform evaporation of the droplets, and do not allow easy estimation of the mass loss to extract information about the drying kinetics. Therefore, most of the experimental studies have been performed using the alternative glass filament method. This approach consists of suspending a drop of a millimeter diameter at the tip of a glass filament and visualizing its shape evolution under controlled experimental conditions. Using this procedure, for instance, the group of Prof. X.D. Chen and his collaborators[32–35] was able to evaluate the impact of solution composition and environmental conditions (temperature, relative humidity) on the development of the drying process stages and the solubility of the dry particles. The same method was used to study the evaporation of still unexplored products, such as noni juice and noni acacia[34] (Figure 5.1).

In a few words, pilot spray-drying systems represent a useful tool to link the functional properties of the particles to their final shape, and the glass-filament experiments have been shown to be quite valuable for characterizing the droplet-to-particle stages at the macro-scale with an engineering approach. However, from a scientific point of

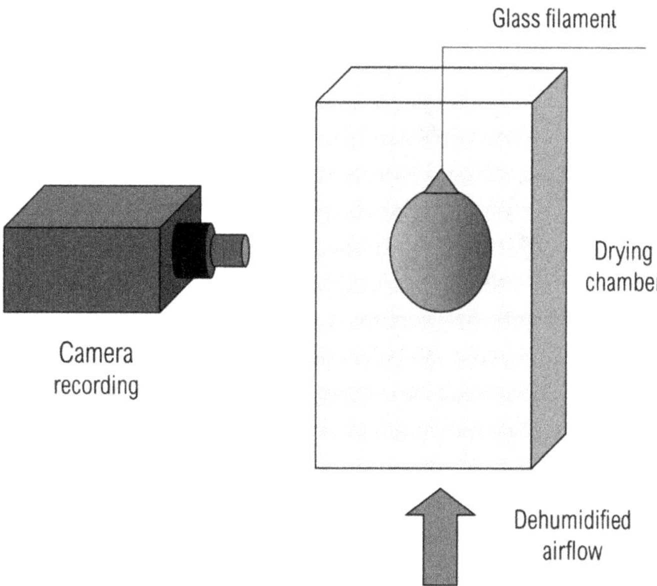

Figure 5.1 Schematic representation of the main components of a typical glass filament apparatus.

view, many crucial open questions remain unanswered. For example, full knowledge of the role of milk solute components (i.e. lipids, proteins) in the sol–gel transition, and therefore of the signature of the characteristics at the micro-scale in the final particle shape, is far from being achieved. Unluckily, the rich and complex composition of milk, and its three-phase nature, make this investigation quite complicated. Therefore, the current investigations on the evaporation of dairy products at the laboratory scale are mainly focusing on simplified models, such as dairy protein solutions. In recent years, this choice contributed to highlighting the considerable interdisciplinary nature of milk drying studies, involving physics, chemistry and biology in addition to process engineering, and to benefiting from the introduction of experimental methods already reliable for other biological systems. The main goal is the gradual and extensive analysis of the mechanisms leading the drying process, and the understanding of the separate role of each component in the development of such mechanisms, in order to tune and control the properties of the dry particles.

5.2.2 *Single-droplet approach in simplified models of milk: the case of dairy protein mixes*

The drying of a colloidal droplet is characterized by phenomena of mass and energy transfers that govern the sol–gel transition and determine the final particle morphology.[36,37] In the case of a sessile or a pendant droplet, from the early stage of the process, internal flows occur to replenish with solvent the regions where the evaporation is more significant. These flows are mainly of two different natures (Figure 5.2A): (i) the Marangoni–Bénard flows originating from chemical and/or thermal gradients,[38,39] which continuously mix the drying solution and simultaneously favor the molecular migration at the air–liquid

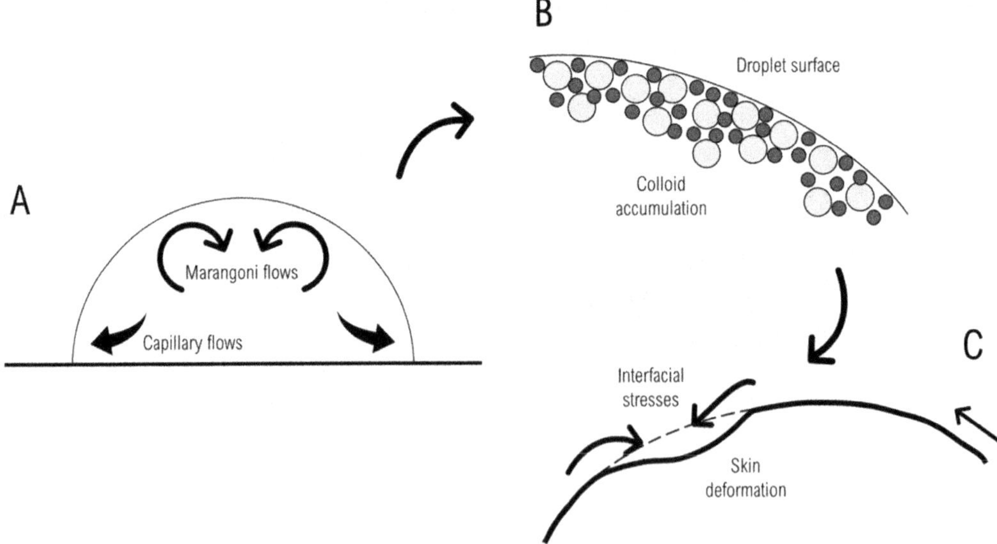

Figure 5.2 Main stages of drying a colloidal droplet: (A) onset of the internal flows after deposit on the substrate; (B) colloid migration and accumulation at the air–liquid interface (skin formation); (C) skin deformation under the action of interfacial stresses when the sol–gel transition is significantly advanced.

interface, and (ii) the capillary flows that enhance the segregation of the solute towards the drop borders.[40,41] When a critical concentration of colloidal particles is reached in proximity to the drop surface, a typical sol–gel transition phenomenon takes place and the formation of a gelling skin is observed (Figure 5.2B). Such a thickening layer can deform, bend and buckle under the action of the internal stress induced by the evaporation process until complete solidification (Figure 5.2C).[42–44] The mechanical response of the droplet skin strongly depends on the evolution of its structural characteristics (e.g. porosity, elasticity, deformability) that, in turn, depend on the physico-chemical properties of the colloidal components. Therefore, in a very schematic way, exploring the mechanisms of skin formation and its final structure would provide essential information about powder composition and functional properties.

Concerning milk, its two main protein components, i.e. whey proteins and caseins, are a glaring example of biocolloids exhibiting different molecular structure and organization. Whey proteins have a rigid globular structure ($d \approx 10$–30 nm),[45] while caseins are mainly organized in the form of micelles, thus exhibiting a sponge-like deformable shape ($d \approx 100$–300 nm).[46,47] It is evident that whey proteins (WP) and casein micelles (CM) show completely different characteristics concerning size, structure, mechanical properties and surface charge. Thus, it would be extremely important to investigate their separate roles in the evaporation process and in dry particle formation, since both WP and CM are crucial ingredients of dairy powders and especially of infant milk formulas. The study of the drying dynamics in mixes of dairy proteins is part of the wider context related to evaporation in binary colloidal solutions. In particular, the challenging investigation of the skin formation is closely linked to the observation of the stratification at the air–liquid interface in dispersions of polydisperse colloids.[48,49] In fact, in addition to the required knowledge of the driving forces that lead to the molecular accumulation at the surface of colloidal droplets and films, it is equally fundamental to shed light on the mechanisms of organization and re-distribution at the micro-scale, occurring when the interfacial layer is sufficiently packed and gelled. For example, are colloids with different sizes and physical properties randomly distributed in the dry particle skin, or can we observe an ordered packed structure depending on component characteristics? In this regard, the discoveries associated with the so-called small-on-top theory represent an absolute breakthrough.[50] Indeed, both experimental and theoretical works recently showed that the skin structure in binary colloidal dispersions is strongly affected by the evaporation rate. In the early stage of the drying process, the deposition and the organization at the molecular scale are mainly driven by Brownian forces and the smaller particles have a more intense motion compared to large ones (Figure 5.3). When the two types of colloids sufficiently differ in terms of size (size ratio ≈ 5) and the evaporation rate is high, the significant osmotic pressure developing in proximity to the air–liquid interface governs the stratification and leads to the accumulation of the smaller colloids on the external part of the gelling shell (small-on-top).[51] On the other hand, when the process is slow enough, such a size-depending stratification is not observed, and it is not unlikely for the presence of large colloids to be detected even in the outer skin layer.

Therefore, it begs the question: is the drying process at the laboratory scale fast enough to observe a protein stratification? Are the evaporation dynamics comparable to those taking place in industrial facilities?

In the last decade, a systematic investigation of the drying dynamics in single protein droplets (WP and CM) was carried out by different research groups. For example, the team of Food Processing Engineering of Wageningen (the Netherlands) observed the evaporation process in WP sessile droplets deposited on a hydrophobic substrate (polypropylene membrane) under controlled environmental conditions (temperature $T = 20$–$80°C$, relative humidity RH $\approx 0\%$, airflow ≈ 0.2 m/s).[52] In parallel, the researchers of the laboratory of

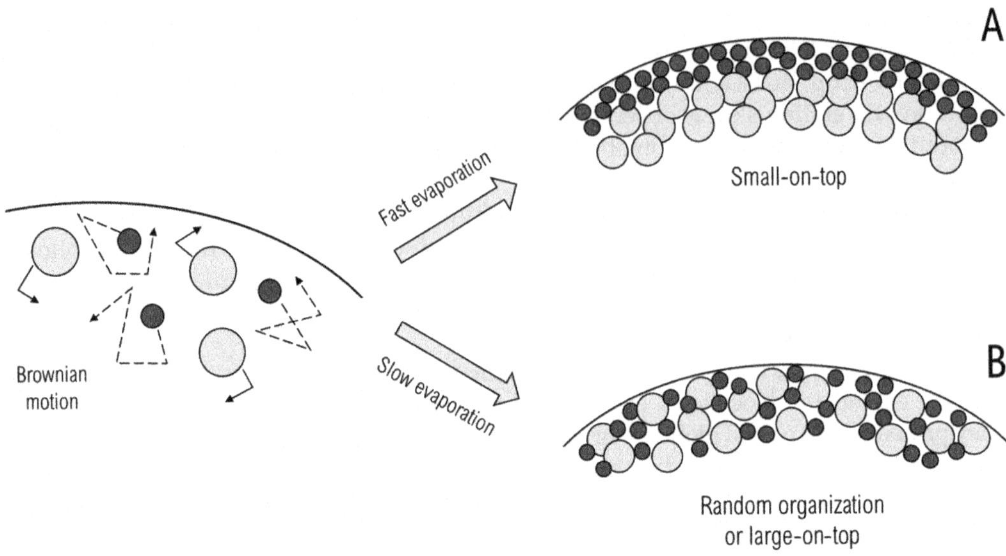

Figure 5.3 Structure of the drying droplet skin at the molecular scale. The early stage of the colloid re-distribution is driven by typical Brownian motion. On the other hand, the final structure depends on the drying rate: (A) small-on-top in case of fast evaporation, (B) random distribution or large-on-top if the evaporation rate is slow.

Science and Technology of Milk and Eggs of Rennes (France) proposed in recent years a thorough investigation of the evaporation in WP and CM droplets. Notably, they highlighted the different stages characterizing the drying process by visualizing the morphological evolution of pendant droplets of such dairy proteins deposited on hydrophobic substrates under constant temperature (T = 20°C) and relative humidity (RH < 3%). These typical examples of single-droplet setups, together with the already mentioned experimental approaches of the model spray-drying systems and the glass-filament procedure, stressed a profound difference in behavior between WP and CM. Indeed, irrespective of the experimental approach and the environmental conditions, the dry WP particles always exhibit a spherical hollow structure,[43] whereas the CM ones display a wrinkled deformed shape (Figure 5.4).[53]

Such morphological characteristics reflect, in a certain way, the physico-chemical properties of the proteins and, thus, the signature of the molecular scale on the final shape of the particles. The evidence of the impact of the colloidal properties on the evaporation dynamics recently led to the challenging open question of the behavior of WP and CM when they are simultaneously present in a drying droplet. From this perspective, the experiments realized by Lanotte et al. represent a scientific novelty in dairy research.[54] Using the pendant configuration similar to the case of pure WP and CM droplets, the authors show that the shape of dry particles of WP/CM mixes is similar to the one of the most represented protein in the sample. However, an elemental analysis performed by scanning electron microscopy (SEM) on the skin of the dry particles revealed the over-representation of the smaller whey protein molecules on the external part of the skin shell, in agreement with the cited small-on-top theory. These promising preliminary results underlined the necessity of a further accurate investigation at the micro-scale to explore the mechanisms of skin formation and the evolution of its viscoelastic properties throughout the evaporation process.

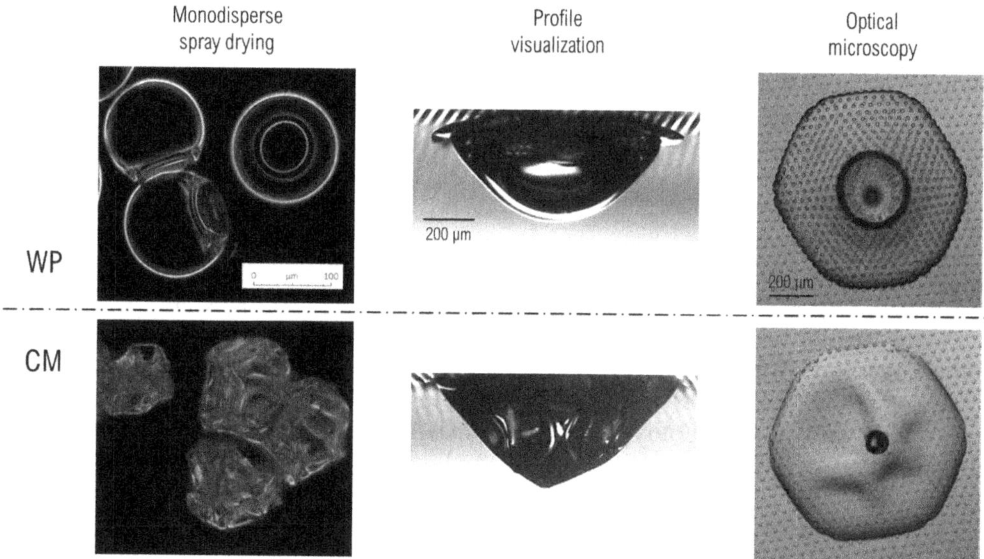

Figure 5.4 Comparison between WP and CM dry particles obtained using different experimental approaches. On the left: WP and CM powders produced by monodisperse spray drying and observed in phase contrast by optical microscopy. In the middle: profile visualization of pendant WP and CM droplets deposited on a patterned hydrophobic support. On the right: WP and MC sessile droplets on patterned hydrophobic support observed in top view by optical microscopy.

5.3 *Microscopy and microfluidic opportunities for studying the drying of dairy protein droplets*

In this last section, we aim at evaluating innovative experimental scenarios related to the study of evaporation in binary dairy protein solutions, from the investigation of the dynamics of skin formation to the evaluation of the evolution of surface rheological properties during the process. Notably, we explore the opportunity of using microscopy and microfluidic approaches that were shown to be efficient for other colloidal systems, in order to develop a strategy for the full investigation of the dairy protein drying process at lab scale. Moreover, the optimization of reliable experimental tests could prove to be crucial in view of the even more complex and challenging study of milk drying, i.e. considering even the presence of fat globules.

5.3.1 *Drying-induced internal flows and colloid stratification*

As already mentioned in this chapter, the evaporation of a colloidal droplet, and therefore also that of WP/CM dispersions, are characterized by the evolution of internal flows. For a better understanding of the stratification of proteins throughout the droplet evaporation, it would be crucial to estimate both the direction and intensity of such flows depending on the distance from the air–liquid interface. For this purpose, the particle-tracking method was shown to be very efficient and easy to use. This technique is frequently used to observe the evaporation of sessile droplets, and it consists of monitoring the movement of tracers, normally micrometric beads, suspended in samples of different types at different distances from the substrate.[17,55,56] The observation of the tracer trajectories is usually combined with the estimation of the flow strength by particle image velocimetry (PIV). Coupling particle tracking and PIV allows then a sort of three-dimensional reconstruction to be obtained of the drying-induced internal dynamics with time. This procedure has been used to investigate different fluid systems, such as colloidal dispersions, binary or ternary liquids and suspensions. Therefore, the current knowledge could be employed for the characterization of the phenomena occurring in mixtures of dairy proteins and, at a later time, in whole milk.

The time-dependent development of the internal flows in drying droplets of WP/CM mixes strongly affects the accumulation and the re-distribution of the proteins at the air–liquid interface. In the wake of the outcomes related to the small-on-top theory in model colloidal systems and of the first experimental investigation of dairy proteins, it would be interesting to provide evidence of the mechanisms of stratification and of the final structure of the skin. Recently, the observation by SEM of the section of a dry skin layer consisting of silica colloids of different sizes confirmed that the selective organization is affected by the component size and the key role played by the time scale of the event. It should be possible to carry out similar investigations on the shells of WP/CM dry particles to identify the impact of the relative percentage of these biocolloids in the initial samples, and their size and deformability on the organization of the skin at the micro-scale. In addition to these post-drying observations, it would be possible to evaluate the selective behavior of both WP and CM during skin formation by fluorescence techniques. Indeed, reliable protocols have been proposed in the literature for colloid labeling, and they could be extremely useful in order to detect the separate role of each protein in the layer formation at the droplet surface by confocal and two-photon excitation microscopy.[57]

5.3.2 Skin mechanical properties

The continuous protein migration and re-distribution towards the droplet surface is linked to the evolution with time of the skin mechanical properties. Here, we propose some transversal techniques that could be useful to characterize the skin mechanical behavior in correspondence to the different stages of the drying process.

- *Beginning of the drying process: protein migration and re-organization at the interface → interfacial dilational rheology*

  Interfacial dilational rheology allows the evaluation of the surface modification due to evolving mechanical forces (e.g. shear and/or dilational stresses) in complex systems with interfacial layers, such as foams, emulsions, colloidal dispersions, polymer solutions and suspensions. In particular, the so-called interfacial dilational rheology, which normally consists of observing the surface changes in compressing/expanding droplets or bubbles, is a powerful tool to estimate the mechanical properties of liquid interfaces.[58,59] Such oscillating drop methods, originally conceived for the measurement of interfacial tension, are nowadays used to measure the evolution of the viscoelastic properties for a wide range of expansion/compression frequencies. Therefore, dilational rheology is promising in the case of dairy protein mixes and, successively, of dairy products in general to evaluate the link between the mechanical properties of the skin layer and the molecular absorption and re-arrangement at the interface during the early stage of the evaporation process.
- *Sol–gel transition during the intermediate stage of the evaporation: skin formation → atomic force microscopy*

One of the most difficult challenges regarding the characterization of the skin during evaporation is undoubtedly exploring and understanding its mechanical properties after the onset of the sol–gel transition at the interface. In this regard, scanning probe microscopy (SPM) comes to the rescue thanks to the impressive advances of recent years in the investigation of the structure of multiple solid and gel-like samples. Notably, atomic force microscopy (AFM) was shown to be an efficient tool for observing soft materials, such as food, biofilms, polymer layers and protein structures.[60,61] Indeed, the high resolution associated with the AFM methodology (of the order of a few nanometers) allows the three-dimensional reconstruction of the observed surface and, at the same time, the estimation of its peculiar mechanical properties (e.g. Young and effective modulus).[62] The classic pointed cantilever extremity, which represented an obstacle to the investigation of soft or gelling interfaces by AFM, is nowadays often replaced by a less intrusive spherical one. Therefore, the AFM approach would be potentially decisive in shedding light on the evolution of the rheological properties of the gelling interface in drying droplets and films of dairy protein dispersions and, in general, of fluid dairy products from the occurrence of the sol–gel transition to the achievement of complete solidification.

- *Complete solidification: mechanical properties of the dry skin → indentation tests*

  AFM tests would be very useful when exploring limited (i.e. a few mm^2) and possibly irregular dry surfaces, as in the case of microdroplets where the interfacial skin is often subjected to buckling phenomena. However, if the experiments are performed on films, i.e. larger and less deformable surfaces, micro-indentation tests on their solid surface can represent an appropriate alternative.[63,64] Indeed, indenters are characterized by the same operating principles as SPM setups; they consist of specific submillimetric tips applying a force to the explored materials and detecting their mechanical response. Recently, Le Floch-Fouéré et al. used this technique to investigate the mechanical properties of dry films of WP and CM dispersions, and they compared the viscoelastic behavior of the dairy samples to that of model colloidal systems in order to develop a theoretical model on the evaporation of such proteins.[65]

5.3.3 Skin microscopic structure

The observation of the rheological behavior of the interfacial skin in drying droplets represents indirect proof of its organization at the molecular scale. In addition to the already mentioned procedures, some groups proposed in the literature a microfluidic approach for evaluating the porosity and the resistance to stress of the dry material deriving from the evaporation of colloidal silica particles. Such methods could be reemployed to better understand the impact of protein composition and overall concentration on the final structure of dairy protein dry surfaces. Potentially, they could be a useful tool even when investigating more complex dairy systems. The key point characterizing such a microfluidic approach is the possibility of observing a so-called monodirectional drying process, which favors the interpretation of the phenomena and the eventual development of a theoretical model.[66] Allain and Limat,[67] for example, studied the directional evaporation of colloidal suspensions introduced in Hele-Shaw cells (see Figure 5.5A for a schematic description). Notably, they observed the formation of regular crack patterns propagating from the air–liquid contact line to the internal part of the sample during the drying process. Moreover, starting from the specific crack formation, the authors deduced a simple

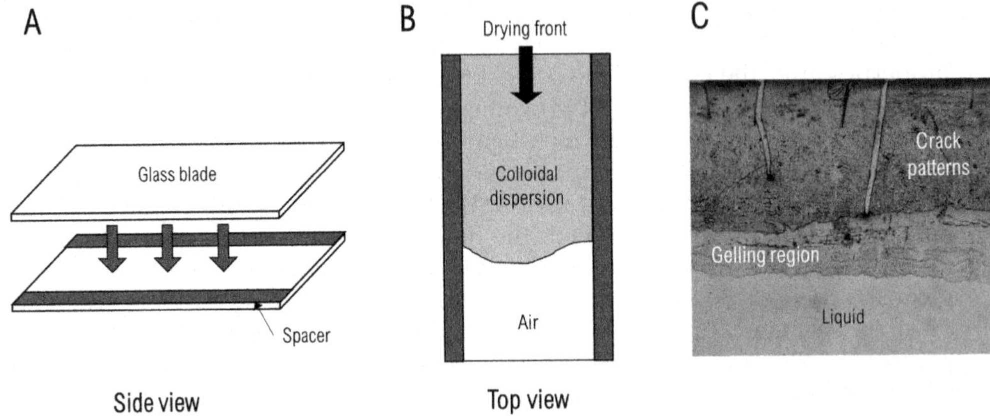

Figure 5.5 Schematic description of a Hele-Shaw cell for directional drying studies. (A) The flow cell consists of two glass blades separated at a controlled distance by a spacer; (B) the colloidal dispersion is introduced in the set up, and the drying starts from the extremity occupied by the liquid; (C) when the evaporation is advanced, patterns of cracks propagate from the initial air–liquid contact line to the internal part of the sample (image from a preliminary test on WP solutions).

model to predict the internal stress induced by evaporation and, thus, to characterize the mechanical properties of the dry material. Using the same basic idea, other researchers carried out works on the directional drying of colloidal solutions in standard laboratory Pasteur pipettes.[68] The authors provided evidence of the link between the formation and growth of a compact porous structure at the tip of the tubes and the overall drying kinetics by coupling the visualization of this cap-like structure development with the measurement of the mass loss throughout the experiment. Most of all, this technique can take advantage of reliable models to estimate the porosity of the dry material forming at the extremity of the pipettes.

References

1. Roos, Y. H. Importance of glass transition and water activity to spray drying and stability of dairy products. *Le Lait* **82**(4), 475–484 (2002).
2. Bhandari, B. R. & Howes, T. Implication of glass transition for the drying and stability of dried foods. *J. Food Eng.* **40**(1–2), 71–79 (1999).
3. Chuy, L. E. & Labuza, T. P. Caking and stickiness of dairy-based food powders as related to glass transition. *J. Food Sci.* **59**(1), 43–46 (1994).
4. Özkan, N., Withy, B. & Dong Chen, X. Effects of time, temperature, and pressure on the cake formation of milk powders. *J. Food Eng.* **58**(4), 355–361 (2003).
5. Lloyd, R. J., Dong Chen, X. & Hargreaves, J. B. Glass transition and caking of spray-dried lactose. *Int. J. Food Sci. Technol.* **31**(4), 305–311 (1996).
6. Fitzpatrick, J. J. et al. Glass transition and the flowability and caking of powders containing amorphous lactose. *Powder Technol.* **178**(2), 119–128 (2007).
7. Min, D. B., Lee, S. H., Lindamood, J. B., Chang, K. S. & Reineccius, G. A. Effects of packaging conditions on the flavor stability of dry whole milk. *J. Food Sci.* **54**(5), 1222–1224 (1989).
8. Coulter, S. T. & Jenness, R. Packing dry whole milk in inert gas. *Minnesota Technical Bulletin 167*, 34 (1945).
9. Chever, S. et al. Agglomeration during spray drying: Physical and rehydration properties of whole milk/sugar mixture powders. *LWT Food Sci. Technol.* **83**, 33–41 (2017).
10. Gaiani, C., Schuck, P., Scher, J., Desobry, S. & Banon, S. Dairy powder rehydration: Influence of protein state, incorporation mode, and agglomeration. *J. Dairy Sci.* **90**(2), 570–581 (2007).

11. Selomulya, C. & Fang, Y. Food powder rehydration. In *Handbook of Food Powders* 379–408 (Elsevier, 2013). doi:10.1533/9780857098672.2.379.
12. Sharma, A., Jana, A. H. & Chavan, R. S. Functionality of milk powders and milk-based powders for end use applications: A review. *Compr. Rev. Food Sci. Food Saf.* **11**(5), 518–528 (2012).
13. Thomas, M. E. C., Scher, J., Desobry-Banon, S. & Desobry, S. Milk powders ageing: Effect on physical and functional properties. *Crit. Rev. Food Sci. Nutr.* **44**(5), 297–322 (2004).
14. Fox, P. F., Uniacke-Lowe, T., McSweeney, P. L. H. & O'Mahony, J. A. Physical properties of milk. In *Dairy Chemistry and Biochemistry* 321–343 (Springer International Publishing, 2015). doi:10.1007/978-3-319-14892-2_8.
15. Christie, W. W. The composition and structure of milk lipids. In *Developments in Dairy Chemistry—2* (ed. Fox, P. F.) 1–35 (Springer Netherlands, 1983). doi:10.1007/978-94-010-9231-9_1.
16. Kinsella, J. E. & Morr, C. V. Milk proteins: Physicochemical and functional properties. *C R C Crit. Rev. Food Sci. Nutr.* **21**, 197–262 (1984).
17. Lanotte, L., Laux, D., Charlot, B. & Abkarian, M. Role of red cells and plasma composition on blood sessile droplet evaporation. *Phys. Rev. E* **96**(5–1), 053114 (2017).
18. Tarasevich, Y. Y. & Pravoslavnova, D. M. Segregation in desiccated sessile drops of biological fluids. *Eur. Phys. J. E* **22**(4), 311–314 (2007).
19. Okubo, T. Sedimentation and drying dissipative structures of green tea. *Colloid Polym. Sci.* **285**(3), 331–337 (2006).
20. Kim, J. Y. & Weon, B. M. Evaporation of strong coffee drops. *Appl. Phys. Lett.* **113**(18), 183704 (2018).
21. Brutin, D., Sobac, B., Loquet, B. & Sampol, J. Patterns formation in drying drops of blood. *J. Fluid Mech.* **667**, 85–95 (2011).
22. Annarelli, C. C., Fornazero, J., Bert, J. & Colombani, J. Crack patterns in drying protein solution drops. *Eur. Phys. J. E* **5**(5), 599–603 (2001).
23. Adhikari, B., Howes, T., Bhandari, B. R. & Truong, V. Experimental studies and kinetics of single drop drying and their relevance in drying of sugar-rich foods: A review. *Int. J. Food Prop.* **3**(3), 323–351 (2000).
24. Wu, W. D., Patel, K. C., Rogers, S. & Chen, X. D. Monodisperse droplet generators as potential atomizers for spray drying technology. *Dry. Technol.* **25**(12), 1907–1916 (2007).
25. You, X. et al. Dairy milk particles made with a mono-disperse droplet spray dryer (MDDSD) investigated for the effect of fat. *Dry. Technol.* **32**(5), 528–542 (2014).
26. Sadek, C. et al. To what extent do whey and casein micelle proteins influence the morphology and properties of the resulting powder? *Dry. Technol.* **32**, 1540–1551 (2014).
27. Schutyser, M. A. I., Perdana, J. & Boom, R. M. Single droplet drying for optimal spray drying of enzymes and probiotics. *Trends Food Sci. Technol.* **27**(2), 73–82 (2012).
28. Perdana, J., Fox, M. B., Schutyser, M. A. I. & Boom, R. M. Single-droplet experimentation on spray drying: Evaporation of a sessile droplet. *Chem. Eng. Technol.* **34**(7), 1151–1158 (2011).
29. Zang, D. et al. Acoustic levitation of liquid drops: Dynamics, manipulation and phase transitions. *Adv. Colloid Interface Sci.* **243**, 77–85 (2017).
30. Yarin, A. L., Brenn, G., Kastner, O. & Tropea, C. Drying of acoustically levitated droplets of liquid–solid suspensions: Evaporation and crust formation. *Phys. Fluids* **14**(7), 2289 (2002).
31. Nuzzo, M., Millqvist-Fureby, A., Sloth, J. & Bergenstahl, B. Surface composition and morphology of particles dried individually and by spray drying. *Dry. Technol.* **33**(6), 757–767 (2015).
32. Chen, X. D. & Lin, S. X. Q. Air drying of milk droplet under constant and time-dependent conditions. *AIChE J.* **51**(6), 1790–1799 (2005).
33. Fu, N., Woo, M. W. & Chen, X. D. Colloidal transport phenomena of milk components during convective droplet drying. *Colloids Surf. B Biointerfaces* **87**(2), 255–266 (2011).
34. Zhang, C. et al. A study on the structure formation and properties of noni juice microencapsulated with maltodextrin and gum acacia using single droplet drying. *Food Hydrocoll.* **88**, 199–209 (2019).
35. Lallbeeharry, P. et al. Effects of ionic and nonionic surfactants on milk shell wettability during co-spray-drying of whole milk particles. *J. Dairy Sci.* **97**(9), 5303–5314 (2014).
36. Ferrari, G., Meerdink, G. & Walstra, P. Drying kinetics for a single droplet of skim-milk. *J. Food Eng.* **10**(3), 215–230 (1989).

37. Nešić, S. & Vodnik, J. Kinetics of droplet evaporation. *Chem. Eng. Sci.* **46**(2), 527–537 (1991).
38. Nguyen, V. X. & Stebe, K. J. Patterning of small particles by a surfactant-enhanced Marangoni-Bénard instability. *Phys. Rev. Lett.* **88**(16), 164501 (2002).
39. Hu, H. & Larson, R. G. Analysis of the effects of Marangoni stresses on the microflow in an evaporating sessile droplet. *Langmuir* **21**(9), 3972–3980 (2005).
40. Deegan, R. D. et al. Capillary flow as the cause of ring stains from dried liquid drops. *Nature* **389**(6653), 827–829 (1997).
41. Deegan, R. D. et al. Contact line deposits in an evaporating drop. *Phys. Rev. E* **62**(1 Pt B), 756–765 (2000).
42. Pauchard, L. & Allain, C. Mechanical instability induced by complex liquid desiccation. *C. R. Phys.* **4**(2), 231–239 (2003).
43. Sadek, C. et al. Shape, shell, and vacuole formation during the drying of a single concentrated whey protein droplet. *Langmuir* **29**(50), 15606–15613 (2013).
44. Brutin, D., Sobac, B. & Nicloux, C. Influence of substrate nature on the evaporation of a sessile drop of blood. *J. Heat Transf.* **134**(6), 061101 (2012).
45. Walstra, P., Walstra, P., Wouters, J. T. M. & Geurts, T. J. *Dairy Science and Technology* (CRC Press, 2005). doi:10.1201/9781420028010.
46. Bouchoux, A., Gésan-Guiziou, G., Pérez, J. & Cabane, B. How to squeeze a sponge: Casein micelles under osmotic stress, a SAXS study. *Biophys. J.* **99**(11), 3754–3762 (2010).
47. Dalgleish, D. G. & Corredig, M. The structure of the casein micelle of milk and its changes during processing. *Annu. Rev. Food Sci. Technol.* **3**, 449–467 (2012).
48. Trueman, R. E. et al. Autostratification in drying colloidal dispersions: Experimental investigations. *Langmuir* **28**(7), 3420–3428 (2012).
49. Sung, P.-F., Wang, L. & Harris, M. T. Deposition of colloidal particles during the evaporation of sessile drops: Dilute colloidal dispersions. *Int. J. Chem. Eng.*, 1–12 (2019).
50. Fortini, A. et al. Dynamic stratification in drying films of colloidal mixtures. *Phys. Rev. Lett.* **116**(11), 118301 (2016).
51. Liu, W., Midya, J., Kappl, M., Butt, H.-J. & Nikoubashman, A. Segregation in drying binary colloidal droplets. *ACS Nano* **13**(5), 4972–4979 (2019).
52. Bouman, J., Venema, P., de Vries, R. J., van der Linden, E. & Schutyser, M. A. I. Hole and vacuole formation during drying of sessile whey protein droplets. *Food Res. Int.* **84**, 128–135 (2016).
53. Sadek, C. et al. Buckling and collapse during drying of a single aqueous dispersion of casein micelle droplet. *Food Hydrocoll.* **52**, 161–166 (2016).
54. Lanotte, L., Boissel, F., Schuck, P., Jeantet, R. & Le Floch-Fouéré, C. Drying-induced mechanisms of skin formation in mixtures of high protein dairy powders. *Colloids Surf. Physicochem. Eng. ASP* **553**, 20–27 (2018).
55. Li, Y. et al. Gravitational effect in evaporating binary microdroplets. *Phys. Rev. Lett.* **122**(11), 114501 (2019).
56. Kim, H. et al. Controlled uniform coating from the interplay of Marangoni flows and surface-adsorbed macromolecules. *Phys. Rev. Lett.* **116**(12), 124501 (2016).
57. Guyomarc'h, F., Jemin, M., Le Tilly, V., Madec, M.-N. & Famelart, M.-H. Role of the heat-induced whey protein/κ-casein complexes in the formation of acid milk gels: A kinetic study using rheology and confocal microscopy. *J. Agric. Food Chem.* **57**(13), 5910–5917 (2009).
58. Lucassen-Reynders, E. H., Benjamins, J. & Fainerman, V. B. Dilational rheology of protein films adsorbed at fluid interfaces. *Curr. Opin. Colloid Interface Sci.* **15**(4), 264–270 (2010).
59. Ravera, F., Loglio, G. & Kovalchuk, V. I. Interfacial dilational rheology by oscillating bubble/drop methods. *Curr. Opin. Colloid Interface Sci.* **15**(4), 217–228 (2010).
60. Crockett, R. et al. Imaging of the surface of human and bovine articular cartilage with ESEM and AFM. *Tribol. Lett.* **19**(4), 311–317 (2005).
61. Shimoni, E. Using AFM to explore food nanostructure. *Curr. Opin. Colloid Interface Sci.* **13**(5), 368–374 (2008).
62. Dimitriadis, E. K., Horkay, F., Maresca, J., Kachar, B. & Chadwick, R. S. Determination of elastic moduli of thin layers of soft material using the atomic force microscope. *Biophys. J.* **82**(5), 2798–2810 (2002).

63. Pauchard, L., Abou, B. & Sekimoto, K. Influence of mechanical properties of nanoparticles on Macrocrack Formation. *Langmuir* **25**(12), 6672–6677 (2009).
64. Léang, M., Giorgiutti-Dauphiné, F., Lee, L.-T. & Pauchard, L. Crack opening: from colloidal systems to paintings. *Soft Matter* **13**(34), 5802–5808 (2017).
65. Le Floch-Fouéré, Cécile, Lanotte, L., Jeantet, R. & Pauchard, L. The Solute mechanical properties impact on the drying of dairy and model colloidal systems. *Soft Matter* **15**(30), 6190–6199 (2019).
66. Dufresne, E. R. et al. Dynamics of fracture in drying suspensions. *Langmuir* **22**(17), 7144–7147 (2006).
67. Allain, C. & Limat, L. Regular patterns of cracks formed by directional drying of a collodial suspension. *Phys. Rev. Lett.* **74**(15), 2981–2984 (1995).
68. Gauthier, G., Lazarus, V. & Pauchard, L. Alternating crack propagation during directional drying. *Langmuir* **23**(9), 4715–4718 (2007).

Index

Printed in the United States
By Bookmasters